Membrane Technology for Air Cleaning

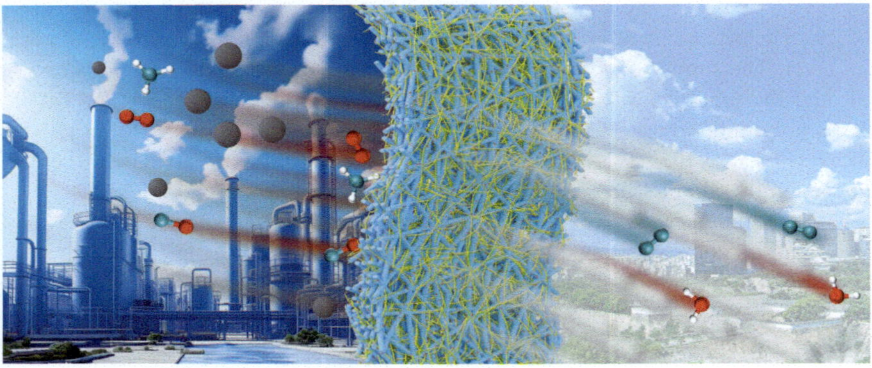

Zhaoxiang Zhong · Weihong Xing

Membrane Technology for Air Cleaning

Fabrication, Characterization and Application

Chemical Industry Press

Springer

Zhaoxiang Zhong
Nanjing Tech University
Nanjing, Jiangsu, China

Weihong Xing
Jiangsu University
Zhenjiang, Jiangsu, China

ISBN 978-981-96-1143-0 ISBN 978-981-96-1144-7 (eBook)
https://doi.org/10.1007/978-981-96-1144-7

Jointly published with Chemical Industry Press
The print edition is not for sale in China (Mainland). Customers from China (Mainland) please order the print book from: Chemical Industry Press.

The original submitted manuscript has been translated into English. The translation was done using artificial intelligence. A subsequent revision was performed by the author(s) to further refine the work and to ensure that the translation is appropriate concerning content and scientific correctness. It may, however, read stylistically different from a conventional translation.

Translation from the Chinese language edition: "Gas Cleaning Membrane Materials and Technology" by Zhaoxiang Zhong and Weihong Xing, © Chemical Industrial Press 2022. Published by Chemical Industrial Press. All Rights Reserved.

This Springer imprint is published by the registered company Springer Nature Singapore Pte Ltd.
The registered company address is: 152 Beach Road, #21-01/04 Gateway East, Singapore 189721, Singapore

Preface

Clean air is among the crucial essences to human health and well-being. One can live without food up to 40 days, without water up to 6 days, and without air up to 6 minutes. Air pollution refers to the contamination of indoor and outdoor air by any chemical, physical or biological agent that modifies the natural characteristics of the atmosphere. They can cause respiratory and other diseases and are important sources of morbidity and mortality. The latest data released by the World Health Organization (WHO) show that almost all of the global population (~99%) breathe air that exceeds WHO guideline limits and contains high levels of pollutants, with low- and middle-income countries suffering from the highest exposures.

In addition to common sources of air pollution such as household combustion devices, motor vehicles, industrial facilities and forest fires, in recent decades, the emission of pollution generated by industry, power plants, car transport, and agricultural and municipal waste becomes more intensified. Pollutants of major public health concern include particulate matter (PM), carbon monoxide, ozone, nitrogen dioxide and sulfur dioxide. When it involves acid gases (SO_2, NO_X) and volatile organic compounds (VOCs), mainly halogen-derived hydrocarbons and aromatic compounds, it becomes exceptionally hazardous.

Therefore, it is important to develop effective technology to control ultrafine dust emissions and achieve efficient and clean utilization of resources. Membrane technology provides an effective approach for PM2.5 control, with the advantages of high separation efficiency and low energy consumption. Especially for the purification of high-temperature gases, it can maximize the utilization of the physical sensible heat of the gas and realize energy recovery. Therefore, in the past ten years, membrane-based separation technology has become one of the best choices for air purification in many countries. With the combination with catalytic reaction, functional nanomaterials, and intelligent manufacturing technology, membrane technology can not only efficiently remove PM through surface filtration but can also realize the synchronous removal of other gaseous pollutants such as VOCs, NOx, and SO_2, significantly broadening the applications in the fields of gas-solid phase reaction processes, industrial flue gas purification processes, and human health and safety protection.

Recent progress on membrane materials and membrane processes has vigorously promoted the application of membrane technology in traditional industries. Integrated application of membrane materials and membrane technology is of important strategic significance to improving energy and resource utilization efficiency and reducing environmental pollution for a green and sustainable future. This book summarizes the latest progress on the fabrication of membrane materials for gas cleaning and their practical application with the emphasis on air cleaning.

Nanjing, China

Zhaoxiang Zhong
Weihong Xing

Contents

Authors and Contributors

About the Authors

Dr. Zhaoxiang Zhong is a distinguished professor at the College of Chemical Engineering, Nanjing Tech University, China, and concurrently holds the position of the Director of the Discipline Development Department at Nanjing Tech University. He was a visiting scholar at Monash University in Australia from 2012 to 2013. Dr. Zhong has published over 200 peer-reviewed papers and been honored with the National Outstanding Youth Science Fund of China, and the position of Chief Scientist of a National Key R&D Program. Dr. Zhong also directs the Gas Purification Membrane and Equipment Engineering Laboratory for the Petroleum and Chemical Engineering Industry and contributes to the academic community as an editorial board member for journals such as Membranes, Chinese Journal of Chemical Engineering, and Advanced Membranes. His leadership extends to professional committees, where he serves as the Director of the Special Separation Membrane Professional Committee and the Deputy Director of the Hydrophobic Membrane Professional Committee at the Membrane Industry Association of China. Additionally, he is the Deputy Secretary-General of the Porous Materials Branch of the Chinese Materials Research Society and the Deputy Director of the Filtration and Separation Professional Committee of the Chemical Industry and Engineering Society of China. Dr. Zhong's research is centered on the development and application of functional membrane materials.

Dr. Weihong Xing is currently a professor at College of Chemical Engineering, Jiangsu University. Also, she serves as the President of Jiangsu University and the director of National Engineering Research Center for Special Separation Membrane, Nanjing Tech University, China. She received her PhD in chemical engineering from Nanjing Tech University (2002), China. She has published four monographs, more than 300 peer-reviewed papers. Her research achievements have won five National Science and Technology Awards, a Ho Liang Ho Li Foundation Science and Technology Progress Award, a National Innovation Pioneer Award, a Hou Debang Chemical Science and Technology Achievement Award, and some other prestigious awards. She also serves as the Vice Chairman of the Membrane Industry Association of China, the Director of the Female Workers in Science and Technology Committee of the Chemical Industry and Engineering Society of China, and the Founding and Executive Editor of the journal Advanced Membranes. She has been elected as a national outstanding professional and technical talent and enjoys special government allowances from the State Council of China. Her research interests include membrane materials, separation engineering, and membrane reactors.

Contributors

Shasha Feng Nanjing Tech University, Nanjing, China

Qilin Gu Nanjing Tech University, Nanjing, China

Feng Zhang Nanjing Tech University, Nanjing, China

Junwei Wu Jiangsu Jiulang High Tech Co., Ltd, Nanjing, China

Feng Han Jiangsu Jiulang High Tech Co., Ltd, Nanjing, China

Qun Zhou Jiangsu Jiulang High Tech Co., Ltd, Nanjing, China

Yutang Kang Nanjing Tech University, Nanjing, China

Nicholas Ze-Xian Low Nanjing Tech University, Nanjing, China

Shipeng Wan Nanjing Tech University, Nanjing, China

Qiuyue Wang Nanjing Tech University, Nanjing, China

Yiqing Zeng Nanjing Tech University, Nanjing, China

Dong Zou Nanjing Tech University, Nanjing, China

Ming Zhou Nanjing Tech University, Nanjing, China

Chapter 1
Introduction

1.1 Definitions and Terms

1.1.1 Definition of Gas Cleaning Membranes

A gas-cleaning membrane mainly refers to a separation membrane material used to remove solid or gaseous pollutants from gases. Gas-cleaning membranes are mainly divided into porous membranes and dense membranes according to pore size. The pore size of porous membranes is about 0.1–10 μm, mainly used for the removal of ultrafine particles in gases and the coordinated management of multiple pollutants (such as dust and NO_x); the pore size of dense membranes is below 1 nm, mainly used for the separation and recovery of volatile organic compounds (VOCs), SO_2 and other aspects [1].

According to the usage temperature, gas cleaning membranes are divided into low and medium-use membranes and high temperature-use membranes. Low and medium-use membranes can directly remove solid impurities in gases below 260 °C. The main material is polymer material and they are mostly prepared by methods such as stretching. A typical low and medium-use membrane is a porous PTFE membrane prepared by polytetrafluoroethylene (PTFE) through processes such as expansion, calendering, and bidirectional stretching [2]. High temperature-use membranes are mainly used for flue gas cleaning at temperatures above 260 °C, with research in this area dating back to the 1980s. At present, the main industrially produced separation materials in high temperature-use membranes are metal materials and inorganic non-metal materials, etc. Common metal membranes are mainly prepared by sintering metal powders such as iron-aluminum alloy, Hastelloy alloy, 310S stainless steel, Inconel 601 alloy [3]. The main inorganic non-metallic materials include porous silicon carbide materials, porous cordierite materials, alumina silicate fiber composite ceramic filter materials, alumina/mullite fiber composite ceramic filter materials, etc. High-temperature-use membranes, with silicon carbide ceramic membranes as a typical example, are produced through a series of processes such as stirring and

Z. Zhong and W. Xing, *Membrane Technology for Air Cleaning*,
https://doi.org/10.1007/978-981-96-1144-7_1

mixing of silicon carbide raw materials and additives, granulation, molding, drying, primary sintering, coating, secondary sintering, etc., to finally obtain silicon carbide ceramic membrane products, which have been successfully used in demonstration projects in fields such as coal chemical industry [4]. Whether in low and medium temperature or high temperature gas cleaning systems, membranes are playing an increasingly important role in addressing the source control of air pollution [5].

1.1.2 Terminology

(1) Particulate Matter

Solid, liquid, or mixed solid and liquid particulate matter suspended in the air, such as dust, smoke, fog, and microorganisms, etc. [6].

(2) $PM_{2.5}$

$PM_{2.5}$ refers to particulate matter in the ambient air with an aerodynamic equivalent diameter less than or equal to 2.5 μm.

(3) Dust

Tiny solid particles suspended in the air, generally produced by the mechanical crushing of solid particles.

(4) Smoke

Tiny solid particles suspended in flue gas, generally produced by the condensation of gas or steam, with a particle size usually smaller than dust.

(5) Aerosol

Aerosol refers to a gaseous dispersion system composed of solid or liquid particles suspended in a gas medium, with a particle size between 0.01 and 10 μm.

(6) Oil-based Aerosol

Aerosols suspended in the air produced by oil smoke, oil mist, asphalt smoke, coke oven smoke, diesel exhaust.

(7) Non-oil-based Aerosol

Aerosols suspended in the air produced by coal dust, cement dust, construction site dust, acid mist, and inorganic salts, etc.

(8) Volatile Organic Compounds (VOCs)

Organic compounds that participate in atmospheric photochemical reactions, or organic compounds determined according to relevant regulations [7].

(9) Hygroscopic Dust

Refers to particulate matter that absorbs environmental moisture in the air, resulting in an increase in moisture content.

(10) Non-hygroscopic Dust

Refers to particulate matter that does not change its moisture content when left in the air for a long time.

(11) Inhalable Particles

Inhalable particles refer to particles that can easily enter the respiratory tract through the breathing process.

(12) Separation Membrane

The separation membrane is a thin layer attached to the outer surface of the support, with fine pore channels and surface filtration properties.

(13) Membrane Area

The effective area of the outer surface of the organic or inorganic support that performs separation [8].

(14) Gas Permeation Flux

The gas permeation flux is the amount of gas passing through per unit time and unit membrane area under given operating conditions, $m^3/(m^2\ h)$.

(15) Gas Permeance

Gas permeance is a way of characterizing the gas permeability of porous materials, also known as the relative permeability coefficient, $m^3/(h\ kPa\ m^2)$ [9].

(16) Filtration Speed

Filtration speed refers to the speed of the gas passing through the membrane filter element, m/min.

(17) Actual Gas Volume

The actual gas volume refers to the actual gas volume flow rate at the inlet and outlet of equipment such as air cleaning membrane, m^3/h [10].

(18) Standard Gas Volume

The standard gas volume refers to the gas volume flow rate at the inlet and outlet of equipment such as air cleaning membrane at a temperature of 273 K and a pressure of 101,325 Pa, m^3/h.

(19) Filter Pressure Drop

The filter pressure drop refers to the pressure drop of the gas passing through the filter element under standard conditions, with clean air as the filtering medium and a filter wind speed of 1 m/min, Pa.

(20) Separation Efficiency (R)

Separation efficiency refers to the percentage of dust retained when a dust-containing airflow passes through the separation membrane to the amount of dust at the inlet, % [11].

(21) Air Cleaning Efficiency

Air cleaning efficiency refers to the air pollutant cleaning speed and ability of an air purifier to a quantified pollution source in a certain indoor space and a certain working time [12].

(22) Clean Air Delivery Rate (CADR)

The clean air delivery rate refers to the total amount of clean air output through the purifier per unit time, m^3/h.

(23) High Efficiency Particulate Air Filter (HEPA)

A high efficiency particulate air filter refers to a filter that, for small particles of size 0.3 μm, such as dust, pollen, mold, bacteria, etc., measured by mass concentration, has a filtration efficiency of not less than 99.97%.

(24) Quality Factor (Q_f)

The quality factor refers to the filtration performance of different filtering materials [13], expressed as follows:

$Q_f = -\ln(1-R)/\Delta P$, where R is the separation efficiency, %; ΔP is the filtration pressure drop of the membrane material under a certain gas flow rate, Pa.

1.2 Classification of Gas Cleaning Membranes

According to the targeted pollutants, gas cleaning membranes are mainly divided into three types: gas dust removal membranes (also known as gas–solid separation membranes), gaseous pollutant cleaning membranes (also known as gas separation membranes), and gas–solid pollutant synergistic cleaning membranes. Among them, gas dust removal membranes are further divided into low and medium gas dust removal membranes and high temperature gas dust removal membranes [14].

1.2.1 Gas Dust Removal Membranes

1.2.1.1 Low- and Medium-Temperature Use Dust Removal Membranes

Low- and medium-temperature dust removal membranes usually refer to organic composite membrane used in gas cleaning systems with working temperatures below

260 °C. The organic composite membrane consists of a support layer and a separation layer. The support layer serves to increase the strength of the membrane material, and the materials of the support layer mainly include polyphenylene sulfide (PPS), polyimide fiber (P84), polyester (PET), glass fiber, and polytetrafluoroethylene (PTFE), etc.; the separation layer is laminated on the support layer, with a thickness generally ranging from several hundred nanometers to several micrometers, and the preparation methods mainly include thermal stretching, impregnation, and electrospinning. A common membrane material is PTFE membrane, capable of sustained operation at temperatures around 200 °C. PTFE resin is pretreated and expanded at high temperatures, then obtained into fiber-like materials with diameters ranging from several hundred nanometers to several micrometers through processes such as rolling and stretching, and finally processed into industrial and civilian gas-cleaning membrane products [15]. Membrane materials generally prepared by stretching or electrospinning also include polylactic acid (PLA), polyamide (PA), polyacrylonitrile (PAN), polyester (PET), polymethyl methacrylate (PMMA), polyetherimide (PEI), polyethylene oxide (PEO), polyvinyl alcohol (PVA), polyvinylpyrrolidone (PVP), cellulose acetate (CA), and polysulfone (PSU), etc.

The first patent on PTFE membranes was applied by W. L. Gore & Associates, Inc. in 1976. After a long period of technology accumulation, it has long been in a dominant position in the field of PTFE membrane manufacturing technology. Its products have been widely used in coal gasification, waste incineration, waste pyrolysis, regeneration of black metal melting, precious metal recovery, polycrystalline silicon production, fluidized bed catalyst purification, boiler devices, chemical manufacturing, and fermentation processes and other industrial fields [16, 17]. The development and research of organic composite membranes for water treatment is earlier than that for gas cleaning. The Membrane Science and Technology Research Institute of Nanjing Tech University is one of the earliest institutions in China to engage in the preparation and application research of gas cleaning membranes. Jiangsu Jiulang High-tech Co., Ltd. has successfully developed high-performance amphiphobic PTFE membrane after years of tackling key problems, and has built a large-scale production line for gas-cleaning membranes. It has promoted more than 100 projects in industrial gas cleaning field and has taken the lead in developing civilian gas cleaning membrane products.

In recent years, nanofiber membranes prepared by electrospinning technology have received increasing attention. The fiber diameter of the membrane materials prepared by this method is between a few nanometers and a few hundred nanometers. The gas-cleaning membrane constructed by nanofibers has a high porosity and high specific surface area on the one hand, and on the other hand, the fiber diameter is comparable to the average free path of the gas (about 66 nm), which greatly reduces the filtration resistance and has become one of the important development directions for the preparation of high-efficiency and low-resistance gas filters [13].

Ultra-high molecular weight polyethylene (UPE) membranes have also entered the high-efficiency gas cleaning market. Apart from exhibiting efficient, low-resistance filtration performance, excellent chemical, surface, and physical characteristics, these membranes also demonstrate precise particle efficiency. UPE membranes

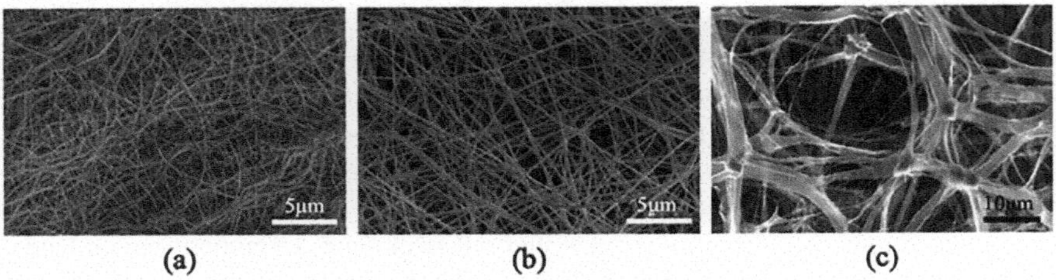

Fig. 1.1 SEM images of low and medium gas cleaning membranes. **a** Biaxially stretched PTFE membrane; **b** electrospun PAN membrane; and **c** UPE membrane

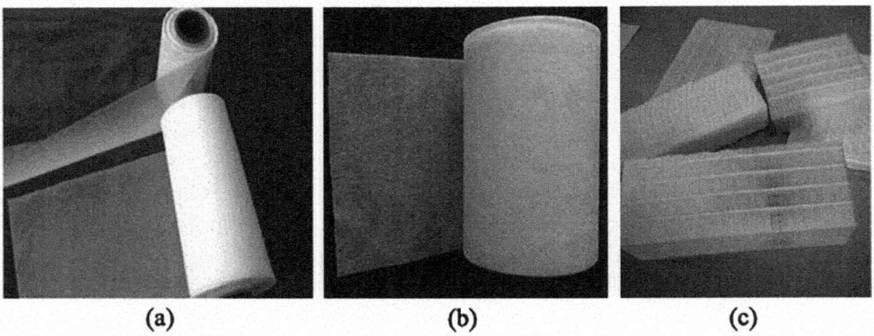

Fig. 1.2 Physical photos of gas cleaning membrane products. **a** Biaxially stretched PTFE membrane; **b** electrospun PAN membrane; and **c** UPE membrane

are environmentally friendly products because their waste will not cause adverse effects on the environment when disposed of by combustion due to their chemical composition of carbon and hydrogen. Because of its good wear resistance and better bonding with the substrate, UPE membranes can be combined with ultra-fine glass fibers or high-strength PET when used as air filter materials, and the products can be processed into pleated filters or pleated filter elements, and have gradually entered the gas cleaning market. However, their resistance to high temperatures and chemical corrosion is weaker compared to expanded polytetrafluoroethylene membranes. Currently, major manufacturers include Lydall Filtration, Millipore, and Gore in the United States. Figures 1.1 and 1.2 are the SEM photos and physical photos of biaxially stretched PTFE membranes, electrospun PAN membranes, and UPE membranes, respectively.

1.2.1.2 High Temperature-Use Dust Removal Membranes

In the industrial processes of energy, metallurgy, chemical, cement and the like, plenty of flue gas with high-temperature will be produced. These flue gases contain harmful components such as dust, nitrogen oxides, and sulfur oxides. Direct discharge will cause a large amount of heat loss and serious environmental pollution. High-temperature gas dust removal is a technology that directly carries out gas–solid separation under high-temperature (> 260 °C) conditions to achieve gas cleaning. It can efficiently harness the sensible heat, latent heat, kinetic energy, and other valuable resources present in the gas. The core of high-temperature gas cleaning is the material, which directly affects whether the dust removal equipment can operate efficiently and stably and determines the service life of the entire filter. Therefore, the development of high-temperature dust removal membrane materials is of great significance [18].

The development and research of high-temperature dust removal membranes began in the 1970s. The U.S. Department of Energy initiated the development and research of high-temperature gas filtration technology using inorganic membranes as the filtering medium. Subsequently, developed countries such as Germany, Japan, and the UK also carried out similar research. The main purpose of developing this technology at that time was to commercialize the "coal clean combustion combined cycle power generation technology", which is known as a cross-century new technology, to solve the wear problem of gas turbine blades, extend the life of gas turbine blades, and improve work efficiency. After a period of development, by the mid-1990s, significant progress had been made in high-temperature gas filtration technology. Firstly, the development of advanced high-performance inorganic membrane filtration materials laid the foundation for the industrial application of high-temperature gas filtration technology; secondly, the improvement of high-temperature dust removal process and technology, such as high-temperature sealing and filter element regeneration technology, etc., greatly promoted the industrial application of high-temperature gas filtration technology [19, 20].

The main features of high-temperature gas cleaning membranes are high-temperature resistance, corrosion resistance, thermal shock resistance, and long service life. The membrane material can directly trap dust in flue gas above 300 °C, and the dust concentration of the outlet gas is less than 5 mg/m^3. High-temperature-use dust removal membranes are mainly divided into porous ceramic membranes and porous metal membranes. Porous ceramic membranes have gradually developed from alumina and cordierite materials to large-pore silicon carbide material. The Pall Schumacher was one of the earliest companies that developed silicon carbide ceramic composite membranes. The preparation process mainly involves cold isostatic pressing, and its typical products are Dia-Schumalith F-20 and Dia-Schumalith F-40. The Dia-Schumalith F-20 product is mainly made of silicon carbide particles with an average particle size of 180 μm, clay as a binder, a support layer is prepared by high-temperature sintering, and then silicon carbide fine particles with an average particle size of 22 μm are used as the membrane layer. The porosity of the support body reaches 37%, the flexural strength reaches more than 30 MPa, the pore size of the membrane layer is 10–20 μm, and the use temperature can reach above 1000 °C [21]. Since the 1990s, universities and research institutes such as Nanjing

Tech University, University of Science and Technology of China, Tsinghua University, Hainan University, Shandong Industrial Ceramic Design Institute, and Institute of Metal Research, and Chinese Academy of Sciences have conducted in-depth research on high-temperature gas filtration membrane materials and technology. They have successfully developed alumina, mullite and cordierite, silicon carbide and other porous ceramic membranes and performed industrial pilot-scale applications [22]. In recent years, the high-temperature dust removal silicon carbide membrane independently developed by Nanjing Tech University has good stability. It uses in-situ reaction sintering to prepare a double-layer structure membrane. The inner layer is a support body with a larger average pore size to ensure the strength of the filter tube. The separation layer with a smaller average pore size is prepared on the outer surface of the support (as shown in Fig. 1.3). The pore size of the support body is relatively large, the porosity is generally greater than 40%, the average pore size is 40–60 μm, the average pore size of the separation layer can be adjusted within the range of 1–10 μm, and the appropriate membrane pore size can be selected according to the actual dust particle size [4, 18, 23–27], the structural parameters of the silicon carbide membrane are shown in Table 1.1.

The most significant advantage of porous metal membrane filters lies in their excellent temperature resistance and mechanical properties. The resilience and thermal

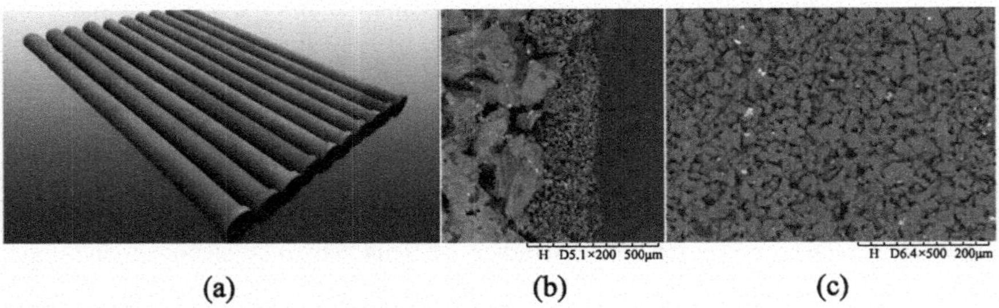

Fig. 1.3 Silicon carbide ceramic membrane (**a**), and their SEM images of cross-section (**b**) and surface micrograph (**c**)

Table 1.1 Typical parameters of two high-temperature gas dust removal films

Parameter	Silicon carbide membrane	Titanium aluminum alloy membrane
Support porosity/%	30–50	30–45
Support pore size/μm	40–60	20–30
Membrane layer porosity/%	45–50	30–40
Membrane layer pore size/μm	1–10	0.5–10
Membrane layer thickness/μm	50–200	100–200

Fig. 1.4 Fe_3Al high-temperature gas cleaning membranes

conductivity inherent in metal materials provide exceptional heat and shock resistance. Additionally, porous metal membrane materials exhibit favorable processability and weldability. However, there are still some problems with porous metal membranes in high-temperature gas filtration: metal filter materials generally show a decrease in strength with increasing temperature at high temperatures. Therefore, there is a maximum temperature limit in use. In recent years, countries around the world have vigorously carried out research on high-performance metal filter materials, among which Fe–Al intermetallic compounds and 310S stainless steel (high cadmium and nickel content) have attracted much attention for their outstanding high-temperature oxidation resistance and sulfur corrosion resistance. For example, the 310S, Inconel 600 (nickel-based alloy), Fe–Al intermetallic compounds and other sintered metal filter tubes produced by American Mott and Pall companies, which can withstand temperatures of 600–900 °C [28, 29]. Chinese research institutions such as Antai Technology Company, Northwestern Polytechnical University, Chengdu Yitai Technology Company, and others have conducted research and development on advanced metal filtration materials like Fe_3Al and 310S. They have also carried out pilot-scale studies on the high-temperature gas dedusting using sintered 310S metal mesh, achieving successful industrial applications. The product image of Fe_3Al membrane is shown in Fig. 1.4, and its structural parameters are shown in Table 1.1.

1.2.2 Gas Phase Component Treatment Membranes

Gas phase component treatment membrane refers to a type of membrane material that can separate certain components from a mixture of various gas components under certain conditions. This book mainly introduces two types of membrane materials for gas pollutant treatment, namely, desulfurization membrane and VOCs separation membrane.

1.2.2.1 Desulfurization Membranes

Membrane gas absorption (MGA) is a new type of separation technology that combines membrane separation with traditional chemical absorption, mainly using hollow fiber ceramic membranes and small pore ceramic membranes. In the mid-1980s, Qi and Cussler first fully proposed the concept of membrane absorption, replacing traditional packed towers with hollow fiber membrane contactors to study the mass transfer process of acidic gases such as SO_2, H_2S and CO_2. Since then, membrane absorption technology has developed rapidly and has been widely used in large-scale applications such as SO_2 and H_2S acid gas removal, CO_2 capture and resource utilization, ammonia recovery in the synthetic ammonia industry, and removal of volatile organic waste gases.

Membrane absorption combines the compact structure of the membrane and the high-efficiency selection of chemical absorption, and is a new type of membrane technology with great development prospects [30]. Currently, membrane materials employed for separating SO_2 consist predominantly of organic and inorganic membranes. Among these, organic membranes like polypropylene (PP) [31], polyvinylidene fluoride (PVDF) [32], polytetrafluoroethylene (PTFE) [33], and polyetheretherketone (PEEK) [34] stand out as extensively researched options. Inorganic membranes are mainly tubular ceramic membranes such as alumina and zirconia. Compared with traditional packed towers, ceramic membranes have a smaller mass transfer unit height [35]. Figure 1.5 shows the separation membrane product used for the absorption treatment of total sulfur compounds in gaseous pollutants, which is made of tubular alumina microfiltration membrane.

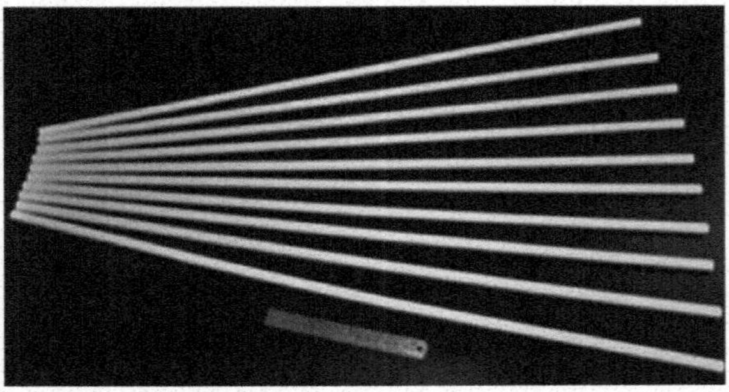

Fig. 1.5 Tubular alumina microfiltration membranes

1.2.2.2 VOCs Removal Membranes

Volatile organic compounds (VOCs), mainly come from industrial waste gases and vehicle exhaust produced by fuel combustion and transportation. VOCs have complex sources and a wide variety and are the main contributors to the formation of fine particulate matter ($PM_{2.5}$), ozone (O_3), and other secondary pollutants, which can cause atmospheric environmental problems such as haze and photochemical smog. Therefore, the removal of VOCs has received increasing attention in recent years. At present, common VOC removal technologies include catalytic combustion, biodegradation, photocatalytic decomposition, plasma destruction, and membrane separation technology [36]. Among them, membrane separation technology, due to its high separation selectivity, easy integration, environmental friendliness, and low energy consumption, is considered as one of the most promising separation technologies and has been widely used in the control of VOCs in the chemical, pharmaceutical, energy, and printing industries [37]. According to different separation mechanisms, membranes can be roughly divided into microporous membranes that follow the molecular sieve separation mechanism and dense membranes that follow the dissolution-diffusion mechanism. The common VOC removal membranes mainly include three types: organic polymer membranes, inorganic composite membranes, and mixed matrix membranes. The organic polymer membrane is mainly PDMS (polydimethylsiloxane), which has a—$Si(CH_3)_2O$—general structure and is also one of the most widely used membrane materials in VOC removal membranes [38]. Inorganic membranes mainly include molecular sieve membranes and MOFs (Metal–Organic Frameworks) membranes [39, 40]. The mixed matrix membrane is a new type of VOC removal membrane that combines the advantages of the first two types of membrane materials. For example, some researchers have hydrophobically modified ZIF-8 particles with 5, 6-dimethylbenzimidazole (DMBIM) and introduced them into the PDMS membrane, preparing a mixed matrix membrane with high separation performance. The selectivity of the PDMS mixed matrix membrane in separating propane/nitrogen increased by 116% compared to pure PDMS, and the permeability increased by 91% [41]. Figure 1.6 is a roll-type PDMS membrane and components used for VOC removal and recovery treatment.

Fig. 1.6 Roll-type PDMS membrane and components for VOCs removal and recovery

1.2.3 Gas–Solid Phase Pollutant Synergistic Cleaning Membranes

According to the separation mechanism and application function of the membrane material, the gas–solid phase pollutant synergistic cleaning membrane is mainly divided into catalytic membranes and adsorptive membranes.

1.2.3.1 Catalytic Membranes

The catalytic membrane refers to a special multifunctional separation membrane material that can simultaneously remove dust in flue gas and catalytically degrade NO_x, SO_x, VOCs and other air pollutants [42, 43]. The catalytic membrane primarily involves embedding the active catalyst precursor into the high-porosity membrane layer or the three-dimensional pore channels of the support using a loading method. Subsequently, active components are obtained through heat treatment, ultimately yielding a separation membrane material with catalytic capabilities. Catalytic membranes are mainly divided into two types, one is using inorganic membranes as carriers, also known as ceramic catalytic membranes (suitable for high-temperature flue gas), which not only have high mechanical strength and good microporous performance, but also have good high-temperature thermal stability, high-temperature corrosion resistance, and lower air resistance, etc. Researchers have carried out a series of studies on VOCs cleaning membranes [44], NO_x-dust cooperative cleaning membrane [45], and achieving significant breakthroughs. At present, this technology has been applied on a small scale of pilot application in China. The denitrification catalyst is evenly distributed in the membrane layer and the support pore channel, and could remove dust and NO_x cooperatively. This type of ceramic catalytic membrane has high filtration accuracy, low ash layer adhesion, good ash cleaning performance, high denitrification efficiency, and the dust content in the filtered gas is less than 2 mg/m^3, and nitrogen oxides are less than 50 mg/m^3. However, there are still problems with the long-term stability of the catalyst in this technology, and it is necessary to further improve the resistance to sulfide poisoning. The other type of catalytic membrane uses organic composite membranes as carriers. The principle is similar to that of ceramic catalytic membranes, and the catalyst is loaded on the fiber to play a catalytic role [46, 47]. It later developed into separating the catalyst from the membrane and placing the catalyst on the inside of the membrane filter bag. In summary, compared with selective catalytic reduction (SCR), the denitrification efficiency still needs to be improved, and this type of catalytic membrane is only suitable for low and medium catalytic processes, generally below 260 °C. Figure 1.7 is a ceramic catalytic membrane product used for the cooperative removal of dust and nitrogen oxides. While this product has demonstrated successful application cases in glass kiln furnaces, it still requires some time before it can be widely implemented on a large scale.

Fig. 1.7 Dust removal and denitrification integrated ceramic fiber catalytic tubular membrane [48]

1.2.3.2 Adsorptive Membranes

The adsorptive membrane is a special separation membrane material that uses porous materials as carriers, loads high specific surface area adsorbents in three-dimensional pore channels, and adsorbs and removes gaseous pollutant molecules when the gas passes through the microscopic pore channels. As the carrier of the adsorptive membrane, it not only needs to have a high porosity but also a coherent three-dimensional pore channel. The commonly used porous carriers are mainly divided into inorganic and organic materials. The inorganic materials are silicon carbide, alumina, activated carbon, etc. [49, 50]; organic materials generally use medium and low-temperature dust removal membrane materials, such as PTFE, PEI, PAN, etc. [51, 52]. The selection of adsorbent in the preparation of adsorptive membranes is essential. Usually, the adsorbent is chemically grafted to carry functional groups, which can enhance the adsorption capacity for different target pollutants, such as amino chain MIL-53(Al)-NH_2 or UiO-66-NH_2. These functional groups not only have adsorption capacity for formaldehyde but also have strong adsorption capacity for gases such as SO_2 and O_3. After a certain period of adsorption saturation, the harmful gases are desorbed by heating or pressure swing adsorption. Feng et al. [53] combined filter coating technology with a hydrothermal method to prepare UiO-66-NH_2@CNT/PTFE composite membrane, which showed excellent SO_2 adsorption performance at a filtration speed of 2.0 m/min, with the filtration pressure drop maintained at 160 Pa. In addition, the construction of multi-level pores can effectively increase the load of adsorbents, increase the residence time of pollutants in the membrane pores, and further enhance the adsorption capacity for pollutants. Hu et al. [54] filled layered MnO_2 crystals into the pores of PS nanofiber membranes with multi-level pore structures. The mesopores of the MnO_2/PS membrane promoted the contact between formaldehyde gas and MnO_2, achieving a formaldehyde removal efficiency of 88.2% and an airflow resistance of only 82 Pa. Figure 1.8 shows a schematic diagram of gas filtration and adsorption principles.

Currently, this type of membrane material has not been fully commercialized on a large scale, and many studies are still in the laboratory stage, mainly due to the high production cost. In addition, in civilian use, long-term adsorption effects are

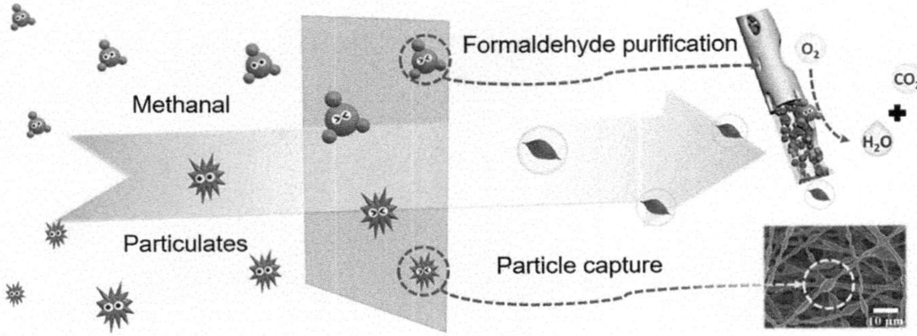

Fig. 1.8 Schematic diagram of gas filtration and adsorption principles of MnO_2/PS membrane [54]

unstable, and the problem of secondary desorption of harmful substances urgently needs to be solved.

In the future, adsorptive membranes will mainly be used for indoor formaldehyde removal. Traditional methods usually use activated carbon layer + HEPA. Still, this method also has the problem of secondary release of formaldehyde pollution in activated carbon, so finding more efficient adsorption materials is of great significance. Adsorptive membranes have multi-level sub-micron channels, which can quickly filter gases, not only effectively intercept $PM_{2.5}$, but also have an efficient adsorption effect on formaldehyde. Therefore, this multifunctional adsorptive membrane material shows good application prospects in gas cleaning filter materials and household screens.

1.3 Principle of Gas–Solid Separation

In the process of gas–solid separation, the filtration forms of dust-containing gas mainly include internal filtration and surface filtration. Internal filtration, also known as deep filtration, is the process of dust-containing gas passing through clean filter material. At this time, the main role of filtration is the structure of the filter material itself; when the filter material reaches a certain dust capacity, the dust layer formed on the surface of the filter material will play a major filtration role for the dust-containing gas. For thick and fluffy filter layers with a large porosity, deep filtration is more obvious; for thin and tight filter layers with a smaller porosity, it mainly manifests as surface filtration. The principle of membrane gas–solid separation is mainly surface filtration, dominated by the controllable structure of the separation layer.

1.3.1 Gas Filtration Principle

Inorganic membranes (ceramic, metal) for filtering dust in high-temperature flue gas and organic membranes (fiber) for purifying gas all belong to the same filtration theory. In the realm of filtration theory, both inorganic membranes (like ceramic and metal) utilized for filtering dust in high-temperature flue gas and organic membranes (such as nanofiber membranes) employed for purifying gas operate within the same framework. In the analysis of filtration mechanisms, it is often assumed that within a single collector, each collision between dust particles and the collector leads to the separation of dust from the airflow. There are many mechanisms that drive dust to collide with the dust collector, and the common separation mechanisms include direct interception, inertial collision, diffusion effect, gravitational effect, electrostatic effect, etc. When the airflow carries particles past a single collector, the possible collection mechanisms are shown in Fig. 1.9. As can be seen from Fig. 1.9: when the airflow bypasses the collector, the streamline deflects; at this time, larger particles continue to move forward in a straight line due to inertia. Therefore, they will deviate from the streamline and hit the collector, which is called the "inertial collision" effect; at the same time, some particles with smaller inertia still follow the streamline, if at this time the particle radius is smaller than the distance from the streamline to the collector surface, they will be collected, which is called the "direct interception" effect; finer particles in the airflow are hit by gas molecules and do not follow the streamline evenly, but make Brownian motion in the airflow, due to this irregular thermal motion, particles may collide with the dust collector and be collected, which is called the "diffusion effect". In addition, if there is an external force such as gravity, electrostatic force, etc., particles will settle and may also hit the collector, which is called "gravity effect" and "electrostatic effect", respectively.

The above discussion is about the collection mechanism of isolated dust collectors. For filters, another important mechanism for particle collection is sieving (Fig. 1.10), that is, dust particles are too large to be collected through specific holes and channels.

1.3.1.1 Inertial Collision

Inertial collision is the most important of all collection mechanisms, especially for particles with $d_p > 1$ μm. Among various effects, the dominant one is the inertia of particles, which depends on the mass and speed of particles, so in the case of neglecting gravity and electrostatic forces, all particles within this limit trajectory can be collected by inertial collision. For a cylinder:

$$\eta_I = D_I / d_f \tag{1.1}$$

where η_I is the collection efficiency of inertial collision; D_I is the range size of the limit trajectory far from the capture group; d_f is the diameter of the capture group. Assuming the particle size d_p is much smaller than d_f, the particle can be regarded

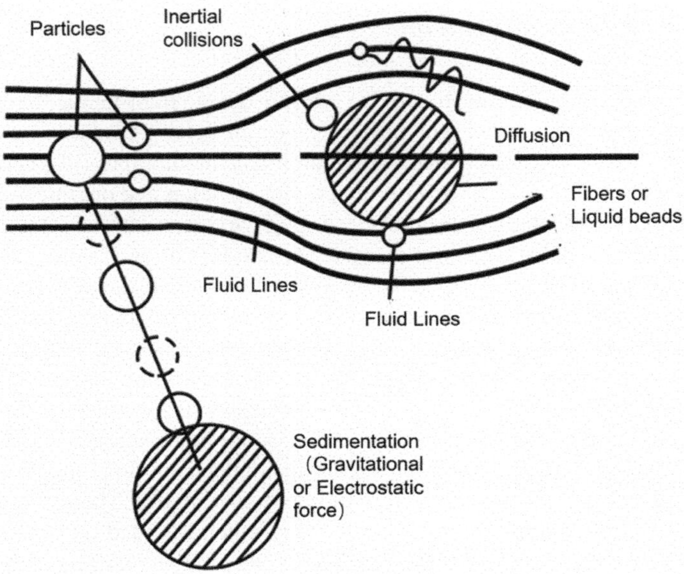

Fig. 1.9 Dust collection mechanism

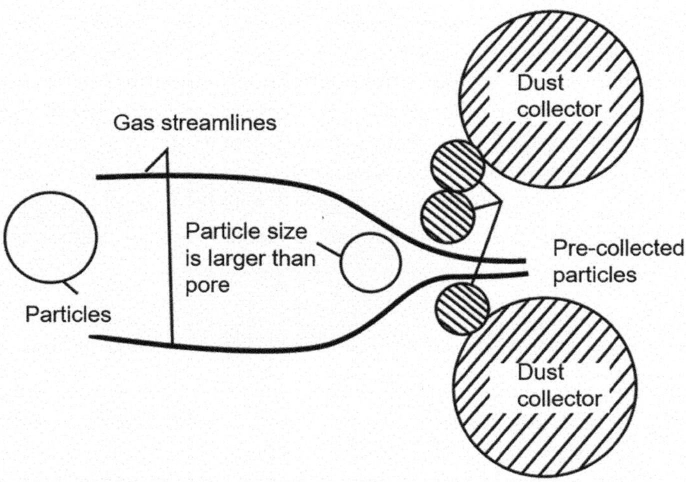

Fig. 1.10 Dust collection mechanism—sieving

as having no size but only mass, thus obtaining various formulas for calculating η_I, see Table 1.2.

Table 1.2 Formulas for the calculation of inertial collision capture efficiency η [55]

Collector	Flow state	Formula	No
Cylinder	Laminar flow ($\mathrm{Re} \leq 1$)	$\eta_I = 1 - \frac{1.2}{\mathrm{Re}^{0.2} Stk^{0.54}} + \frac{0.36}{\mathrm{Re}^{0.4} Stk^{1.08}}$	(1.2)
	Potential flow ($\mathrm{Re} \to \infty$)	$\eta_I = \frac{Stk}{Stk+1.5}$	(1.3)
	Transition zone ($\mathrm{Re} = 10$)	$\eta_\mathrm{I} = \frac{Stk^3}{Stk^3+0.77Stk^2+0.22}$	(1.4)

Note Re—Reynolds number; *Stk*—Stokes number

Table 1.3 Formulas for calculation of direct interception efficiency η_R [56]

Collector	Flow state	Formula	No
Cylinder	Laminar flow	$\eta_R = \frac{1}{(2.002-\ln \mathrm{Re})}\left[(1+R)\ln(1+R) - \frac{R(2+R)}{2(1+R)}\right]$	(1.5)
		When $R < 0.07$, Re is much less than 0.5, $\eta_\mathrm{R}=\mathrm{R}^2/(2.002\text{-}\ln\mathrm{Re})$	
	Potential flow	$\eta_R = \left(1 + R^2 - \frac{1}{(1+R)}\right)$(If $R < 0.1$, $\eta_\mathrm{R} = 2R$)	(1.6)

1.3.1.2 Direct Interception

The direct interception mechanism assumes that particles have size but no mass. Therefore, particles of different sizes move along the streamline of the airflow. When the airflow encounters the capture body, the streamline deflects, and if the particle size d_p is less than the distance from the streamline to the surface of the capture body, the particle will also be captured. When the ratio of particle size d_p to the size of the capture body R is relatively large, the direct interception effect is more prominent. When calculating the interception efficiency alone, the interception efficiency η_R is not related to the Stokes number *Stk*, but is a function of the airflow streamline and R. The calculation formulas of η_R proposed by various scholars are shown in Table 1.3.

1.3.1.3 Diffusion Effect

As the particle size decreases, the flow rate slows down, the temperature increases, the thermal motion of the particles accelerates, and the probability of collision with the dust collector also increases, and the diffusion effect becomes more significant. Studies have shown that for particles with $d_\mathrm{p} \leq 1\,\mu\mathrm{m}$, the diffusion mechanism plays a dominant role. Under the diffusion effect, particles will be collected by the collector when they are close to it. Hence, the diffusion collection efficiency η_D is generally mainly related to the dimensionless parameters *Pe* and Re. The calculation formulas for collection efficiency under different flow states are shown in Table 1.4.

Table 1.4 Formulas for calculation of collection efficiency η_D under diffusion action [57]

Collector	Flow state	Formula	No
Cylinder	Laminar flow (Re < 1, $Pe \ll 1$)	$\eta_D = \frac{2\pi}{Pe(1.502 - \ln Pe)}$	(1.7)
	Laminar flow (Re < 1, $Pe \geq 1$)	$\eta_D = 2.9L_a^{-\frac{1}{3}} Pe^{-\frac{2}{3}} + 0.624Pe^{-1}$	(1.8)
		$\eta_D = 2^{L_a^{-1}}\left[2(1+x)\ln(1+x) + \frac{1}{(1+x)}\right]$ $x = 1.308\left(\frac{L_a}{Pe}\right)^{\frac{1}{8}}$, $L_a = 2.002 - \ln \text{Re}$	(1.9)
	Laminar flow (Re < 1, Pe is large)	$\eta_D = K^{L_a^{-\frac{1}{3}}} P^{-\frac{2}{3}}$, $K = 1.71 \sim 2.92$	(1.10)
	Potential flow (Re $\gg$ 1, $Pe \gg 1$)	$\eta_D = 2.05L_a^{-0.4}\text{Pe}^{-0.6}$ $\eta_D = KPe^{-\frac{1}{2}}$, $K = 1.57 \sim 3.19$	(1.11) (1.12)
	$0.1 < \text{Re} < 10^4$ $10 < Pe < 10^5$	$\eta_D = \frac{\pi}{Pe}\left(\frac{1}{\pi} + 0.55\text{Re}^{\frac{1}{6}} Pe^{\frac{1}{3}}\right)$ or $\eta_D = \frac{1}{Pe} + 1.727R^{e^{\frac{1}{6}}} Pe^{-\frac{2}{3}}$	(1.13)

1.3.1.4 Gravity Settling

Gravity separation is the most basic type of separation. When particles have a certain size and density, and the flow rate is low, the particles will settle on the dust collector due to gravity, which is the gravity effect. For a horizontal cylindrical collector, Rauz and Wang proposed that the gravity settling collection efficiency η_G should be:

$$\eta_G = G = \frac{u_s}{v_0} = \frac{C_u \rho_p d_p^2 g}{18\mu v_0} \tag{1.14}$$

where u_s is the *Stk* settling velocity, m/s; v_0 is the average flow velocity, m/s; μ is the gas viscosity coefficient, Pa·s; ρ_p is the particle density, kg/m^3; d_p is the particle diameter, m; g is the gravity acceleration, m/s^2; and C_u is the Cummingham correction factor.

As can be seen, only when the particles are large and the gas speed is low, the effect of gravity settling is more significant. Equation (1.14) is the case when the direction of airflow is the same as the direction of gravity. For any horizontally placed cylinder, the value in the above equation should be multiplied by the ratio of the projection area of the cylinder perpendicular to the direction of the airflow to the projection area along the direction of the airflow.

1.3.1.5 Electrostatic Effect

Both general dust and filter materials may carry charges, and when they carry opposite charges, the electrostatic attraction effect becomes apparent, improving the dust filtration efficiency but making dust removal difficult. When no external electrostatic field is applied, this charge will slowly decrease due to factors such as the conductivity of the collection body, the passage of ionized gas molecules, radioactive irradiation, and the settling of charged particles. For particles that obey Stokes' law, the electrostatic force can be written as the dimensionless parameter K_{E}, and the electrostatic collection efficiency η_{E} is a function of K_{E}. Their calculation formulas are listed in Table 1.5.

Table 1.5 Formulas for calculation of electrostatic attraction effects [56]

Charge situation	Parameter	Cylinder	Formula number
Particle charged, collector neutral	F_{EM}	$\eta_{\mathrm{E}} = \dfrac{\left(\frac{\varepsilon_{\mathrm{C}}-1}{\varepsilon_{\mathrm{C}}+1}\right)q^2}{4\varepsilon_0\left(r-\frac{d_{\mathrm{f}}}{2}\right)^2}$	(1.15)
	K_{EM}	$\eta_{\mathrm{E}} = \left(\frac{\varepsilon_{\mathrm{c}}-1}{\varepsilon_{\mathrm{c}}+1}\right)\frac{c_{\mathrm{u}}^2 q}{3\pi u d_{\mathrm{p}} d_{\mathrm{f}}^2 \varepsilon_0 v_0}$	(1.16)
	η_{EM}	Laminar flow: $\eta_{\mathrm{E}} = 2\sqrt{\frac{K_{\mathrm{EM}}}{La}}$	(1.17)
		Potential flow: $\eta_E = (6\pi K_{EM})^{1/3}$	(1.18)
Particle neutral, collector charged	F_{EI}	$\eta_E = \frac{4Q^2}{\varepsilon_0}\left(\frac{\varepsilon_{\mathrm{p}}-1}{\varepsilon_{\mathrm{p}}+2}\right)\frac{\left(\frac{d_{\mathrm{p}}}{2}\right)^3}{r^3}$	(1.19)
	K_{EI}	$\eta_{\mathrm{E}} = \frac{4}{3\pi}\left(\frac{\varepsilon_{\mathrm{p}}-1}{\varepsilon_{\mathrm{p}}+2}\right)\frac{C_{\mathrm{u}} d_{\mathrm{p}}^2 Q_{\mathrm{a}}^2}{d_{\mathrm{f}}^3 \mu v_0 \varepsilon_0}$	(1.20)
	η_{EI}	$\eta_E = \left(\frac{3\pi K_{\mathrm{EI}}}{2}\right)^{\frac{1}{3}}, \frac{2r}{d_{\mathrm{f}} \gg 1}$	(1.21)
Both charged	F_{EL}	$\eta_E = \frac{2Q_{\mathrm{b}}}{\varepsilon_0 r}$	(1.22)
	K_{EC}	$\eta_E = \frac{4Q_b q C_u}{3\pi \mu d_p d_f v_0 \varepsilon_0}$	(1.23)
	η_{EL}	$\eta_E = -\pi K_{EC}$	(1.24)

Note Q represents the charge on the collector, C; Q_{a} represents the charge per unit length on the collector, C; Q_{b} represents the charge per unit area on the collector, C; q represents the charge on the particle, C; ε_0 represents the permittivity of free space, $\varepsilon_0 = 8.85 \times 10^{-12}$ C/(V m); ε_p represents the permittivity of the particle, C/(V m); ε_c represents the permittivity of the collector, C/(V m); d_{f} represents the diameter of the collector, m; d_{p} represents the diameter of the particle, m; r represents the distance between the particle and the collector, m; and C_{u} represents the Cummingham correction factor

1.3.1.6 Synergistic Effects of Various PM Capture Mechanisms

In the practical filtration process, many parameters can affect the separation characteristics of particles, such as the separation forces involved in different separation mechanisms, the physical properties of gas–solid mixtures, fluid dynamics characteristics, particle size and distribution, and particle density, etc. Various collection mechanisms often coexist, and the collection efficiency of the synergistic effect can be approximately written as:

$$\eta_0 = 1 - (1 - \eta_I)(1 - \eta_D)(1 - \eta_R)(1 - \eta_G)(1 - \eta_E) \tag{1.25}$$

For various specific situations, different scholars have also proposed some different calculation methods.

(1) The combination of inertial collision and direct interception always coexists. For the cylinder under laminar flow conditions ($Re = 0.2$):

$$\eta_{IR} = 0.16\left[R + (0.5 + 0.8R)\,Stk - 0.1052(Stk)^2\right] \tag{1.26}$$

where η_{IR} represents the combined efficiency of direct interception and inertial collision. The calculation results of this formula can be seen in Fig. 1.11.

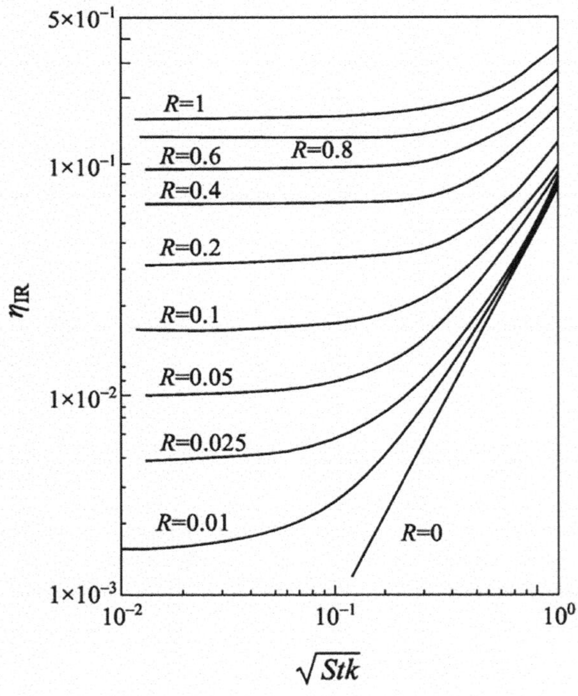

Fig. 1.11 Combined efficiency of inertial collision and direct interception

(2) For the combined effect of diffusion and direct interception of the cylinder, Langmuir proposed:

$$\eta_{DR} = \frac{1}{L_a\left[(1+2x)\ln(1+2x) - \frac{x(2+2x)}{1+2x}\right]} \tag{1.27}$$

where x can be obtained from the following formula:

$$x\left(x - \frac{R}{2}\right)^2 = 0.28L_a Pe^{-1} \tag{1.28}$$

Friedlander and Pasceri proposed a semi-empirical formula for laminar flow:

$$\eta_{DR} = 6\mathrm{Re}^{\frac{1}{6}} Pe^{-\frac{2}{3}} + 3\mathrm{Re}^2 Pe^{\frac{1}{2}} \tag{1.29}$$

Billings later pointed out that in some experiments, the dust removal efficiency of solid particles is higher than the above calculations, and when collecting droplets, it is more in line with the above formula, which may be due to the existence of electrostatic effects.

(3) For the combined effect of inertia, interception, and diffusion of the cylinder, Davies suggested that it is only necessary to add a $1/Pe$ item in the Stk item of formula (1.26), that is:

$$\eta_{IRD} = 0.16\left[R + (0.5 + 0.8R)\left(Stk + \frac{1}{Pe}\right)\right] - 0.16\left[0.1052R\left(Stk + \frac{1}{Pe}\right)^2\right] \tag{1.30}$$

1.3.2 Factors Affecting Gas Filtration

1.3.2.1 The Impact of Particulate Matter

The impact of particulate matter on gas filtration is mainly reflected in the shape, size, and concentration of dust particles. The shapes of particulate matter are divided into regular and irregular shapes. The surface of regular-shaped dust particles is smooth, with a small specific surface area, and is not easily intercepted when passing through the filter material; irregular-shaped particles not only have a rough surface, but also a larger specific surface area, and are easily intercepted when passing through the filter material, but they can cause wear on the filter bag or tubular membrane.

The impact of dust particle size on gas filtration is mainly the impact of the dust rejection process and dust collector. Generally, the larger the dust particle size, the slower the rate of pressure drop increase, and the higher the rejection rate; and the smaller the dust particle size, the faster the pressure drop increases, and the lower the rejection rate. This is because pore blockage occurs at the initial stage of filtration, and then particles are deposited on the surface of the filter material. Large particles are not easy to aggregate on the surface and are easy to remove. The pressure drop will slowly rise and then stabilize over a short period of time; as the dust particle size decreases, particles are more likely to enter the filter material channel and cause blockage, increasing the pressure drop. At the same time, the smaller the dust particle size, the greater the filter cake pressure drop, because the smaller the particle size, the smaller the voids in the filter cake [58].

The impact of dust concentration on gas filtration is mainly reflected in filtration pressure drop, filtration efficiency, wear of dust collector, and dust removal cycle. Within a single back-blow cycle, the filtration pressure drop basically increases linearly with time. As the inlet concentration rises, the filtration pressure drop also escalates. Higher dust concentrations lead to the deposition of a thicker filter cake on the membrane surface within a given timeframe, consequently accelerating the rate at which the pressure drop intensifies. Typically, during the initial stages of filtration, a higher inlet concentration correlates with an increased likelihood of small particles penetrating the filter material. This results in a lower filtration efficiency as smaller particles find pathways through the material, affecting the overall effectiveness of the filtration process. However, as time goes on, a filter cake is formed on the surface of the filter material. This layer of filter cake plays an actual filtration role, and the retention efficiency slowly rises and remains stable.

1.3.2.2 The Impact of Gas Properties

Industrial gases generally contain a certain amount of water vapor. In typical application scenarios such as coal-fired power plants, high-humidity dust will affect the structure of the filter cake by affecting the inter-particle forces, affecting the stable operation of the dust removal process and limiting the service life of the filter [59]. Therefore, studying the impact of humidity on inter-particle forces and dust removal performance is of significant importance for reducing the energy consumption of dust collectors. Typically, the size of inter-particle forces can be influenced by changing the environmental humidity, thereby altering the structure of the filter cake accumulation and reducing the resistance of the filter cake. Particles with stronger hydrophilicity exhibit a faster rate of change in filter cake resistance in response to environmental humidity variations. Conversely, particles with stronger hydrophobicity showcase a slower rate of change in filter cake resistance concerning shifts in environmental humidity levels [60, 61].

Corrosive gases contained in the industrial exhaust, such as SO_2 and alkali metal vapors, can corrode the membrane, resulting in a decrease in filtration efficiency, a reduction in filtration precision, and a shortening of the material's service life. Figure 1.12 shows the corrosion of the particle surface of the alumina ceramic membrane when filtering industrial kiln flue gas. As can be seen, after a long period of gas corrosion, many tiny particles appear on the surface of the alumina, and the degree

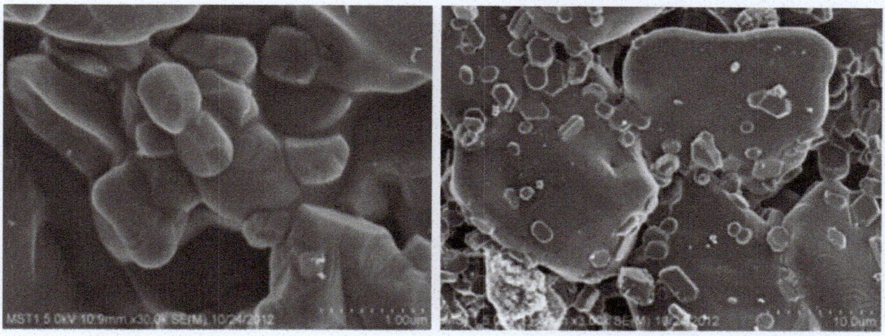

(a) (b)

Fig. 1.12 Microscopic morphology of ceramic material before (**a**) and after (**b**) corrosion by industrial kiln flue gas [62]

of neck binding is significantly reduced, resulting in a decrease in the mechanical strength of the ceramic membrane material [62].

1.3.2.3 Impact of Operating Conditions

An increase in gas temperature will promote an increase in the diffusion coefficient of particles and an increase in viscosity, thereby reducing the settling efficiency of large particles that rely on inertia and gravity effects, and small particles will also reduce the volume flow rate of gas and increase filtration resistance due to the intensification of Brownian motion [62, 63]. According to Darcy's law, an increase in pressure drop leads to an increase in the driving force for nitrogen permeation, so the gas flow rate will increase; while an increase in temperature leads to an increase in nitrogen viscosity, hence the permeation resistance of nitrogen increases, and the gas flow rate will decrease.

Figure 1.13 shows the effect of temperature on the gas permeance. As the temperature increases, the gas permeance tends to decrease. Due to the increase in temperature, the gas viscosity increases, hence the nitrogen permeance decreases. Generally, the larger the pore size of the gas–solid separation membrane, the greater the impact of temperature on the gas permeance. This phenomenon arises due to the relationship between temperature and molecular movement. Higher temperatures lead to larger molecular free paths, increasing the likelihood of molecular collisions with the pore wall. Consequently, at elevated temperatures, the probability of gas molecules passing through the pore decreases compared to lower temperatures, impacting the gas flow rate.

Gas-containing dust will form a layer of filter cake on the surface of the membrane during the filtration process. When the filter cake exceeds the limit, it will cause deformation, damage, a decrease in gas volume, and a decrease in the efficiency of the filter. Therefore, it is necessary to clean the filter cake to restore the filtration

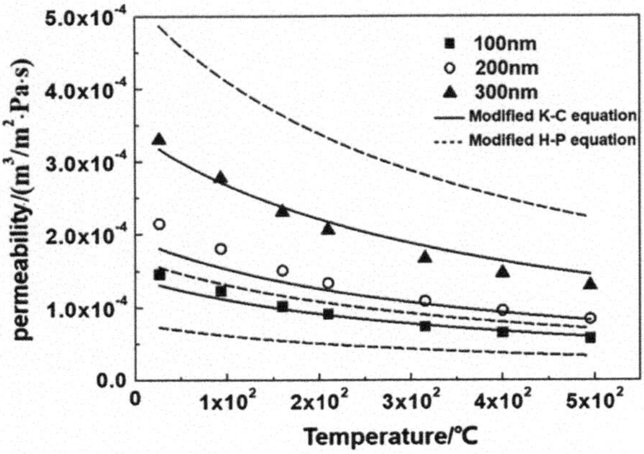

Fig. 1.13 The effect of temperature on gas permeance [62]

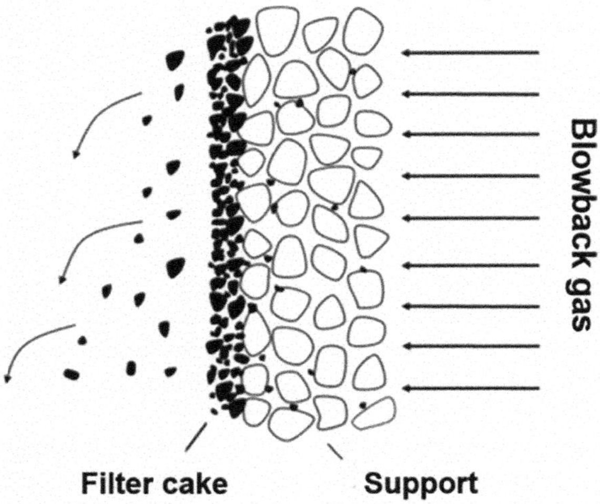

Fig. 1.14 Schematic diagram of back-blowing ash removal

performance [64, 65]. In such scenarios, integrating a back-blowing system within the dust collector is necessary for effective dust removal. This system incorporates a switching device that halts the filtration airflow. Utilizing external power generates a reverse airflow with adequate momentum. This reverse airflow induces vibrations or expansions and contractions on the surface of the filter material, effectively dislodging and peeling off the accumulated dust [66]. The process of back-blowing ash removal is shown in Fig. 1.14. When high-pressure gas travels through the pore channels, entering the component from the permeation side, it generates an impactful force. If this force surpasses the adhesion force holding the filter cake on the tubular membrane surface or the cohesion between the particles in the filter cake, the filter cake detaches from the membrane surface. This mechanism completes the cleaning process by dislodging and removing the accumulated filter cake [67, 68].

References

1. Xing WH, Gu XH. High performance membrane materials and membrane technology. Chemical Industry Press. 2017.
2. Hao XM. Polytetrafluoroethylene microporous membrane and fiber. Chemical Industry Press. 2000.
3. Tang LP, Wang JZ. Metal fiber porous materials: pore structure and properties. Metallurgical Industry Press. 2016.
4. Han F, Zhong ZX, Yang Y, et al. High gas permeability of SiC porous ceramics reinforced by mullite fibers. J Eur Ceram Soc. 2016;36:3909–17.
5. Du LH. $PM_{2.5}$ and gas cleaning technology. Chemical Industry Press. 2019.
6. National Safety Production supervision administration. Respiratory protection equipment—self-priming filtering anti-particulate respirator: GB 2626—2006. China Standard Press. 2006.
7. National Market Supervision Administration. Control standard for volatile organic compounds unorganized emission: GB/T 37822—2019. China Environment Press. 2019.
8. China National Standardization Management Committee. Silicon carbide high temperature ceramic filter elements: GB/T 32978—2016. China Standard Press. 2017.
9. China National Standardization Management Committee. Intermetallic compound membrane filters: GB/T 35250—2017. China Standard Press. 2018.
10. General Technical Specifications for Bag Dust Removal Engineering: HJ 2020. 2012.
11. Shi J, Yuan Q. Membrane technology handbook. Chemical Industry Press. 2001.
12. China National Standardization Management Committee. Air Purifiers: GB/T 18801-2015. China Standard Press. 2016.
13. Li P, Wang CY, Zhang YY, et al. Air Filtration in the free molecular flow regime: a review of high-efficiency particulate air filters based on carbon nanotubes. Small. 2014;10(22):4543–61.
14. Xu NP. Design, preparation and application of ceramic membrane materials for application process. Science Press. 2017.
15. Hu M, Zhong ZX, Xing WH. Research progress on the application of nanofiber membranes in air purification. Chem Ind Prog. 2018;37(4):1305–13.
16. Astakhov EY, Shutov AA. Porous polytetrafluoroethylene film. Phys Lett. 2007;33:228–30.
17. Feng SS, Zhong ZX, Drioli E, et al. Progress and perspectives in PTFE membrane: preparation, modification, and applications. J Membr Sci. 2018;549(1):332–49.
18. Han F, Zhong ZX, Xing WH, et al. Preparation and characterization of SiC whisker-reinforced SiC porous ceramics for hot gas filtration. Ind Eng Chem Res. 2015;54:226–32.
19. Zhang F, Han F, Zhong ZX, et al. Research progress on ceramic membranes for high-temperature exhaust gas cleaning. In: Proceedings of the fourth national conference on the application of membrane separation technology in the metallurgical industry. 2014.
20. Heidenreich S. Hot gas filtration: a review. Fuel. 2013;104:83–94.
21. Heidenreich S, Wolters C. Hot gas filter contributes to IGCC power plant's reliable operation. Filtr Sep. 2004;41(5):22–4.
22. Fu XN, Yao G, Liu M, et al. Application of porous ceramic materials in high-temperature gas dry dust removal. Environ Eng. 2012;30(3):49–54.
23. Wei W, Zhang WQ, Jiang Q, et al. Preparation of non-oxide SiC membrane for gas cleaning by spray coating. J Membr Sci. 2017;540:381–90.
24. Yang Y, Han F, Xu WQ, et al. Low-temperature sintering of porous silicon carbide ceramic support with SDBS as sintering aid. Ceram Int. 2017;43(3):3377–83.
25. Yang Y, Xu WQ, Zhang F, et al. Preparation of highly stable porous SiC membrane supports with enhanced air purification performance by recycling NaA zeolite residue. J Membr Sci. 2017;541:500–9.
26. Han F, Xu CN, Wei W, et al. Corrosion behaviors of porous reaction-bonded silicon carbide ceramics incorporated with CaO. Ceram Int. 2018;44(11):12225–32.
27. Xu CN, Xu C, Han F, et al. Fabrication of high performance macroporous tubular silicon carbide gas filters by extrusion method. Ceram Int. 2018;44(15):17792–9.

28. Dennis R, McMahon T, Dorchak T, et al. Department of energy's high temperature and high-pressure particulate clean up program for advanced coal-based power systems. High Temp Gas Cleaning. 1999;2(4):303–6.
29. Alvin MA. Hot gas filter development and performance. High Temp Gas Cleaning. 1992;21(4):455–7.
30. Zhao SF, Feron PM, Deng LY, et al. Status and progress of membrane contactors in post-combustion carbon capture: a state-of-the-art review of new developments. J Membr Sci. 2016;511:180–206.
31. Demontigny D, Tontiwachwuthikul P, Chakma A. Using polypropylene and polytetrafluoroethylene membranes in a membrane contactor for CO_2 absorption. J Membr Sci. 2006;277(1–2):99–107.
32. Atchariyawut S, Feng C, Wang R, et al. Effect of membrane structure on mass-transfer in the membrane gas-liquid contacting process using microporous PVDF hollow fibers. J Membr Sci. 2006;285(1–2):272–81.
33. Chen S, Li S, Chien R, et al. Effects of shape, porosity, and operating parameters on carbon dioxide recovery in polytetrafluoroethylene membranes. J Hazard Mater. 2010;179(1–3):692–700.
34. Li S, Pyrzynski TJ, Klinghoffer NB, et al. Scale-up of PEEK hollow fiber membrane contactor for post-combustion CO_2 capture. J Membr Sci. 2017;527:92–101.
35. Han SX, Gao XY, Fu KY, et al. Application of hydrophobic single-tube ceramic membrane contactor in SO_2 absorption. J Chem Ind Eng. 2017;68:2415–22.
36. Drioli E, Stankiewicz AI, Macedonio F. Membrane engineering in process intensification: an overview. J Membr Sci. 2011;380(1):1–8.
37. Ambrosi A, Cardozo NSM, Tessaro IC. Membrane separation processes for the beer industry: a review and state of the art. Food Bioprocess Technol. 2014;7(4):921–36.
38. Barrer RM, Chio HT. Solution and diffusion of gases and vapors in silicone rubber membranes. J Polym Sci Polym Symp. 2007;10(1):111–38.
39. Barrer RM. Zeolites and their synthesis. Zeolites. 1981;1(3):130–40.
40. Li J, Sculley J, Zhou H. Metal-organic frameworks for separations. Chem Rev. 2011;112(2):869–932.
41. Yuan JW, Li QQ, Shen J, et al. Hydrophobic-functionalized ZIF-8 nanoparticles incorporated PDMS membranes for high-selective separation of propane/nitrogen. Asia Pacific J Chem Eng. 2017;12(1):110–20.
42. Sandra F, Ballestero A, Nguyen VL, et al. Silicon carbide-based membranes with high soot particle filtration efficiency, durability and catalytic activity for CO/HC oxidation and soot combustion. J Membr Sci. 2016;501:79–92.
43. Mirvakili A, Bahrani S, Jahanmiri A. An environmentally friendly configuration for ammonium nitrate decomposition. Ind Eng Chem Res. 2013;52:13276–87.
44. Li C, Zhang F, Feng SF, et al. SiC@TiO_2/Pt catalytic membrane for collaborative removal of VOCs and nanoparticles. Ind Eng Chem Res. 2018;57(31):10564–71.
45. Chen JH, Pan B, Wang B, et al. Hydrothermal synthesis of a Pt/SAPO-34@SiC catalytic membrane for the simultaneous removal of NO and particulate matter. Ind Eng Chem Res. 2020;59:4302–12.
46. Zhao Y, Low ZX, Feng SS, et al. Multifunctional hybrid porous filters with hierarchical structures for simultaneous removal of indoor VOCs, dusts and microorganisms. Nanoscale. 2017;9:5433–44.
47. Li DY, Gu C, Han F, et al. Catalytic performance of hybrid Pt@ZnO NRs on carbon fibers for methanol electro-oxidation. Chin J Chem Eng. 2017;25(12):1871–6.
48. Fulikang Technology Co., Ltd. Dust removal and denitration integrated ceramic fiber catalytic membrane tube. [2021-07-01]. www.cat-filter.com/product/.
49. Li LL, Zhang F, Zhong ZX, et al. Novel synthesis of a high performance Pt/ZnO/SiC filter for the oxidation of toluene. Ind Eng Chem Res. 2017;56(46):13857–65.
50. Yang S, Zhu Z, Wei F, et al. Enhancement of formaldehyde removal by activated carbon fiber via in situ growth of carbon nanotubes. Build Environ. 2017;126:27–33.

51. Feng S, Li X, Zhao S, et al. Multifunctional metal organic framework and carbon nanotube-modified filter for combined ultrafine dust capture and SO_2 dynamic adsorption. Environ Sci Nano. 2018;5(12):3023–31.
52. Park KT, Hwang J. Filtration and inactivation of aerosolized bacteriophage MS_2 by a CNT air filter fabricated using electro-aerodynamic deposition. Carbon. 2014;75:401–10.
53. Feng SS, Li XY, Zhao SF, et al. Multifunctional metal organic framework and carbon nanotube-modified filter for combined ultrafine dust capture and SO_2 dynamic adsorption. Environ Sci-Nano. 2018;5(12):3023–31.
54. Hu M, Kang WM, Zhong ZX, et al. Porphyrin-functionalized hierarchical porous silica nanofiber membrane for rapid HCI gas detection. Ind Eng Chem Res. 2018;57(34):11668–74.
55. Zhou ZQ. Experimental study on the combustion of coal powder particles in airflow: collection of engineering thermophysics papers. Tsinghua University Press. 1985.
56. Tan TY. Industrial ventilation dust removal technology. China Architecture & Building Press. 1984.
57. Chemical Engineering Handbook Editorial Committee. Chemical engineering handbook: volume 5. Chemical Industry Press. 1989.
58. Liu W. Preparation and gas-solid separation performance study of silicon carbide porous ceramics. Nanjing: Nanjing University of Technology; 2014.
59. Li N, Yang FS, Zhang WD. Study on the effect of environmental humidity on the dust removal performance of PTFE coated filter material. Forum on Environmentally Friendly Chemical Engineering Technology. 2015.
60. Li N. Study on the effect of environmental humidity on the dust removal performance of different nature dust. Beijing University of Chemical Technology. 2013.
61. Yuan XL. Study on the effect of environmental humidity on the filtration performance of PTFE coated filter material. Beijing University of Chemical Technology. 2015.
62. Zhu JJ. Study on the high temperature stability of ceramic membrane materials. Nanjing University of Technology. 2013.
63. Luax S, et al. Hot gas filtration with ceramic filter elements. Process of the 12th International conference on FBC. ASME; 1993. pp. 1241–1250.
64. Heidenreich S, Haag W, Salinger M. Next generation of ceramic hot gas filter with safety fuses integrated in venturi ejectors. Fuel. 2013;108:19–23.
65. Zhong ZX, Li X, Xing WH, et al. Study on the dust removal performance of porous ceramic membrane. Environ Sci Technol. 2013;36(6):155–8.
66. Ji ZL, Ding FX, Meng XB. Transient flow outside ceramic filter tube. J Chem Ind. 2000;51(2):165–8.
67. Guo JG, Ji ZL. Performance study of ceramic filter pulse backblow system. Power Eng. 1999;19(5):394–8.
68. Ji ZL, Meng XB, Shi MX. Flow characteristics of ceramic filter tube pulse backblow process. Filtr Sep. 2001;11(2):1–4.

Chapter 2
Global Air Pollution Status and Related Standards

2.1 Current Situation and Causes of Air Pollution

2.1.1 Current Status of Air Pollution

According to the definition by the International Organization for Standardization (ISO), air pollution typically refers to the phenomenon where certain substances enter the atmosphere due to human activities or natural processes, reach a sufficient concentration, persist for a sufficient amount of time, and thereby harm human comfort, health, and welfare or the environment [1].

According to estimates by the World Health Organization, the global population prematurely dying from air pollution each year has exceeded 7 million. Air pollution causes comprehensive harm to the human body. In addition to common pulmonary and cardiovascular diseases (such as lung cancer, hypertension, heart disease, and stroke), air pollution can also affect all human organs, causing diabetes, dementia, intellectual decline, neonatal diseases, etc. At the same time, air pollution can further worsen the symptoms of those infected with the novel coronavirus.

Developed countries have significantly improved air quality through years of efforts. However, the air quality in most developing countries has deteriorated significantly due to large-scale urbanization and economic development that are heavily reliant on fossil fuels. In 2019, over 90% of the global population lived in areas where the $PM_{2.5}$ concentration exceeded 10 μg/m^3 (Global Air Quality Guidelines, 2005). According to real-time data on global air quality, there are very few areas in the world that can control the $PM_{2.5}$ concentration to within 5 μg/m^3. The World Health Organization states that premature deaths caused by air pollution are mainly concentrated in low- and middle-income countries. The higher the exposure to air pollution, the greater the impact on human health, especially for patients with chronic diseases, the elderly, children, and pregnant women.

A research report released by the Energy Policy Institute at the University of Chicago (EPIC) states that particulate air pollution remains the greatest external risk

Z. Zhong and W. Xing, *Membrane Technology for Air Cleaning*,
https://doi.org/10.1007/978-981-96-1144-7_2

to global human health. Bangladesh, India, Pakistan, China, Nigeria, and Indonesia are among the countries with the severe air pollution, still suffering from three-quarters of the global impact of air pollution. PM $_{2.5}$ pollution index in South Asia has risen by more than 50% since the beginning of this century, mainly due to rapid industrialization and population growth. India's capital, New Delhi, remains the world's most polluted metropolis, with an annual average particulate concentration exceeding 126.5 μg/m^3, far exceeding the air standard recommended by the World Health Organization (5 μg/m^3). The report also shows that since 2013, about 59% of the increase in global air pollution has come from India. According to monitoring by the Indian National Clean Air Program, the annual average PM2.5 concentration in New Delhi in 2022 was 99.7 μg/m^3, more than twice the safety limit of 40 μg/m^3 set by India.

China is the largest developing country in the world, and it began to vigorously develop its coal economy in the 1980s and 1990s. In order to maintain the sustainable and healthy development of the economy and society, a series of resolute and decisive prevention and control measures have been takenen. Since 2013, when the State Council of China issued the "Air Pollution Prevention and Control Action Plan" (referred to as "Ten Measures on Air"), clear air quality goals have been proposed, and various departments have successively introduced relevant laws and policies to fully support the work of air pollution prevention and control [2]. Among them, the new "Environmental Air Quality Standard" and the emission standards for industries such as thermal power, steel, cement, chemical industry, and non-electric coal-fired boilers are the most powerful policies to encourage enterprises to implement flue gas treatment projects. According to the "National Ecological Environment Quality Report" released by the Ministry of Ecology and Environment in 2020 [3], the average proportion of excellent and good days in 337 prefecture-level and above cities in China was 87.0%, an increase of 5.0 percentage points year-on-year. The environmental air quality of 202 cities met the standard, accounting for 59.9% of all prefecture-level and above cities, an increase of 45 year-on-year. The annual average concentration of $PM_{2.5}$ was 33 μg/m^3, a decrease of 8.3% year-on-year; the annual average concentration of PM $_{10}$ was 56 μg/m^3, a decrease of 11.1% year-on-year. It is worth mentioning that the average concentration of PM $_{2.5}$ in China has decreased by 42.3% from 2013 to 2021. If this trend continues, the average life expectancy of the Chinese can be expected to increase by 2.2 years.

2.1.2 *Hazards of Air Pollution*

The harm of air pollution to human health can be divided into two types: acute harm and chronic harm [4]. Acute harm is caused by a sharp increase in the concentration of air (specifically outdoor environmental air) pollutants in a short period of time (such as severe smog), and the human body inhales a large amount of contaminants; the main manifestations are respiratory and eye irritation symptoms, cough, chest pain, difficulty breathing, sore throat, headache, vomiting, heart function disorders, lung function failure, acute attacks of chronic cardiovascular and cerebrovascular diseases, etc.

Chronic harm mainly includes:

(1) Long-term stimulation leads to chronic inflammation of the eyes and respiratory system, such as conjunctivitis, pharyngitis, tracheitis, etc. Severe cases can cause chronic obstructive pulmonary disease (COPD), and even lead to pulmonary heart disease.
(2) The body's immune function declines. In areas with severe air pollution, the salivary lysozyme and secretory immunoglobulin A (IgA) content in residents significantly decreases, and other immune indicators also decrease.
(3) Exacerbates chronic cardiovascular and cerebrovascular diseases.
(4) Exacerbates allergic reactions or allergic diseases. Some pollutants in the air have sensitizing effects, which can exacerbate diseases such as asthma and allergic rhinitis.
(5) Increases the risk of lung cancer. The International Agency for Research on Cancer (IARC) has classified air pollutants, including particulates, as Class A carcinogens. Review articles published internationally suggest that although overall, the risk of cancer from air pollution is relatively low, for lung cancer, atmospheric pollutants, especially particulates, often contain carcinogens such as benzo[a]pyrene (BaP) and arsenic, posing a higher risk.

According to statistics on respiratory diseases in China, the incidence of respiratory diseases has rapidly increased in the past decade [5]. In addition, it is found that long-term air pollution is associated with premature birth, low birth weight, and other birth defects. It is clear that the deteriorating atmospheric environment is slowly eroding people's health [6].

2.1.3 Sources of Air Pollutants

Air pollution can be attributed to natural and human causes. Some changes in nature can also cause changes in the composition of air. For example, volcanic eruptions can spew large amounts of dust and carbon dioxide into the atmosphere, causing smog and toxic gases in surrounding areas. However, generally speaking, the impact of such natural changes is local and short-term.

Human production and living activities are the main causes of air pollution, mainly in the following aspects [7]:

(1) Industrial pollution: Industrial pollution is the main source of air pollution, including the combustion of fuels, mainly the large amount of harmful substances emitted during the combustion of coal, oil, and biomass fuels. Combustion flue gas contains a large amount of dust, sulfur dioxide, nitrogen oxides, organic compounds, etc.; dust and exhaust gases are emitted during the production process, with power plants, steel mills, petrochemical plants, cement plants, etc. causing the most severe air pollution.

(2) Residential pollution: A large number of domestic stoves and heating boilers consume a lot of coal, and the combustion of coal releases a large amount of dust, sulfur dioxide, carbon monoxide and other harmful substances into the atmosphere, thus causing pollution. In addition, during home decoration, materials such as wood panels, paints, and floors will emit formaldehyde, toluene and other VOC gases. Once these gases enter the atmospheric environment, they can cause air pollution to a certain extent.

(3) Transportation pollution: The use of transportation tools such as cars, trains, airplanes, and ships all require coal or oil as power materials, inevitably emitting a large number of pollutants, such as carbon monoxide, sulfur dioxide, nitrogen oxides, and hydrocarbons.

(4) Construction dust pollution: With the continuous acceleration of urbanization in developing countries, the scale of construction projects is getting larger and larger, resulting in increasingly serious environmental pollution problems caused by construction. The pollution of the atmosphere during construction mainly comes from dust pollution.

(5) Agricultural process pollution: When pesticides are applied in the field, a portion of the pesticides will escape into the atmosphere in the form of aerosols, and the residues on the crops or adhering to the surface of the crops can volatilize into the atmosphere. The pesticide components that enter the atmosphere can be absorbed by suspended particles and transported to various places with the air flow, causing atmospheric pesticide pollution. Agricultural process pollution also includes pollution caused by straw burning and other activities.

The sources and proportions of air pollutants in China are shown in Fig. 2.1.

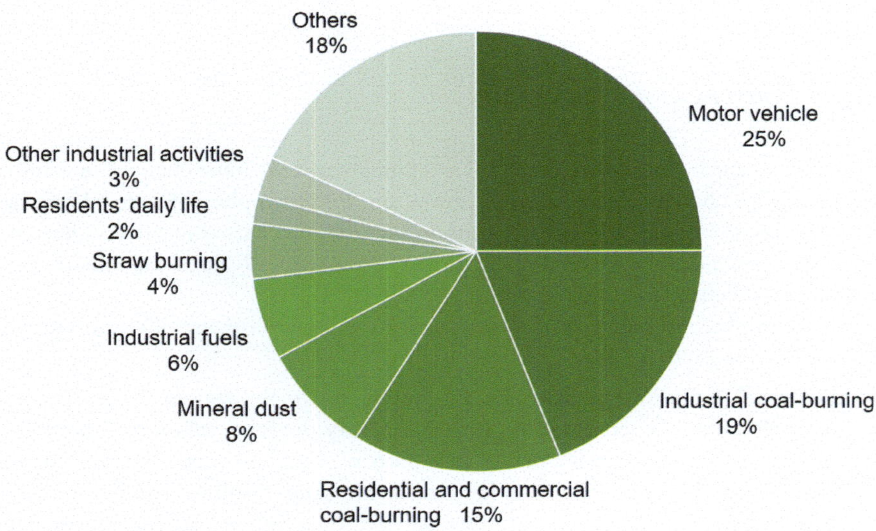

Fig. 2.1 Sources and proportions of air pollutants in China [8]

2.2 Main Types of Air Pollutants

According to their generation ways, air pollutants can be divided into primary pollutants and secondary pollutants. Primary pollutants refer to pollutants directly emitted from pollution sources, such as sulfur dioxide, nitrogen dioxide, carbon monoxide, particulates, etc., which can be further divided into reactive and non-reactive substances. The former are unstable and often react with other substances in the atmospheric environment, or act as catalysts to promote reactions between other pollutants; the latter do not react or react slowly. Secondary pollutants refer to new air pollutants with completely different physical and chemical properties from primary pollutants, formed by the interaction of primary pollutants in the atmosphere through chemical reactions or photochemical reactions. Their toxicity is stronger than that of primary pollutants. The most common secondary pollutants include sulfuric acid and sulfate aerosols, nitric acid and nitrate aerosols, ozone, photochemical oxidants, and many different lifespan active intermediates (also known as free radicals), such as $HO_2\cdot$and HO [9].

According to their sources, air pollutants can be divided into natural pollutants and anthropogenic pollutants. Among them, those that cause public nuisance are often anthropogenic pollutants, which mainly come from fuel combustion and the production process of large-scale industrial and mining enterprises. Anthropogenic pollutants include particulates, sulfur oxides, nitrogen oxides, carbon monoxide, hydrocarbons, and other harmful substances. Among them, particulates refer to liquid and solid substances in the atmosphere. Sulfur oxides are a general term for sulfur-containing oxides, including sulfur dioxide, sulfur trioxide, disulfur trioxide, and sulfur monoxide, etc. Nitrogen oxides are a general term for nitrogen-containing oxides, which can be subdivided into nitrous oxide, nitric oxide, nitrogen dioxide, and dinitrogen trioxide, etc. Hydrocarbons are compounds formed by carbon and hydrogen elements, such as methane and ethane, etc. Other harmful substances including heavy metals, fluorine-containing gases, chlorine-containing gases, and so on [10].

Air pollutants can be broadly divided into two categories according to their state of existence: aerosol state pollutants and gaseous state pollutants. Aerosol state pollutants mainly include dust, smoke, fly ash, and fog, while gaseous pollutants mainly include nitrogen oxides, sulfur oxides, hydrocarbons, halogen compounds, etc. [11, 12].

2.2.1 Aerosol State Pollutants

Aerosol state pollutants refer to suspended bodies composed of solid or liquid particles suspended in the atmosphere (particles with a diameter of 0.01–10 μm). Fine solid particles are the main components of aerosol state pollutants. Particulates can be divided into primary and secondary particulates. Primary particulates are particles that cause pollution directly released into the atmosphere from pollution sources, including natural and anthropogenic sources. Natural sources include ground dust, earthquake ash, sandstorms, and biological particles such as pollen and spores; anthropogenic sources are mainly solid particles produced in the process of production, fuel combustion, construction, and transportation, such as dust, coal smoke, fly ash, etc. Secondary particulates are particles formed through photochemical oxidation reactions, catalytic oxidation reactions or other chemical reactions between certain pollutant gas in the atmosphere (such as sulfur dioxide, nitrogen oxides, hydrocarbons, etc.) or these components with normal components in the atmosphere (such as oxygen), such as sulfur dioxide transformed into sulfate [13].

2.2.2 Gas Phase Pollutants

Gas phase pollutants are gaseous pollutants that exist in a molecular state at normal temperature and pressure. This includes inorganic gas phase pollutants and volatile organic pollutants. Common inorganic gas phase pollutants mainly include CO, SO_2, NO_2, NH_3 H_2S, etc. [14], which are mainly related to the industrial emissions from power plants, steel, cement, chemical and other industries. Volatile organic pollutants are mainly formaldehyde, benzene, toluene and other organic gases, and their sources are mainly indoor decoration materials, such as wood panels, adhesives, etc. [15] In air pollution control, primary pollutants, including sulfur oxides, nitrogen oxides, carbon oxides and organic compounds and secondary pollutants, including sulfuric acid smoke and photochemical smoke, have gained wider attentions [16].

2.3 Air Quality Standards

Environmental air quality standards are limit regulations made for the allowable content of various pollutants in the atmosphere aiming at protecting human health and the ecological environment. They are the basis for specifying air pollution control plans and air pollutant emission standards, as well as the legal basis for environmental management. According to the air quality standards, air pollution control technical standards can be formulated for production, design and management personnel.

Countries around the world have carried out active exploration and systematic research in the formulation, revision and implementation of environmental air quality standards. At present, the main air quality monitoring indicators in various countries include PM2.5, PM_{10}, NO_2, SO_2, CO, and O_3. In 2013, the World Health Organization classified PM2.5 as a carcinogen. In recent years, the standard for atmospheric particulate matter has become a hot topic of concern in many countries. Monitoring technology and the assessment of $PM_{2.5}$ have attracted global attention. In addition, the monitoring and assessment of toxic substances and heavy metal pollutants in the air is an important direction for future development.

2.3.1 United Nations World Health Organization (WHO)

In 2005, the United Nations World Health Organization formulated the "Air Quality Guidelines" (AQGs), setting guideline values and three graded transitional period target values for the annual and daily concentrations of $PM_{2.5}$ and PM_{10}. Among them, the guideline values are the most stringent, mainly based on the more reasonable and less health-threatening particulate matter limit standards derived from research, which are difficult to achieve immediately, even for these developed countries. Therefore, three graded transitional period target values were set for $PM_{2.5}$ and PM_{10}, which could be helpful in assessing the progress made in efforts to reduce particulate matter concentrations.

In 2021, the World Health Organization (WHO) updated the "Global Air Quality Guidelines" for the first time, tightening almost all air quality levels with the aim of reducing key air pollutant levels to protect human health. In addition to the standard values, the new version also provides four phased target values (Interim Target) for $PM_{2.5}$, aimed at helping regions with severe air pollution assess the progress of gradually reducing population exposure to particulate matter. The target values for the first to fourth stages are 35, 25, 15, and 10 $\mu g/m^3$, respectively, and each increase in stage can reduce the mortality rate by about 7%.

2.3.2 United States

In 1971, the US government was the first to establish particulate matter monitoring standards, setting the first and second standards for Total Suspended Particles (TSP). As research on particulate matter pollution deepened, it was found that the smaller the particle size, the greater the harm to human health. Therefore, the US Environmental Protection Agency lowered the particle size of the monitored particulate matter to PM_{10} in 1987 and issued the $PM_{2.5}$ monitoring standard in 1997. In 2006, the United States issued new air particulate matter quality standards based on the original standards.

2.3.3 European Union

In the 1980s, the European Union began to focus on air particulate matter monitoring. In 2005, it established the PM_{10} standard, but 23 of its member countries exceeded the PM_{10} concentration standard. In 2008, the European Union passed a new air quality law (2008 /50/EC), began to supervise the implementation of air quality standards strictly, and severely punished violations. In May of the same year, the European Union issued the "Regulation on European Air Quality and Clean Air Law", which stipulated the target concentration limit, exposure concentration limit, and reduction target value for $PM_{2.5}$, and established the $PM_{2.5}$ standard in 2010. This regulation requires that by 2020, EU member states should reduce the average $PM_{2.5}$ content by 20% based on 2010 levels and control the inhalable particulate matter content in urban areas to an annual average concentration of 20 $\mu g/m^3$ or less by 2015. For example, the UK has set the annual average of $PM_{2.5}$ at 25 $\mu g/m^3$ as the target concentration limit for 2020, requiring all administrative regions to reduce the exposure concentration by 15% between 2010 and 2020, while Scotland aims to reach an annual average concentration limit of 12 $\mu g/m^3$ by 2020.

2.3.4 China

The ambient air quality standards in China are established according to the "Environmental Protection Law of the People's Republic of China" and the "Air Pollution Prevention and Control Law of the People's Republic of China". On January 6, 2000, the State Environmental Protection Administration of China issued a notice on the release of the "Amendment to the Ambient Air Quality Standard" (GB3095-1996), mainly modifying some index standards. In February 2012, the Ministry of Environmental Protection of China issued a new standard - "Ambient Air Quality Standard" (GB3095-2012) and proposed a phased implementation plan: Beijing-Tianjin-Hebei, Yangtze River Delta, Pearl River Delta and other key regions and municipalities and provincial capitals in 2012; 113 key cities for environmental protection and national environmental protection model cities in 2013; all cities at or above the prefecture level in 2015; and nationwide implementation of the new standard on January 1, 2016. At the same time, the Ministry of Environmental Protection of China issued the "Technical Regulations for Ambient Air Quality Index (AQI) (Trial)", which stipulates the grading scheme, calculation method, index release content and related requirements for the ambient air quality index, and is implemented in sync with the "Ambient Air Quality Standard" (GB3095-2012). When the AQI is greater than 50, the pollutant with the highest IAQI is the primary pollutant, and pollutants with an IAQI greater than 100 exceed the standard pollutants.

2.4 Industrial Atmospheric Pollutant Emission Standards

In China, "Comprehensive Emission Standards for Atmospheric Pollutants" (GB 16297-1996) officially came into effect on January 1, 1997, which stipulates the emission limits of 33 atmospheric pollutants and various requirements in the implementation of the standards. In the existing national atmospheric pollutant emission standard system, according to the principle of non-overlapping implementation of comprehensive emission standards and industry-specific emission standards, thermal power, steel, cement, petroleum refining, and other industries with special emission limits for atmospheric pollutants have national standards, and the special emission limits for atmospheric pollutants are implemented. Table 2.1 gives the main atmospheric pollutant emission limits in the "Comprehensive Emission Standards for Atmospheric Pollutants".

In 2013, the State Council of China issued the "Air Pollution Prevention and Control Action Plan," which required that during the "Twelfth Five-Year Plan" period, special emission limit standards for air pollutants should be formulated for "6 + 1" key industries (fields) including thermal power, steel, petrochemicals, cement, non-ferrous metals, chemical industries, and coal-fired boiler projects, as shown in Table 2.2. This involves multiple revisions of national pollutant emission (control) standards. By the end of 2019, 28 national emission standards for special emission limits of air pollutants had been issued. The requirements for emission indicators in

Table 2.1 Main atmospheric pollutant emission limits stipulated in GB 16297-1996, China

No.	Pollutant	Maximum allowable emission concentration/(mg/m^3)	Unorganized emission monitoring concentration limit/(mg/m^3)
1.	Sulfur Dioxide	1200 (Production of sulfur, sulfur dioxide, sulfuric acid, and other sulfur compounds) 700 (Production of sulfur, sulfur dioxide, sulfuric acid, and other sulfur compounds)	0.50 (Difference in concentration between monitoring point and reference point)
2.	Nitrogen Oxides	1700 (Production of nitric acid, nitrogen fertilizers, and explosives) 420 (Use of nitric acid, etc.)	0.15 (Difference in concentration between monitoring point and reference point)
3.	Particulates	22 (Carbon black dust, dye dust) 80 (Glass wool dust, quartz dust, slag wool dust) 150 (Others)	Invisible to the naked eye 2.0 (Difference in concentration between monitoring point and reference point) 5.0 (Difference in concentration between monitoring point and reference point)
4.	Hydrogen Fluoride	150	0.25
5.	Chromic Acid Mist	0.080	0.0075
6.	Sulfuric Acid Mist	1000 (Explosives factory) 70 (Others)	1.5
7.	Fluorides	100 (General calcium industry) 11 (Others)	20 $\mu g/m^3$ (Difference in concentration between monitoring point and reference point)
8.	Chlorine Gas	85	0.50
9.	Lead and its Compounds	0.90	0.075
10.	Mercury and its Compounds	0.015	0.0015

the newly established standards are much higher than the old standards. During the "Twelfth Five-Year Plan" period, existing enterprises in the thermal power, steel, and petrochemical industries in the main urban areas of 47 cities in key control areas, as well as coal-fired boiler projects, implemented the special emission limit standards for air pollutants. During the "Thirteenth Five-Year Plan" period, the requirements for special emission limits were extended to the urban areas of key control areas. On June 27, 2018, the State Council of China issued the "Three-Year Action Plan to Win the Blue-Sky Defense War", which clearly requires key areas to fully implement the special emission limit standards for air pollutants such as sulfur dioxide, nitrogen oxides, particulate matter, and volatile organic compounds (VOCs). At the same time, local governments have also formulated stricter local standards. The nationwide concerted effort to fight the tough battle against air pollution has begun.

2.4.1 Cement Industry Emission Standards

Since 2013, China has successively introduced relevant policies, proposed emission reduction targets for the cement industry, and formulated new emission standards, "Emission Standards of Air Pollutants for Cement Industry" (GB 4915-2013). This standard stipulates the emission limits, detection, supervision and management requirements of atmospheric pollutants from cement manufacturing enterprises (including independent grinding stations), cement raw material mines, bulk cement transfer stations, cement product enterprises, and their production facilities. The standard stipulates that the nitrogen oxide emission standard (cement kiln and other thermal equipment) is tightened to 400 mg/m^3 (key areas 320 mg/m^3), sulfur dioxide emission standard (cement kiln and other thermal equipment) is tightened to 200 mg/m^3 (key areas 100 mg/m^3), and dust emission standard (cement kiln and other thermal equipment) is tightened to 30 mg/m^3 (key areas 20 mg/m^3), see Table 2.3. Some regions formulate and implement local standards, and local standards are generally higher than national standards, such as the "Emission Standards of Air Pollutants for Cement Industry" (DB 41/1953—2020) issued by the Henan Provincial Department of Ecology and Environment, which stipulates that the emission limits of nitrogen oxides, sulfur dioxide, and dust are 100 mg/m^3, 35 mg/m^3, and 10 mg/m^3 respectively.

2.4.2 Emission Standards for the Waste Incineration Industry

Waste incineration is one of the main methods of waste treatment. The heat generated by waste incineration can be used for power generation and heating, achieving comprehensive utilization of resources. According to incomplete statistics, at the beginning of 2018, the waste incineration power generation capacity exceeded 6.8 million kilowatts in China, annual power generation exceeded 350 billion kilowatt-hours, and annual waste treatment exceeded 105 million tons, accounting for more

Table 2.2 28 special emission standards for air pollutants and main pollutant emission limits

No.	Name	Series number	Pollutant	Maximum allowable emission limit/(mg/m^3)
1.	Emission standards for air pollutants in thermal power plants	GB 13223-2011	Smoke dust Sulfur dioxide Nitrogen oxide	20 50 100
2.	Emission standard for industrial pollutants in iron ore mining	GB 28661-2012	Particulate matter	20 (Key area 10)
3.	Emission standard for air pollutants in iron and steel sintering, pelletizing industry	GB 28662-2012	Particulate matter Sulfur dioxide Nitrogen oxide	50 200 300
4.	Emission standard for air pollutants in iron making industry	GB 28663-2012	Particulate matter Sulfur dioxide Nitrogen oxide	20 100 300
5.	Emission standard for air pollutants in steelmaking industry	GB 28664-2012	Particulate matter	50 (Converter)
6.	Emission standard for air pollutants in steel rolling industry	GB 28665-2012	Particulate matter Sulfur dioxide Nitrogen oxide	20 150 300
7.	Emission standard for air pollutants in ferroalloy industry	GB 28666-2012	Particulate matter	50 (Key area 30)
8.	Emission standard for air pollutants in the coking chemical industry	GB 16171-2012	Particulate matter	Crushing and screening 30 Coaling 50 Coke pushing 50
9.	Emission standard for air pollutants in petroleum refining industry pollutants	GB 31570-2015	Particulate matter Sulfur dioxide Nitrogen oxide	20 100 150
10.	Emission standard for air pollutants in petrochemical industry	GB 31571-2015	Particulate matter Sulfur dioxide Nitrogen oxide	20 100 150
11.	Emission standard for air pollutants in synthetic resin industry pollutants	GB 31572-2015	Total non-methane hydrocarbons Particulate matter	100 30
12.	Emission standard for air pollutants in caustic soda and polyvinyl chloride industry pollutants	GB 15581-2016	Particulate matter Sulfur dioxide Nitrogen oxide Total non-methane hydrocarbons	60 (Calcium carbide crushing) 100 (Incinerator) 200 (Incinerator) 50

(continued)

Table 2.2 (continued)

No.	Name	Series number	Pollutant	Maximum allowable emission limit/(mg/m^3)
13.	Emission standard for air pollutants in nitric acid industry	GB 26131-2010	Nitrogen oxide	300
14.	Emission standard for air pollutants in sulfuric acid industry	GB 26132-2010	Particulate matter Sulfur dioxide	50 860
15.	Emission standard for air pollutants in inorganic chemical industry	GB 31573-2015	Particulate matter Nitrogen oxide Sulfur dioxide	30 200 400
16.	Emission standard for air pollutants in aluminum industry	GB 25465-2010	Particulate matter Sulfur dioxide	50 (Crushing, screening, transshipment) 400 (Clinker firing kiln)
17.	Emission standard for air pollutants in lead and zinc industry	GB 25466-2010	Particulate matter Sulfur dioxide	80 400
18.	Emission standard for air pollutants in copper, nickel and cobalt industries	GB 25467-2010	Particulate matter Sulfur dioxide	80 (Smelt) 400 (Smelt)
19.	Emission standards for air pollutants in magnesium and titanium industries	GB 25468-2010	Particulate matter Sulfur dioxide	50 (Raw material preparation) 400 (Refine)
20.	Emission standards for air pollutants in the rare earth industry	GB 26451-2011	Particulate matter Sulfur dioxide	40 (Decomposition and extraction) 300 (Decomposition and extraction)
21.	Emission standard for air pollutants in vanadium industry	GB 26452-2011	Particulate matter Sulfur dioxide	50 400
22.	Emission standard for air pollutants in tin, antimony and mercury industry	GB 30770-2014	Particulate matter Sulfur dioxide	30 (Smelt) 400
23.	Emission standard for air pollutants in copper, aluminum, lead and zinc recycling industry	GB 31574-2015	Particulate matter Alumina nitride Sulfur dioxide	10 100 100

(continued)

Table 2.2 (continued)

No.	Name	Series number	Pollutant	Maximum allowable emission limit/(mg/m^3)
24.	Emission standard of air pollutants in cement industry	GB 4915-2013	Particulate matter Alumina nitride Sulfur dioxide	30 (Cement kiln) 200 (Cement kiln) 400 (Cement kiln)
25.	Emission standard for air pollutants from boilers	GB 13271-2014	Particulate matter Alumina nitride Sulfur dioxide	30 (Coal-fired boiler) 200 (Coal-fired boiler) 200 (Coal-fired boiler)
26.	Non-organized emission control standard for volatile organic compounds	GB 37822-2019	VOCs	10 (1 h mean concentration at the monitoring point)
27.	Emission standard of air pollutants in pharmaceutical industry	GB 37823-2019	Particulate matter VOCs Ammonia	30 150 30
28.	Emission standard for air pollutants in coatings, inks and adhesives industry	GB 37824-2019	Particulate matter VOCs	20 80

than 35% of the urban waste transportation volume [17]. According to the "13th Five-Year Plan for Biomass Energy Development", by 2020, the urban domestic waste incineration power generation capacity has reached 7.5 million kilowatts; the "Guiding Opinions on the Implementation of the 13th Five-Year Plan for Renewable Energy Development" also clarified the layout of the "13th Five-Year Plan" for waste incineration power generation, clearly laying out 529 waste incineration power generation projects in 30 provinces (municipalities, autonomous regions) and the Xinjiang Production and Construction Corps, with an installed capacity of 10.22 million kilowatts.

With the large-scale construction of waste incineration power plants in China and the increasingly strict national environmental management, the issue of atmospheric pollutant emissions from the waste incineration industry has gradually attracted the attention of the whole society. The flue gas emissions from domestic waste incineration power plants generally have three standards: the Chinese national standard "Domestic Waste Incineration Pollution Control Standard" (GB 18485-2014), the EU standard "EU Industrial Emission Directive" (2010/75/EU) and local standards. Among them, the Chinese national standard must be met, some areas choose the stricter one between the Chinese national standard and the EU standard, and some areas implement it according to the local standard higher than the Chinese national standard. Table 2.4 compares the Chinese national standard, the EU standard, and the Shanghai local standard.

Table 2.3 GB 4915-2013 stipulates the emission limits of atmospheric pollutants for existing and new enterprises

Production process	Production equipment	Particulate matter	Sulfur Dioxide	Nitrogen Oxides (calculated as NO_2)	Fluorides (calculated as total F)	Mercury and its compounds	Ammonia
Mining	Crushers and other ventilation production equipment	20 (Key areas 10)	–	–	–	–	–
Cement Manufacturing	Cement kilns and waste heat recovery systems	30 (Key areas 20)	200 (Key areas 100)	400 (Key areas 320)	5 (Key areas 3)	0.05	10 (Key areas 8)
	Dryers, drying mills, coal mills, and coolers	30 (Key areas 20)	600 (Key areas 400)	400 (Key areas 300)	–	–	–
	Crushers, mills, packaging machines, and other ventilation production equipment	20 (Key areas 10)	–	–	–	–	–
Bulk cement Transit station and Cement products Production	Cement silos and other ventilation production equipment	20 (Key Areas 10)	–	–	–	–	–

Unit: mg/m^3

2.4.3 *Ultra-Low Emission Standards*

In 2014, China was the first to propose "ultra-low emission" standards in the coal-fired boiler industry, that is, coal-fired boilers in power plants use a variety of high-efficiency coordinated pollutant removal integrated system technologies in power generation operations and end treatment processes, making the concentration of atmospheric pollutant emissions basically comply with the emission limits of gas turbine units. The emission concentrations of soot, sulfur dioxide, and nitrogen oxides (reference oxygen content 6%) do not exceed 5, 35 and 50 mg/m^3, which are 75%, 30% and 50% lower than the special emission limits for coal-fired boilers in key areas stipulated in the "Emission Standards for Atmospheric Pollutants from Thermal

Table 2.4 A comparison between Chinese National Standard, EU Standard, and Shanghai Local Standard

No.	Pollutant	Unit	GB 18485—2014		2010/75/EU		DB 31/768 2013	
1.	Particulate matter	mg/m^3	30	Average in 1 h	30	0.5 h 100%	10	Measured average
					10	0.5 h 97%		
			20	Average in 24 h	10	Average in 24 h		
2.	Nitrogen oxides	mg/m^3	300	Average in 1 h	400	0.5 h 100%	250	Average in 1 h
					200	0.5 h 97%		
			250	Average in 24 h	200	Average in 24 h	200	Average in 24 h
3.	Sulfur dioxide	mg/m^3	100	Average in 1 h	200	0.5 h 100%	100	Average in 1 h
					50	0.5 h 97%		
			80	Average in 24 h	50	Average in 24 h	50	Average in 24 h
4.	Hydrogen chloride	mg/m^3	60	Average in 1 h	60	0.5 h 100%	50	1 h Average
					10	0.5 h 97%		
			50	Average in 24 h	10	24 h Average	10	24 h Average
5.	Hydrogen fluoride	mg/m^3	–	–	4	0.5 h 100%	–	–
					2	0.5 h 97%		
			–	–	1	Average in 24 h	–	–
6.	Mercury and its compounds	mg/m^3	0.05	Measured average	0.05	Sampling period 30 min, Max. 8 h	0.05	Measured average
7.	Cadmium, thallium and their compounds	mg/m^3	0.1	Measured average	0.05	Sampling period 30 min, Max. 8 h	0.05	Measured average
8.	Antimony, arsenic, lead, chromium, cobalt, copper, manganese, nickel and their compounds	mg/m^3	1.0	Measured average	0.5	Sampling period 30 min, Max. 8 h	0.5	Measured average

(continued)

Table 2.4 (continued)

No.	Pollutant	Unit	GB 18485—2014		2010/75/EU		DB 31/768 2013	
9.	Dioxin-like	ng-TEQ/m^3	0.1	Measured average	0.1	Sampling period 30 min, Max. 8 h	0.1	Measured average
10.	Carbon monoxide	mg/m^3	100	Average in 1 h	100	0.5 h 100% (day)	100	Average in 1 h
					150	10 min95% (day)		
			80	Average in 24 h	50	Average in 24 h	50	Average in 24 h

Power Plants" (GB 13223-2011), and have fully surpassed the level of the European Union [18].

The 2015 Chinese "Government Work Report" pointed out that the carbon dioxide emissions in 2015 should be reduced by more than 3.1%, the chemical oxygen demand and ammonia nitrogen emissions should be reduced by about 2%, and the sulfur dioxide and nitrogen oxide emissions should be reduced by about 3% and 5% respectively. Deeply implement the action plan for air pollution prevention and control, implement regional joint prevention and control, promote the ultra-low emission transformation of coal-fired power plants, and promote zero growth in coal consumption in key areas [19].

On December 2, 2015, the executive meeting of the State Council of China decided that by 2020, ultra-low emission and energy-saving transformations will be fully implemented for coal-fired units, greatly reducing power generation coal consumption and pollutant emissions [20].

2.4.4 Emission Standards in Other Industries

To prevent air pollution and improve air quality, other industries in China such as petrochemicals, non-ferrous metals, pharmaceuticals, coatings, inks and adhesives, and other industries have successively issued relevant pollutant emission control standards. The following is a brief introduction to the iron and steel industry and the inorganic chemical industry.

2.4.4.1 Iron and Steel Industry

The iron and steel industry are the second largest high-pollution industry after the thermal power industry. The production process of steel rarely uses coal-fired boilers, but the sintering and pelletizing processes of the iron and steel industry produce a large amount of sulfur dioxide, nitrogen oxides, and particulates. In June 2012, the Ministry of Environmental Protection of China issued the "Emission Standard of Air Pollutants for Iron and Steel Sintering and Pelletizing Industry" (GB 28662-2012), which clearly states that production process equipment that produces air pollutants must set up local gas collection systems and centralized removal treatment devices, and the gas should be discharged after meeting the standard. The standard requires that the emission limits of existing enterprises' sintering machines and pellet roasting equipment are 80 mg/m^3 for particulates, 600 mg/m^3 for sulfur dioxide, 500 mg/m^3 for nitrogen oxides, 6.0 mg/m^3 for chlorides, and 1.0 ng-TEQ/m^3 for dioxins. The emission limit for particulates from production equipment, such as the tail of the sintering equipment is 50 mg/m^3. The emission limits for newly built sintering machines and pellet roasting equipment are 50 mg/m^3 for particulates, 200 mg/m^3 for sulfur dioxide, 300 mg/m^3 for nitrogen oxides, 4.0 mg/m^3 for chlorides, and 0.5 ng-TEQ/m^3 for dioxins; the emission limit for particulates from production equipment such as the tail of the sintering machine is 30 mg/m^3 (see Table 2.5).

Table 2.5 Emission limits of air pollutants specified in GB 28662-2012, China

Production process or facility	Pollutant	Existing enterprises Emission limit	New enterprises Emission limit	Key areas Emission limit
Sintering machine pellet roasting equipment	Particulates	80	50	40
	Sulfur dioxide	600	200	180
	Nitrogen oxides (calculated as NO_2)	500	300	300
	Fluorides (calculated as F)	6.0	4.0	4.0
	Dioxins (ng-TEQ/m^3)	1.0	0.5	0.5
Sintering machine tail, belt roasting machine tail and other production equipment	Particulate matter	50	30	20

Unit: mg/m^3

2.4.4.2 Inorganic Chemical Industry

The inorganic chemical industry is an important basic raw material and material industry. China is a major producer, consumer, and exporter of the inorganic chemical industry, with the capacity and output ranking first in the world. While supporting the rapid development of the national economy, this industry also brings serious environmental pollution. To further strengthen the control and management of pollutant emissions from the inorganic chemical industry, in April 2015, the Chinese Ministry of Environmental Protection issued the "Emission Standards for Pollutants in the Inorganic Chemical Industry" (GB 31573-2015) (see Table 2.6), which was implemented from July 1, 2015. This standard has made clear control requirements for water pollution and air pollution. According to the emission characteristics of air pollutants in the inorganic chemical industry, a total of 25 pollutant emission limits have been set, including particulate matter, nitrogen oxides, sulfur dioxide, hydrogen sulfide, chlorine, hydrogen chloride, cyanides, ammonia, sulfuric acid mist, fluorides (chromic acid mist, arsenic and its compounds) and other 10 pollutants, as well as lead, mercury, cadmium, tin, nickel, zinc, manganese, antimony, copper, cobalt, molybdenum, zirconium, thallium, chromic acid mist and other 14 heavy metals and their compounds and arsenic compounds. According to the production process and product category, different emission limits have been set for pollutants, such as the emission limits for hydrogen chloride and chlorine pollutants in the inorganic chloride and chlorate industry are 8 mg/m^3 and 20 mg/m^3, respectively, while the emission limits for hydrogen chloride and chlorine pollutants in other inorganic chemical industries (except for sulfide and sulfate industries, inorganic cyanide industry) are tightened, which are 5 mg/m^3 and 10 mg/m^3, respectively. In addition, it is stipulated that some areas implement the special emission limits for air pollutants in this standard, which particularly restricts the emission of particulate matter, nitrogen oxides, sulfur dioxide, hydrogen sulfide and other pollutants.

2.5 Domestic and Foreign Standards for Civilian Gas Cleaning Products

2.5.1 Chinese Air Purifier Standards

With the continuous improvement of people's living standards, air purifiers have become commonly used household appliances. Currently, the standard system for air purifiers in China can be divided into two types: safety standards and performance standards. Safety standards include GB 4706.1-2005 "Safety of household and similar electrical appliances Part 1: General requirements" and GB 4706.45-2008 "Safety of household and similar electrical appliances Special requirements for air purifiers". Performance standards include GB/T18801-2015 "Air purifiers", GB 21551.1-2008 "Antibacterial, disinfection, purification function general rules

Table 2.6 Air pollutant emission limits stipulated in GB 31573-2015, China

No	Pollutant	Source of pollution control	Limit value	Key area limit value
1	Particulate Matter	All	30	10
2	Nitrogen Oxides	All	200	100
3	Sulfur Dioxide	Sulfur Compounds and their Sulfate Industry, Heavy Metal Inorganic Compound Industry	400	100
		Others	100	
4	Hydrogen Sulfide	Excluding Inorganic Cyanide Compound Industry, Halogen and its Compound Industry	10	5
5	Chlorine Gas	Inorganic Chloride and Chlorate Industry	8	8
		Others	5	5
6	Hydrogen Chloride	Inorganic Chloride and Chlorate Industry	20	20
		Others	10	10
7	Hydrogen Cyanide	Excluding Sulfide and Sulfate Industry, Halogen and its Compound Industry	0.3	0.3
8	Ammonia	Excluding Heavy Metal Inorganic Compound Industry, Halogen and its Compound Industry	20	10
9	Sulfuric Acid Mist	Sulfides and sulfate industry, involving barium, strontium heavy metal inorganic compound industry	20	10
10	Fluorides (calculated as F)	Industry involving cobalt, zirconium heavy metal inorganic compounds	3	3
		Inorganic fluoride industry	6	
11	Chromic acid mist	Chromium and its compound industry	0.07	0.07
12	Arsenic and its compounds	All	0.5	0.5
13	Lead and its compounds	Industry involving lead heavy metal inorganic compounds	2	0.1
		Others	0.1	

(continued)

Table 2.6 (continued)

No	Pollutant	Source of pollution control	Limit value	Key area limit value
14	Mercury and its compounds	All	0.01	
15	Cadmium and its compounds	All	0.5	
16	Tin and its compounds	Industry involving tin heavy metal inorganic compounds	4	
17	Nickel and its compounds	Industry involving nickel heavy metal inorganic compounds	4	
18	Zinc and its compounds	Industry involving zinc heavy metal inorganic compounds	5	
19	Manganese and its compounds	Industry involving manganese heavy metal inorganic compounds	5	
20	Antimony and its compounds	Industry involving antimony heavy metal inorganic compounds	4	
21	Copper and its compounds	Industry involving copper heavy metal inorganic compounds	5	
22	Cobalt and its compounds	Industry involving cobalt heavy metal inorganic compounds	5	
23	Molybdenum and its compounds	Industry involving molybdenum heavy metal inorganic compounds	5	
24	Zirconium and its compounds	Industry involving zirconium heavy metal inorganic compounds	5	
25	Thallium and its compounds	Industry involving thallium, zinc, copper, lead heavy metal inorganic compounds		

Unit: mg/m^3

for household and similar electrical appliances", GB 21551.3-2010 "Antibacterial, disinfection, purification function special requirements for air purifiers of household and similar electrical appliances".

In terms of safety standards, GB 4706.1-2005 stipulates the general provisions for the safety of household and similar electrical appliances, while GB 4706.45-2008 makes requirements for the safety of the classified product of air purifiers.

The entire safety standard system is set up for the safety of air purifier products. The standard mainly regulates the content of air purifier's electric shock prevention, power consumption, overload temperature rise, current leakage and electrical strength, working under humid environment, abnormal operation, stability and mechanical hazards, mechanical strength, structure, components, power connection, grounding measures, creepage distance and electrical clearance, non-metallic materials, radiation toxicity and similar hazards.

Performance standard GB/T GB/T 18801-2015 "Air Purifier" specifies the basic technical indicators of air purifiers as Clean Air Delivery Rate (CADR) and Cumulative Purification Amount (CCM); that is, the purification ability of air purifier products and the sustainability of purification ability. It divides the noise limit of air purifiers into 4 levels, from low to high. Also, it improves the energy efficiency level of air purifiers for different pollutants, including qualified and high-efficiency levels. Moreover, it stipulates the experimental methods for testing the purification ability of air purifier products to remove various target pollutants, including the measurement for cumulative purification amount of particulate matter and formaldehyde; that is, the life time test of air purifiers. In addition, it has formulated the purification ability test and repeatability evaluation for formaldehyde. The performance indicators stipulated in GB/T 18801-2015 are shown in Table 2.7.

Table 2.7 Main performance indicators of air purifiers stipulated in GB/T 18801-2015, China

Division of Particulate Purification Efficiency	
Purification Efficiency Level	Purification Efficiency $\eta_{\text{Particulate}}$/[m^3/(W h)]
High Efficiency Level	$\eta_{\text{Particulate}} \geq 5.00$
Qualified Level	$2.00 \leq \eta_{\text{Particulate}} < 5.00$
Division of Gaseous Pollutant Purification Efficiency	Purification Efficiency $\eta_{\text{Gaseous Pollutant}}$/[m^3/(W h)]
Purification Efficiency Level	
High Efficiency Level	$H_{\text{Gaseous Pollutant}} \geq 1.00$
Qualified Level	$0.50 \leq \eta_{\text{Gaseous Pollutant}} < 1.00$
Noise Standard	
Clean Air Volume/(m^3/h)	Sound Power Level/dB(A)
$Q \leq 150$	≤ 55
$150 < Q \leq 300$	≤ 61
$300 < Q \leq 450$	≤ 66
$Q > 450$	≤ 70

2.5.2 Chinese Cleanroom Grade Standards

Cleanroom is also known as dust-free room or dust-free workshop. The development of cleanroom is closely related to modern industry and cutting-edge technology, and its application in biopharmaceuticals, health care, food and daily chemicals, electronic optics, energy and precision instruments and other industries is quite common and mature. Cleanrooms are divided into several grades, namely $1 \geq 100 \geq 1000 \geq 10{,}000 \geq 100{,}000 \geq 1{,}000{,}000$, with 1 being the highest grade. The most common are from 100 to 100,000, and different grades have different grade standards.

The grade standard of cleanroom is divided by the maximum concentration limit of particles in the air that are greater than or equal to the considered particle size. Dust-free workshops should be tested and examined in terms of air state, static state, and dynamic state, requiring compliance with "Clean Room Design Specifications" (GB 50073-2013, Table 2.8), and "Clean Room Construction and Acceptance Specifications" (GB 50591-2010). The cleanliness and the continuous stability of pollution control are the core standards for inspecting the quality of dust-free workshops. This standard is divided into several grades according to factors such as regional environment and cleanliness.

2.5.3 Mask Protection Level Standards in China

The wearing of masks has become normalized, with the wearers extending from workers in the medical, coal mining, metallurgy, and other fields to everyday life. Masks have become an indispensable item for people who are at home or traveling. The current Chinese mask standards are mainly divided into medical protective masks (GB 19083-2010, YY 0469-2011), labor protection masks (GB 2626-2019), and daily

Table 2.8 Air cleanliness grade specified in GB 50,073-2013, China

Air Cleanliness Grade (N)	Maximum concentration limit of particles greater than or equal to the required particle size/(pc/m^3)					
	0.1 μm	0.2 μm	0.3 μm	0.5 μm	1 μm	5 μm
1	10	2	–	–	–	–
2	100	24	10	4	–	–
3	1000	237	102	35	8	–
4 (Tenth level)	10,000	2370	1020	352	83	–
5 (Hundredth level)	100,000	23,700	10,200	3520	832	29
6 (Thousandth level)	1,000,000	237,000	102,000	35,200	8320	293
7 (Ten thousandth level)	–	–	–	352,000	83,200	2930
8 (Hundred thousandth level)	–	–	–	3,520,000	832,000	29,300
9 (Millionth level)	–	–	–	35,200,000	8,320,000	293,000

protective masks (GB/T 32610-2016). The comparison of mask standards and main technical indicators is shown in Table 2.9.

(1) Standards for Medical Protective Mask

There are two national standards for medical protective masks in China: the People's Republic of China Pharmaceutical Industry Standard "Medical Surgical Masks" YY0469-2011 and "Technical Requirements for Medical Protective Masks" GB 19083-2010 (Table 2.10). The former was issued by the China National Food and Drug Administration on December 31, 2011, and implemented on June 1, 2013. This standard specifies the technical requirements, test methods, marking and usage instructions, and packaging, transportation, and storage requirements for disposable masks worn by clinical medical staff during invasive surgeries and other procedures.

Table 2.9 Comparison of mask standards and main technical indicators

Item	Medical Protective Masks		Labor Protection Masks	Daily Protective Masks
Standard Name	Medical Protective Masks	Medical Surgical Masks	Respiratory Protection Equipment-Self-priming Filtering Anti-particulate Respirator	Technical Specifications for Daily Protective Masks
Standard Number	GB 19083-2010	YY 0469-2011	GB 2626-2019	GB/T 32610-2016
Airflow Resistance	Inhalation resistance less than 343.2 Pa under 85 L/min flow	$\leq$ 49 Pa	Inhalation resistance $\leq$ 350 Pa	Inhalation resistance $\leq$ 175 Pa
			Exhalation resistance $\leq$ 250 Pa	Exhalation resistance $\leq$ 145 Pa
Filtration Efficiency	Level 1 $\geq$ 95%	BFE $\geq$ 95%	KN90 $\geq$ 90%	Level 1 $\geq$ 99%
	Level 2 $\geq$ 99%	PFE $\geq$ 30%	KN95 $\geq$ 95%	Level 2 $\geq$ 95%
	Level 3 $\geq$ 99.7%		KN100 $\geq$ 99.97%	Level 3 $\geq$ 90%
Synthetic Blood Penetration	When 2 mL of synthetic blood is sprayed at the mask at a pressure of 10.7 kPa (80 mmHg), there should be no penetration on the inner side of the mask	After 2 mL of synthetic blood is sprayed at the outer side of the mask at a pressure of 16.0 kPa (120 mmHg), there should be no penetration on the inner side of the mask	–	–

Table 2.10 Technical Requirements for Medical Protective Masks in GB 19083-2010, China

Filter Efficiency Level (Non-oil particle filtration efficiency at a gas flow rate of 85 L/min)					
Level 1	$\geq$ 95%				
Level 2	$\geq$ 99%				
Level 3	$\geq$ 99.97%				
Airflow Resistance: Under a gas flow rate of 85 L/min, the inhalation resistance must not exceed 343.2 Pa					
Ethylene Oxide Residue: Must not exceed 10 μg/g					
Microbial Indicators of Masks					
Total Bacterial Colony Count/ (CFU/g)	Coliforms	Pseudomonas aeruginosa	Staphylococcus aureus	Hemolytic Streptococcus	Total Fungal Colony Count/ (CFU/g)
$\leq$ 200	Not detectable	Not detectable	Not detectable	Not detectable	$\leq$ 100

(2) Standards for Labor Protection Masks

The standard "Respiratory Protection—Self-priming Filtering Anti-particulate Respirator" (GB 2626—2019) for labor protection masks was issued by the National Safety Production Supervision and Administration Bureau on December 31, 2019, and officially implemented on July 1, 2020. It mainly targets industrial dust masks and stipulates the technical requirements, testing methods, and marking of self-priming filtering anti-particulate respirators. According to the filtration efficiency of the mask, this standard divides the masks into three levels, including 90 (KN90, KP90), 95 (KN95, KP95), and 100 (KN100, KP100). Among them, KN masks are only suitable for filtering non-oily particles, while KP masks are suitable for filtering oily and non-oily particles. The main technical indicators of labor protection masks are shown in Table 2.11.

(3) Standard for daily protective masks

Due to the differences in usage environments and main objects to be protected, neither of the first two types of masks can be fully applicable to civilian anti-smog ($PM_{2.5}$). In order to protect people's daily health, the General Administration of Quality Supervision, Inspection and Quarantine of the People's Republic of China and the Standardization Administration of China have specifically issued the "Technical Specifications for Daily Protective Masks" (GB/T 32610—2016) for anti-smog and other daily protective masks. This standard innovatively matches the mask grading with the degree of air pollution, allowing people to selectively wear masks according to the level of pollution, in order to prevent the inhalation of fine particles. The main technical indicators of daily protective masks are shown in Table 2.12.

Table 2.11 Main technical indicators of labor protection masks stipulated in GB 2626-2019, China

Category and level of filter element	Filtration efficiency			
	Tested with Sodium Chloride Particles		Oil particle detection	
KN90	$\geq$ 90.0%		Not applicable	
KN95	$\geq$ 95.0%			
KN100	$\geq$ 99.97%			
KP90	Not applicable		$\geq$ 90.0%	
KP95			$\geq$ 95.0%	
KP100			$\geq$ 99.97%	
Mask Category	Breathing Resistance Requirements			
	Inhalation Resistance/Pa		Exhalation Resistance	
	KN90 and KP90	KN95 and KP95	KN100 and KP100	
Disposable mask, without exhalation valve	$\leq$ 170	$\leq$ 210	$\leq$ 250	Same as inhalation resistance
Disposable mask, without exhalation valve	$\leq$ 210	$\leq$ 250	$\leq$ 300	$\leq$ 150
Replaceable half mask and full face mask	$\leq$ 250	$\leq$ 300	$\leq$ 350	

Table 2.12 Main technical indicators of daily protective masks specified in GB/T 32610-2016, China

Test items		Quality indicators
Formaldehyde content/(mg/kg)		$\leq$ 20
pH value		4.0–8.5
Ethylene oxide residue/(μg/g)		$\leq$ 10
Inhalation resistance/Pa		$\leq$ 175
Exhalation resistance/Pa		$\leq$ 145
Microorganisms	Escherichia coli	Not detectable
	Pathogenic pyogenic bacteria	Not detectable
	Total fungal colony count/(CFU/g)	$\leq$ 100
	Total bacterial colony count/(CFU/g)	$\leq$ 200

Filter efficiency level requirements				
Filter efficiency grading		Level 1	Level 2	Level 3
Filter efficiency/%	Saline medium	$\geq$ 99	$\geq$ 95	$\geq$ 90
	Oily medium	$\geq$ 99	$\geq$ 95	$\geq$ 90

2.5.4 American Mask Standards

According to the "42 CFR Part 84" regulation (Respiratory Protective Devices) of the Department of Health and Human Services (HHS), the National Institute for Occupational Safety and Health (NIOSH) in the United States classifies its certified masks into 9 categories, as shown in Table 2.13.

Based on the filtering characteristics of the middle filter, masks in America are divided into the following three types, including Level 100, indicating a minimum filtration efficiency of 99.97%; Level 99, indicating a minimum filtration efficiency of 99%; and Level 95, indicating a minimum filtration efficiency of 95%. According to the minimum filtration efficiency of the filter material, masks can be divided into the following three levels. N series, where N stands for not resistant to oil, can be used to protect against non-oily suspended particles; P series, where P stands for oil proof, can be used to protect against non-oily and oily suspended particles; R series, where R stands for resistant to oil, can be used to protect against non-oily and oily suspended particles.

Taking the N95 mask as an example, it is one of the nine masks certified by NIOSH. "N" means not suitable for oily particles (the smoke produced by cooking is oily particles, while the droplets produced by people talking or coughing are not oily); "95" means that under the testing conditions specified by NIOSH, the filtration efficiency reaches 95%. N95 is not a specific product name, as long as it meets the N95 standard and passes the NIOSH review, it can be called an "N95 mask".

Table 2.13 NIOSH mask classification and main technical indicators

Serial number	Series	Level	Efficiency/%	Types of Particles that can be Protected Against
1	N Series	N100	≥ 99.97	Non-oily Particles
2		N99	≥ 99	
3		N95	≥ 95	
4	R Series	R100	≥ 99.97	Non-oily Suspended Particles, Oily Suspended Particles
5		R99	≥ 99	
6		R95	≥ 95	
7	P Series	P100	≥ 99.97	Non-oily Suspended Particles, Oily Suspended Particles
8		P99	≥ 99	
9		P95	≥ 95	

Table 2.14 Main technical indicators of EN149: 2001+A1: 2009 masks

No.	Test items		Quality Indicators			
1.	Filter material	Salt (95 L/min)	FFP1: ≤ 20			
			FFP2: ≤ 6			
			FFP3: ≤ 1			
	Penetration Rate/%	Oil (95 L/min)	FFP1: ≤ 20			
			FFP2: ≤ 6			
			FFP3: ≤ 1			
2.	Breathing Resistance/mbar		Level	Inhalation		Exhalation
				30 L/min	95 L/min	160 L/min
			FFP1	≤ 0.6	≤ 2.1	≤ 3.0
			FFP2	≤ 0.7	≤ 2.4	≤ 3.0
			FFP3	≤ 1.0	≤ 3.0	≤ 3.0

2.5.5 *European Mask Standards*

The European personal protective mask standard is EN149: 2001+A1: 2009 "Respiratory protective devices - Filtering half masks to protect against particles - Requirements, testing, marking", established by the European Committee for Standardization (Comité Européen de Normalisation, CEN), see Table 2.14. This standard divides masks into three levels according to filtration efficiency: FFP1, with a minimum filtration effect of > 80%; FFP2, with a minimum filtration effect of > 94%; FFP3, with a minimum filtration effect of > 97%.

The EN149 standard divides particulate protective filter materials into two types: solid particle protection and liquid particle protection, which are tested and graded with NaCl and DOP (paraffin oil) aerosols. According to the penetration rate of the tested particles, the qualified solid particle protective filter materials are further divided into three levels: P1 (FFP1), P2 (FFP2), and P3 (FFP3). Among them, P3 has the best protection performance and P1 has the lowest. Liquid particle protective filter materials are divided into P2 and P3 levels, where with P3 has higher protective performance than P2. Its detection flow rate is 95 L/min, using DOP dusting.

References

1. Ni WQ, Zhao T, Yang CC, et al. Analytical investigation on the situation of air quality in atmospheric environment. Sci Technol Innov. 2020;31(66–67):69.
2. The State Council Information Office of the People's Republic of China. Notice of the state council on issuing the action plan for air pollution prevention and control. State Council Guo Fa [2013] No. 37. (2013-9-10).
3. Ministry of Ecology and Environment of the People's Republic of China. The ministry of ecology and environment released the 2020 national ecological environment Quality Briefing

[EB/OL]. [2021-07-01]. http://www.mee.gov.cn/xxgk2018/xxgk/xxgk15/202103/t20210302_823100.html.
4. Domingo JL, Rovira J. Effects of air pollutants on the transmission and severity of respiratory viral infections. Environ Res. 2020;187: 109650.
5. Xu LP. Modification effects of temperature and gaseous pollutants in the relationship between particulate matter and outpatient visits of pediatric acute respiratory diseases. Taiyuan: Shanxi Medical University; 2020.
6. Kelly F, Fussell J. Global nature of airborne particle toxicity and health effects: a focus on megacities, wildfires, dust storms and residential biomass burning. Toxicol Res. 2020;9(4):331–45.
7. Zhang GC. Analysis and treatment of air pollution in environmental engineering. Huanjingyufazhan. 2019;31(9):31–3.
8. Wang XL, Zou B, Li SX, et al. Design and implementation of air pollution exposure risk prevention and early warning service platform. Adm Tech Environ Monit. 2020;32:1–7.
9. Yang P, Liu J. Research on the spatiotemporal variation of atmospheric pollutants and their intelligent optimization algorithm. Beijing: Science Press; 2017.
10. Zhang XG, Yao WQ, Peng YC. Air pollution control technology. Beijing: Chemical Industry Press; 2020.
11. Feng L, Liao W. Legislation, plans, and policies for prevention and control of air pollution in China: achievements, challenges, and improvements. J Clean Prod. 2017;112:1549–58.
12. Maji KJ, Li VOK, Lam JCK. Effects of China's current air pollution prevention and control action plan on air pollution patterns, health risks and mortalities in Beijing 2014–2018. Chemosphere. 2020;260: 127572.
13. Guo SE, Chi MC, Hwang SL. Effects of particulate matter education on self-care knowledge regarding air pollution, symptom changes, and indoor air quality among patients with chronic obstructive pulmonary disease. Int J Environ Res Public Health. 2020;17(11):4103.
14. Yang LF. Atmospheric pollution control technology. Beijing: China Environmental Publishing House; 2011.
15. Salonen H, Salthammer T, Morawska L, et al. Human exposure to indoor air contaminants in indoor sports environments. Indoor Air. 2020;30(6):1109–29.
16. Huang XH, Han XX, Li SD, et al. Spatial and temporal variations and relationships of major air pollutants in Chinese cities. Res Environ Sci. 2017;30(7):1001–11.
17. Yang W, Zheng RD, Zhang HD, et al. Development process and trend of waste to energy projects in China. Environ Eng. 2020;38(12):124–9.
18. Wang RF. Effects of ultra-low emissions and power substitution on the atmospheric particulate matter. Hefei: University of Science and Technology of China; 2020.
19. People's Network. This year's carbon dioxide emission intensity will be reduced by at least 3.1%. [2021-07-01]. http://www.gov.cn/guowuyuan/2015-03/16/content_2835101.htm.
20. Ministry of Environmental Protection, National Development and Reform Commission. National Energy Administration. Full implementation of ultra-low emission and energy-saving transformation of coal-fired power plants. 2015.

Chapter 3
Structural Characterization and Performance Evaluation of Gas Cleaning Membranes

3.1 Microstructure and Characterization Methods

3.1.1 Microstructure

The microstructure of the gas cleaning membrane mainly includes surface morphology, cross-sectional morphology, three-dimensional structure, etc. By characterizing the microstructure of the gas cleaning membrane, the basic information in terms of the pore structure characteristics, pore distribution, porosity, membrane thickness, and membrane integrity of the membrane material can be grasped. The conventional microstructure characterization methods of membrane materials include scanning electron microscopy, atomic force microscopy, and transmission electron microscopy [1–3].

3.1.1.1 Scanning Electron Microscope (SEM)

The scanning electron microscope uses an electron beam instead of visible light, emits a high-energy electron beam with an electron gun to bombard the sample surface, and collects secondary electron information, which can present a vivid and three-dimensional secondary electron morphology image. For membrane materials with low conductivity, it is usually necessary to spray gold on the membrane to prevent the accumulation of surface charges caused by the low conductivity, which affects the image quality.

(1) Selection and preparation of samples

Common gas cleaning membranes are mainly divided into particle stacking types (such as ceramic membranes and metal membranes) and fiber constructed types (such as biaxially stretched organic membranes, electrospun nanofiber membranes, and ceramic fiber membranes, etc.) in terms of pore structure characteristics. Scanning

Z. Zhong and W. Xing, *Membrane Technology for Air Cleaning*,
https://doi.org/10.1007/978-981-96-1144-7_3

electron microscopy (SEM) characterization of the surface and cross-section of the membrane material can reflect the basic structural characteristics of the membrane material. For particle stacking type membrane materials, the preparation of cross-section samples needs to ensure that the cross-section is as neat as possible, which can be achieved by sample clamps, water jet cutting, etc. For nanofiber membrane materials, the corresponding cross-section can be obtained by means of freeze-fracture. Due to the relatively small thickness of this membrane type, usually between 5 and 125 μm, the scanning quality of the cross-section image can be improved by oblique platform sampling.

(2) SEM imaging analysis

Figures 3.1a, c are the surface and cross-sectional morphology of silicon carbide analyzed by a desktop scanning electron microscope (Hitachi TM-3000), respectively. As can be seen from the figure, the silicon carbide particles are continuous, the neck connection between the particles is obvious, and there are no obvious defects on the membrane surface. The cross-sectional image shows that the membrane thickness is about 200 μm, and the membrane thickness is relatively uniform. There is almost no infiltration of the particles. Figures 3.1b, d are the surface and cross-sectional morphology of polytetrafluoroethylene (PTFE) membrane scanned by a cold field emission scanning electron microscope (Hitachi S-4800). As can be seen from Fig. 3.1, the PTFE membrane is a continuous network structure connected by fibers and nodes, with uniform fiber diameter. The cross-sectional image shows that the membrane layer is well connected with the support layer, and the membrane thickness is smaller than 10 μm. By analyzing the images of the membrane surface, basic information such as the continuity of the membrane surface and the characteristics of the pore structure can be obtained. Analysis of the cross-sectional images can deliver information about the thickness of the membrane layer, transition layer, and support layer. The status of connection between the membrane layer and the support layer, particle infiltration, etc. can be roughly estimated from the continuity between the layers. By using software analysis, information on the surface porosity, particle or fiber size, and pore size of the membrane material can be further obtained. In addition, using an energy-dispersive X-ray spectroscopy (EDX) coupled scanning electron microscope is able to analyze the distribution of elements on the surface and cross-section of the membrane material [4, 5], thereby judging the quality, particle bonding, fouling or corrosion, etc., of the membrane material.

3.1.1.2 Atomic Force Microscope (AFM)

The atomic force microscope studies the surface structure and properties of materials by detecting the extremely weak atomic interaction between the surface of the tested sample and a micro force-sensitive element, which can obtain the three-dimensional structure of the material surface. It has high requirements for the surface roughness of the sample and is generally suitable for the surface structure characterization of membrane materials with nano-scale pore diameters.

Fig. 3.1 SEM images of silicon carbide membrane surface and cross-section (**a**, **c**); SEM images of polytetrafluoroethylene porous membrane surface and cross-section (**b**, **d**)

(1) Selection and preparation of samples

Preparation of powder samples: The adhesive tape method was often used for the preparation of powder samples. First, double-sided adhesive tape needs to be attached to the sample stage, then the powder is sprinkled on the adhesive tape, and the excess powder that is not attached to the adhesive tape is blown off.

Preparation of bulk samples: Solid samples such as glass, ceramics, and crystals need to be polished, according to the surface roughness of the solid samples.

Preparation of liquid samples: The concentration of liquid samples should not be too high, otherwise particle agglomeration can damage the needle tip. (Nanoparticles: the nanopowder needs to be dispersed into the solvent (the thinner, the better), which is then coated onto a mica or silicon slice, by either manual drop coating or spin coating, and dried naturally.)

(2) AFM imaging analysis

Imaging with electron microscopes and optical microscopes is susceptible to electromagnetic diffraction, which makes it difficult to identify three-dimensional structures. Atomic force microscopy can compensate for these shortcomings, with peaks and valleys clearly visible in AFM images [6, 7]. In the meantime, light or electricity has minor effect on its imaging, and it can be used to measure the true morphology of the sample surface and to directly observe the fine structure of various uneven sample surfaces. Figure 3.2 shows the surface characterization of gas separation membranes

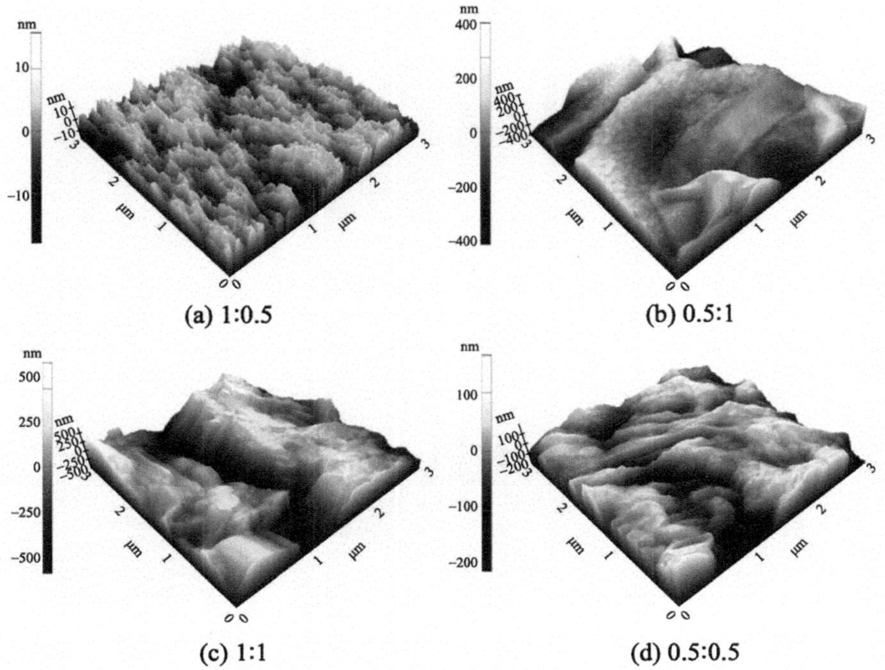

Fig. 3.2 Atomic force microscope images of CO_2/CH_4 gas separation membranes obtained at different ratios of 1, 3-cyclohexanedimethanamine to isophthaloyl chloride [8]

measured by the Nanoman AFM system, Veeco model AFM [8], from which the effect of different ratios of 1,3-cyclohexanedimethanamine to isophthaloyl chloride on the surface roughness of the prepared membranes can be seen. The white area is the highest position, and the black area is the lowest position.

3.1.1.3 Transmission Electron Microscopy (TEM)

Although TEM has a resolution of 0.2 nm, its electron beam needs to penetrate the sample, which is thus strictly limited to the thickness of the test sample. The thickness of the observation area of the sample needs to be controlled below 100 nm. The sample preparation technology, to some extent, determines the depth and breadth of the application of transmission electron microscopy. Therefore, transmission electron microscopy is often used to analyze nanofiber membrane materials with multi-level pore structures and nanofiber membrane materials with catalytic or adsorption functions [9]. The common sample preparation method is the physical peeling of samples, which are dispersed in solution and then characterized by transmission electron microscopy. For thicker materials, samples can be prepared by chemical

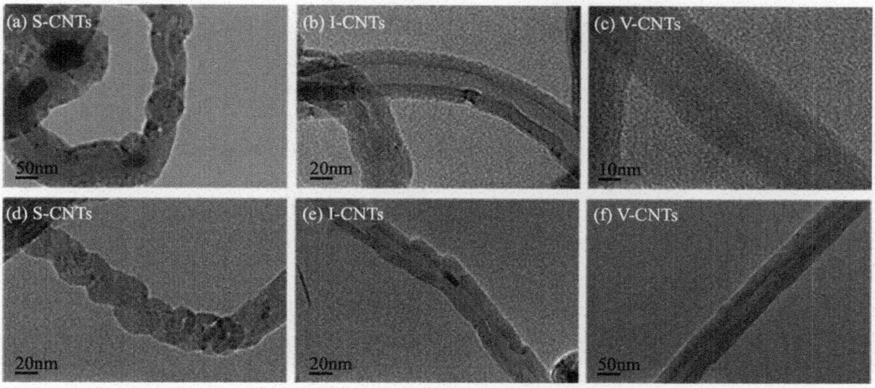

Fig. 3.3 TEM characterization of carbon nanotube membranes [10]

thinning, ultrathin slicing, ionic thinning, etc. Figure 3.3 shows the TEM characterization results of the basic structures of three types of carbon nanotube membranes [10], where S-CNTs are helical, I-CNTs interwoven, and V-CNTs vertical. The figure shows the basic structure, wall thickness, tube diameter, and other fundamental information of carbon nanotubes.

3.1.2 Membrane Pore Size and Distribution

The measurement of membrane pore size includes geometric and physical methods. The geometric method is direct observation of the geometric structure of the membrane pores with a transmission or scanning electron microscope, and determination of the surface pore size and pore distribution through software analysis. Scanning electron microscopy can intuitively provide images of the membrane surface morphology, but it cannot observe the pore structure inside the membrane. Transmission electron microscopy can observe the pore structure of the membrane after the sample is ultra-thinly sliced. The geometric method is suitable for symmetric membrane materials with small thickness and consistent pore structure in the cross-sectional direction, such as nanofiber membranes prepared by electrospinning, but it is not suitable for asymmetric membrane materials.

The physical method is to calculate the pore size of a membrane by measuring the physical effects related to the membrane pores. This pore size is often defined as the physical pore size. Common measurement methods include the bubble point method, mercury intrusion method, and gas adsorption–desorption method [11–13]. There are open and closed pores in a particle stacking membrane. Different measurement methods used for the same membrane sample can result in certain variations in the average pore size obtained, because they depend on the different physical effect.

3.1.2.1 Bubble Point Method

Assuming that the pores of a porous membrane are filled with a liquid of known surface tension, e.g., a capillary with a radius of r is wetted by a liquid with a surface tension of σ, the liquid phase pressure P_1 and the gas phase pressure P_2 of the capillary reach a static balance, wherein the relationship between P_1 and P_2 can be represented by the Washburn equation. There is a relation between the pressure, required for the gas to pass through the membrane pore, and the radius of the membrane pore, as follows:

When the forces reach a balance, the capillary force equals the weight of the liquid column, resulting in:

$$\Delta P = P_1 - P_2 = \frac{2\sigma \cos\theta}{r} \tag{3.1}$$

where θ is the contact angle between the liquid phase and the capillary (°); ΔP is the additional pressure difference in the capillary (MPa); σ is the surface tension of the liquid (N/m); and r is the radius of the capillary (m).

The liquid in the capillary will be expelled, when the pressure difference between both sides of the membrane is greater than $\frac{2\sigma \cos\theta}{r}$. The bubble point method is used to measure the pore size based on this principle.

Given the known surface tension and contact angle of the gas and liquid, the pressure P, corresponding to the gas passing through the membrane pore and generating bubbles on the membrane surface, can be used to calculate the pore radius using Eq. (3.1). In the experiment, the pore radius calculated from the pressure, corresponding to the first bubble present on the membrane surface, is taken as the maximum radius of the membrane. The pore radius calculated from the pressure, corresponding to the largest number of bubbles, is regarded as the minimum pore size of the membrane. The average pore size of the membrane can be calculated from the maximum and minimum pore sizes.

The experimental setup of the bubble point method is shown in Fig. 3.4.

During the experiment, the membrane should be completely wetted by the liquid otherwise, it will cause errors. Hydrophilic membranes use water as the wetting

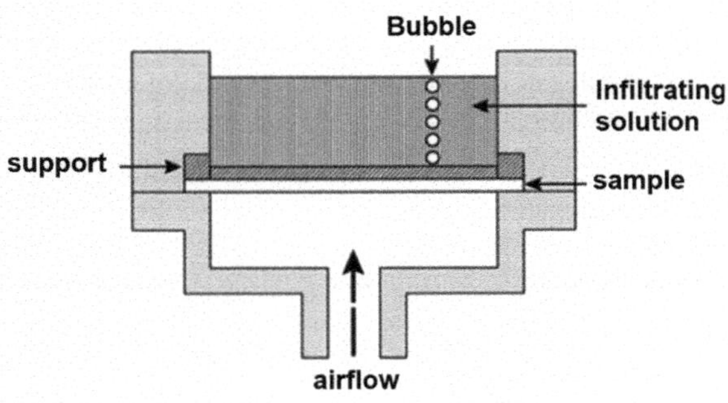

Fig. 3.4 Schematic diagram of the experimental setup for bubble point method [5]

liquid, while hydrophobic membranes often use alcohol as the wetting liquid that evaporates more slowly. The bubble pressure method is suitable for measuring the pore size of membranes ranging from 0.1 to 10 μm.

3.1.2.2 Mercury Intrusion Method

The basic principle of the mercury intrusion method is similar to that of the bubble pressure method. The mercury intrusion method is suitable for measuring larger membrane pore diameters. Mercury usually cannot wet many solid substances. In the case of non-wetting, the contact angle is greater than 90°, whereby the surface tension prevents the liquid from entering the pores. Therefore, exerting an external pressure is needed to make mercury enter the membrane pores. The greater the external pressure, the smaller the pore radius that mercury can enter. The Laplace or Washburn equation can be used when the force reaches a balance:

$$-2\pi r\sigma \cos\theta = 2\pi rp \tag{3.2}$$

then:

$$r = \frac{2\sigma \cos\theta}{p} \tag{3.3}$$

The contact angle of mercury on polymers is normally 141.3°, and the interfacial tension of mercury/air is 0.48 N/m. The above formula can be simplified to:

$$r = \frac{7492}{p} \tag{3.4}$$

where p is the applied pressure (bar); and r is the smallest pore diameter that mercury can enter under a given pressure p (nm). As the pressure increases, the small pores are gradually filled until all the pores are filled. In the meantime, the volume of mercury injected reaches the maximum. The pore size distribution can be determined from the variation curve of volume as a function of pressure, according to the cumulative volume of mercury entering the membrane sample under each pressure. The membrane pore size can be calculated from the pore size distribution curve.

The disadvantages of the mercury intrusion method are as follows:

(1) The cost of the mercury intrusion instrument is high, and the measurement pressure is high;
(2) In the calculation, it is assumed that the contact angle of mercury on the membrane is 141.3°, but the contact angle of mercury on the membrane can actually vary with the type of membrane samples. Moreover, mercury is inevitably contaminated during the test, causing changes in surface tension and contact angle, and thus making errors;

(3) The smaller the pore size, the higher the pressure required to press mercury into the pores, which may inevitably cause compressive deformation of the samples;
(4) It cannot distinguish between open pores and closed pores.

3.1.2.3 Gas Adsorption–Desorption Method

The gas adsorption–desorption method for measuring pore size and its distribution is based on the capillary condensation phenomenon and the Kelvin equation, which can reflect the information of pores in the range of 3–60 nm. Inert gases (such as N_2) are usually used as adsorbates. The adsorption amounts during adsorption and desorption of N_2 at constant temperature, but changing relative vapor pressures are measured separately to obtain the adsorption–desorption isotherm. The pore size distribution is calculated from the experimental data, based on the Kelvin equation. The adsorption–desorption method can be affected by the support when measuring the pore structure of the support membrane, and is thus often used for the measurement of unsupported membranes.

$$\ln \frac{p}{p_0} = -\frac{2\sigma V_m}{\gamma_k RT} \cos \varphi \tag{3.5}$$

where p/p_0 is the relative pressure of the gas on the curved liquid surface; σ is the surface tension of the adsorbate liquid (10^{-5} N/cm); V_{m} is the molar volume of the adsorbed liquid (mL/mol); φ is the contact angle between the crescent surface and the solid wall, taken as 0 when the liquid can wet the surface; R is the gas constant; T is the adsorption temperature (K); and γ_{k} is the Kelvin radius (m).

3.1.3 Membrane Porosity

Porosity is generally defined as the percentage of the pore volume of a porous membrane accounting for the apparent volume of the membrane, that is,

$$\varepsilon = \frac{V_{pore}}{V_{membrane\ apparent}} \tag{3.6}$$

In the application and research of gas cleaning membrane materials, porosity is an important and commonly used indicator, which can be used to evaluate the filtration performance and separation ability of membrane materials. At present, the porosity of inorganic gas cleaning membrane materials is generally above 40%, the porosity of organic membrane materials is generally greater than 70%, with its highest value exceeding 90%. The specific examples of several commonly used test methods for membrane porosity are as follows [14–16].

3.1.3.1 Weighing Method

The weighing method determines the pore volume of a membrane V_{pore} based on the weight change before and after the membrane is wetted with a proper liquid (such as water). The skeleton volume of the membrane $V_{\mathrm{membrane\ skeleton}}$ can be obtained through the density of the raw material of the membrane and the dry membrane weight, then the porosity of the membrane is:

$$\varepsilon = \frac{V_{pore}}{V_{membrane\ apparent}} = \frac{V_{pore}}{V_{pore} + V_{membrane\ skeleton}} = \frac{m_{wet\ weight} - m_{dry\ weight}}{m_{wet\ weight} - m_{floating\ weight}} \tag{3.7}$$

where $m_{wet\ weight}$ is the mass of the membrane after complete wetting; $m_{dry\ weight}$ is the mass of the dry membrane in the air; $m_{floating\ weight}$ is the mass of the fully wetted membrane in the wetting liquid.

3.1.3.2 Density Method

The density method generally only requires the density of the raw material of a membrane ($\rho_{\mathrm{membrane\ material}}$) and the apparent density of the membrane $\rho_{\mathrm{membrane\ apparent}}$, to calculate the porosity (ε). The apparent density ($\rho_{\mathrm{membrane\ apparent}}$) can be obtained from the apparent volume and mass.

$$\begin{aligned}\varepsilon &= \frac{V_{pore}}{V_{membrane\ apparent}} = \frac{V_{membrane\ apparent} - V_{membrane\ skeleton}}{V_{membrane\ apparent}} \\ &= \frac{\rho_{membrane\ apparent} - \rho_{membrane\ skeleton}}{\rho_{membrane\ apparent}}\end{aligned} \tag{3.8}$$

3.1.3.3 Mercury Intrusion Method

The mercury intrusion method is using pressure to force mercury into the various structures of the "pores" in a membrane, whereby the pore volume and sizes of the membrane can be calculated based on the pressure and volume of the injected mercury. The disadvantage of this method is that a large pressure is needed to force mercury into the micropores, making it more suitable for analyzing rigid materials. However, most membrane materials are elastic and can easily deform or "collapse" during the mercury injection process, thereby resulting in significant errors.

3.1.4 Membrane Thickness

The thickness of gas cleaning membranes significantly affects the gas permeability and gas cleaning performance. One of the development directions of gas cleaning membranes is to develop membranes with small thickness and good retention performance. The measurement of membrane thickness is an important step in the membrane preparation process, which can provide feedback on the preparation method, preparation times of membrane materials, the sintering temperature and program of ceramic membranes, the stretching ratio of organic membranes, and the electrospinning time, etc.

Mechanical measurement and electron microscope measurement are the most commonly used methods for measuring membrane thickness [17]. The mechanical measurement method determines the membrane thickness using a thickness tester [18], wherein the force applied to the sample by the measuring surface should be between 0.5 and 1.0 N, the diameter of the measuring surface should be between 2.5 and 10 mm, and the measurement result has an accuracy of 0.1 μm. The measuring surfaces of the sample and the measuring instrument should be free of oil stains, dust, and other contaminants. The formula for calculating the thickness of gas cleaning membranes, observed by the mechanical measurement method or the electron microscope method, is as follows:

$$T = \frac{\sum T_i}{i} \tag{3.9}$$

where T is the average thickness (μm or nm); T_i is the actual thickness at a specific location (μm or nm); i is the measurement times of the actual thickness at a specific location. In addition, a scanning electron microscope can also be used to observe the thickness of the membrane layer [19], each sample should have at least two electron microscope images scanned at different locations. Two or more images need to be taken at different locations, and at least two local membrane thicknesses should be measured for each image when the uniformity of the membrane layer needs to be examined. However, the observation of the thickness of the membrane layer by a scanning electron microscope can produce a large error and often varies with the magnification because there is no clear boundary between the membrane layer and the support layer. Figure 3.5 shows the thickness of the silicon carbide membrane layer measured by a scanning electron microscope, which is about 125 μm [4].

Fig. 3.5 Measurement of silicon carbide membrane thickness by scanning electron microscope [4]

3.2 Membrane Material Stability

3.2.1 Mechanical Properties

Tensile strength and elongation at break are generally the main indicators for evaluating the mechanical properties of low-temperature gas cleaning membranes [20]. Tensile strength refers to the maximum tensile stress recorded when the tensile specimen is stretched to fracture. The elongation at break is the percentage of the increment of the marked distance to the marked distance of the unstretched specimen, when the specimen is stretched to fracture. For high-temperature gas cleaning membranes, the mechanical properties are usually evaluated by tensile strength and flexural strength tests.

3.2.1.1 Tensile Strength

Tensile strength refers to the ratio of the tensile force, when a membrane breaks, to the original cross-sectional area of the fracture surface of the membrane material [21–23], namely, the fracture stress. Its calculation formula is:

$$\sigma = \frac{F}{A} \tag{3.10}$$

where σ is the stress (MPa); F is the corresponding load measured (N); and A is the original cross-sectional area of the specimen (mm^2).

3.2.1.2 Elongation at Break

Elongation at break is an indicator of the flexibility and elasticity of a membrane material [24–26], the greater the elongation at break, the better the flexibility and elasticity. Elongation at break, expressed as ε, can represent the elongation deformation ability of the membrane material under the maximum load. The sample can be clamped under pre-tension, or loosely clamped (without tension). The resulting elongation should not exceed 2%, when the sample is clamped under pre-tension. The calculation is as follows:

Pre-tension clamped sample:

$$\varepsilon = \frac{\Delta L}{L_0} \times 100\% \tag{3.11}$$

where ε is the elongation at break (%); ΔL is the elongation at break when the sample is clamped under pre-tension (mm); and L_0 is the gap length (mm).

Loosely clamped sample:

$$\varepsilon = \frac{\Delta L' - L_0'}{L_0 + L_0'} \times 100\% \tag{3.12}$$

where ε is the elongation at break (%); $\Delta L'$ is the elongation at break when the sample is loosely clamped (mm); L_0' is the elongation when the loosely clamped sample reaches the specified pre-tension (mm); and L_0 is the gap length (mm).

3.2.1.3 Flexural Strength

The flexural strength of a membrane material mainly depends on its inherent ability to resist fracture [27]. Flexural strength is usually evaluated by three-point flexural testing or four-point flexural testing. The three-point bending method is easy to operate, but it can only measure a small part of the local stress of the sample, and the strength measured is often much larger than the four-point flexural strength. Therefore, the four-point bending method is advocated in the property characterization of gas cleaning membrane material. Before the flexural strength test, the width (b) and thickness (d) of the membrane material should be measured at the closest location to the midpoint, with the span set at 30 mm and the beam rate set at 0.5 mm/min. The number of the test samples should not be fewer than 10. The formulas of the three-point bending method and the four-point bending method for calculating the flexural strength are as follows:

Three-point bending method:

$$\sigma_f = \frac{3FL}{2bd^2} \tag{3.13}$$

where σ_f is the flexural strength (MPa); F is the maximum load (N); L is the lower span of the fixture (mm); b is the width of the sample (mm); and d is the height (thickness) of the sample parallel to the loading direction (mm).

Four-point bending method:

$$\sigma_f = \frac{3Fa}{bd^2} \tag{3.14}$$

where σ_f is the flexural strength (MPa); F is the maximum load (N); a is the length of the bending arm of the sample (mm); b is the width of the sample (mm); and d is the height (thickness) of the sample parallel to the loading direction (mm).

The three-point flexural strength is one of the main technical indicators for evaluating the strength of inorganic membrane materials. For ceramic membranes used for gas cleaning, the three-point flexural strength needs to reach higher than 15 MPa.

3.2.2 *Wetting and Non-wetting Properties*

The gas cleaning system is complex, where the membrane material often needs to be suitable for high humidity systems and systems containing liquid phase aerosol pollutants. The wetting and non-wetting properties of the membrane material significantly affect the formation of the filter cake layer, the setting of the back-blowing cycle, the evaluation of back-blowing efficiency, the regenerability of the membrane, and the service life.

The contact angle test of a liquid on the membrane surface is one of the main methods for evaluating the wetting and non-wetting properties of the membrane material [28–31]. The tested membrane material sample needs to remain flat and be placed horizontally, without wrinkles and twists. A droplet of $1-2\ \mu L$ is then hanged at the end of the needle, and contacts the membrane surface. The contact angle is measured after the droplet drops within (60 ± 10) s, and calculated as follows:

$$\theta = 2 \times \tan^{-1}(H/R) \tag{3.15}$$

where θ is the contact angle between the droplet and the sample surface; H is the height of the droplet image (mm); and R is the half-height of the droplet image (mm).

The membrane sample is then moved to make the next droplet drop on a new test location of the membrane. Measurements for 10 times on the same sample are needed to obtain an average of the contact angles:

$$\theta = \frac{\theta_{sta,1} + \theta_{sta,2} + \cdots + \theta_{sta,10}}{10} \tag{3.16}$$

where $\theta_{sta,1}$, $\theta_{sta,2}$ and $\theta_{sta,10}$ are the contact angles of the membrane tested for 10 times; and θ is the average contact angle of the membrane material.

3.2.3 Temperature Resistance Characteristics

The flue gas emitted from industrial fields, such as coal chemical industry, mine hot furnace, ferroalloy, and phosphorus chemical industry, not only contains many fine particles and harmful chemical substances, but also often has residual heat. The temperature of the flue gas is generally above 300 °C, and can reach up to 800–1000 °C [32]. Therefore, the membrane materials, used for high-temperature flue gas cleaning, need to have good heat resistance [33, 34].

Heat resistance mainly refers to the ability to resist thermal damage of membrane materials, when working in a hot environment for a long time, which is usually expressed as heat resistance temperature. The limit of the operating temperature of polymer membranes is the glass transition temperature of the membrane material. However, the actual operating temperature of a membrane in practical applications is far lower than the glass transition temperature of the membrane material, due to the influence of environmental factors. At present, the heat resistance of the membrane is often tested by Differential Scanning Calorimetry (DSC) [35–37], through recording the DSC curve of the sample during the programmed heating, cooling, or constant temperature process. The DSC method for measuring the heat resistance of low and medium gas cleaning membrane materials is as follows [38]: the temperature is raised from room temperature to 300 °C with a heating rate of 10 °C/min, under the atmosphere of N_2 with a gas flow rate of 15 mL/min. The DSC method for measuring the heat resistance of high-temperature gas cleaning membrane materials is as follows [39]: the temperature is raised from room temperature to 1600 °C with a heating rate of 10 °C/min, under the atmosphere of N_2 with a gas flow rate of 15 mL/min.

3.2.4 High-Temperature Thermal Expansion Performance

In terms of high-temperature gas cleaning, the practical application of membrane materials is often accompanied by a thermal shock environment of extremely high-temperature (flue gas environment) and cooling (back-blowing gas). Therefore, the membrane materials need to have good thermal expansion resistance. The most effective way to improve the thermal expansion resistance of membrane materials is to choose materials with low thermal expansion coefficients for membrane preparation. In addition, additives or sintered new phases need to have a thermal expansion coefficient similar to that of the membrane materials [40, 41].

High-temperature thermal expansion usually refers to the phenomenon that the volume increases as the temperature rises and decreases as the temperature drops under constant external pressure [42]. Thermal expansion is related to physical properties of the membrane material, such as temperature, heat capacity, binding energy, and melting point. Material composition and organization, phase change, and the characteristics of different phases, etc., are the main factors affecting the expansion

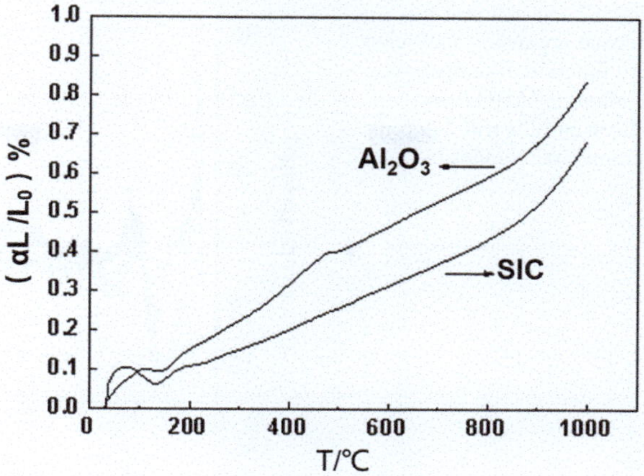

Fig. 3.6 Thermal expansion curves of alumina (Al_2O_3) and silicon carbide (SiC) ceramic membranes measured by dilatometer [44]

properties of the material. The measurement methods of thermal expansion mainly include optical method, electrical method, and mechanical method. Currently, the thermal expansion coefficient is usually used to characterize the expansion or contraction properties of the membrane material due to temperature changes [43], which is expressed as follows:

$$\frac{\Delta L}{L_0} = \left(\frac{\Delta L}{L_0}\right)_\alpha + A \tag{3.17}$$

where $\frac{\Delta L}{L_0}$ is the thermal expansion tested within a specified temperature range; $\left(\frac{\Delta L}{L_0}\right)_\alpha$ is the thermal expansion value measured by a dilatometer within a specified temperature range; and A is the correction constant.

Figure 3.6 shows the thermal expansion curves of alumina and silicon carbide ceramic membranes. Both the alumina and silicon carbide are in a state of thermal expansion, with temperatures above 100 °C. The degree of thermal expansion of the alumina materials is greater than that of the silicon carbide, and the higher the temperature, the greater the degree of expansion of the materials. Figure 3.7 shows the relationship between the thermal expansion coefficients of alumina and silicon carbide ceramic membranes and temperature. The average thermal expansion coefficients of the two materials are $8.49 \times 10^{-6}/K$ and $6.75 \times 10^{-6}/K$, respectively. Silicon carbide is superior to alumina, in terms of thermal expansion properties [44].

3.2.5 Thermal Shock Resistance

Membrane materials, used for gas cleaning under high temperature conditions, must be able to withstand the shock, caused by the pulse jet of extremely fine dust particles,

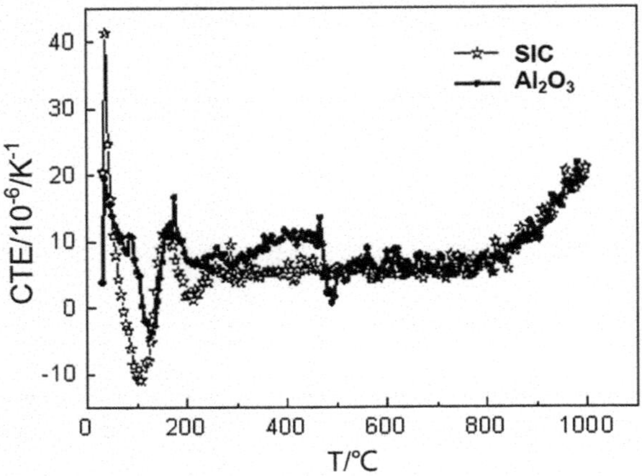

Fig. 3.7 Variations of thermal expansion coefficient as a function of temperature of alumina (Al_2O_3) and silicon carbide (SiC) porous ceramic membranes [44]

and the changes in thermal stress, resulting from a rapid increase in temperature or cooling. Thermal shock resistance is the ability of gas cleaning membrane materials withstanding a rapid temperature change, to some extent, without causing structural damage to the membranes, which is also known as thermal impulse resistance or thermal stability. The ability to resist thermal shock is one of the key factors determining the service life of ceramic membranes.

Currently, thermal shock resistance is characterized by the thermal cycles, that the sample undergoes, and the methods for evaluating its thermal shock damage. Each thermal cycle that the sample undergoes includes two stages. In the first stage, the entire sample or only a part of it (for example, one side) is heated to the initial temperature T_i. During this heating period, the heating rate will not cause excessive stress. Thermal shock occurs in the second stage, where the initial temperature T_i quickly changes to the final temperature T_f. If $T_i > T_f$, thermal shock is caused by cooling; If $T_i < T_f$, thermal shock is caused by heating. In the case of thermal shock caused by cooling, the sample is first heated to the initial temperature T_i in a preheating furnace and held for 10–30 min. The final temperature T_f is achieved by quickly moving the sample to a low-temperature furnace, natural cooling in a room temperature environment, forced air cooling, or quenching in a water bath (or other baths) at T_f temperature. In the case of thermal shock caused by heating, the sample temperature changes rapidly from T_i to T_f, which can be achieved by moving the sample to a high-temperature furnace.

The method used to evaluate the degree of thermal shock damage is to measure the residual strength of the sample after thermal shock [45]. However, there is a large statistical deviation for the parameter of residual strength. Other evaluation methods include visual inspection, mass loss, changes in elastic modulus, etc. [46].

The sample is heated from room temperature to 1000 °C or 1100 °C (or other test temperatures) at a specified rate in a heating furnace, held for 30 min, and then quenched in flowing water or air at 5–35 °C. The processes of rapid cooling and heating are repeated, whereby the degree of damage is evaluated by a percentage of the changes in flexural strength before and after thermal shock [47, 48]. The formula for calculating the change in flexural strength is as follows:

$$R_r = \frac{R_a}{R_h} \times 100\% \tag{3.18}$$

where R_r is the retention rate of flexural strength (%); R_a is the flexural strength of the sample after thermal shock (MPa); and R_b is the flexural strength of the sample before thermal shock (MPa).

The R_r value undergoes a minor change, when ΔT is in the range of 0–1000 °C, indicating that the thermal shock damage of the material is small due to the good thermal shock resistance in this range of temperature difference. Figure 3.8 shows the changes in flexural strength of the silicon carbide support during the thermal shock test. As observed in Fig. 3.8, the flexural strength of the silicon carbide support tested by the air cooling method is higher than that tested by the water cooling method. The heat release of the water cooling method is intense. As a result, the material is more prone to produce micro-cracks during thermal expansion and contraction. The strength of the support almost remains stable, after 20 cycles of water cooling and air cooling. The large-pore silicon carbide support, prepared by this method, still maintains a flexural strength of higher than 15 MPa after 60 cycles of air cooling, indicating that silicon carbide material has very good thermal shock resistance [49].

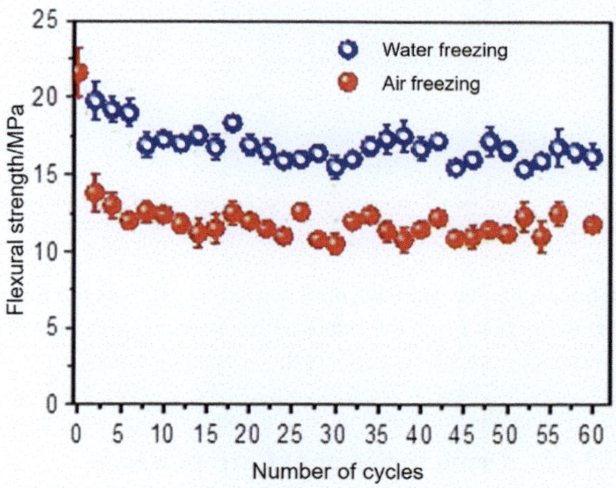

Fig. 3.8 Changes in fracture strength of the support during cold and hot circulation (30–800 °C)

3.2.6 Corrosion Resistance

Common pollutants in industrial exhaust gases mainly include smoke, particulate matter, sulfur dioxide, nitrogen oxides, fluorides, sulfides, and toluene, etc. The corrosive components, such as sulfur, alkali, and chlorine, carried in the dust-containing gas, can affect the stability of membrane materials. Therefore, the membrane material needs to have good corrosion resistance to ensure the normal continuous operation of the membrane in industrial gas cleaning processes. At present, the corrosion resistance of the membrane material is represented by the change rate of its mass and thickness after being corroded by a certain concentration of hydrochloric acid and sodium hydroxide, etc. [50, 51]. The degree of uniform corrosion can be expressed by a mass indicator and depth indicator, etc., as detailed below.

3.2.6.1 Mass Indicator of Corrosion Rate

The mass indicator of the corrosion rate includes a weight loss and weight increase per unit time and per unit area. The weight loss is the difference between the mass of the sample before corrosion and the sample mass after removing the corrosion products. The weight increase is the difference between the mass of the sample with corrosion products after corrosion and the mass before corrosion.

Weight loss:

$$v^- = \frac{w_0 - w_1}{St} \tag{3.19}$$

Weight increase:

$$v^+ = \frac{w_2 - w_0}{St} \tag{3.20}$$

where w_0 is the mass before corrosion (g); w_1 is the mass after removing the corrosion products (g); w_2 is the mass with corrosion products after corrosion (g); S is the area of the sample (m^2); and t is the corrosion time (h).

3.2.6.2 Depth Indicator of Corrosion Rate

The depth indicator of the corrosion rate (v_h) refers to the difference between the thickness of the sample before corrosion (H_0) and after corrosion (H_1) per unit time, that is:

$$v_h = \frac{H_0 - H_1}{t} \tag{3.21}$$

When comparing the corrosion degree of materials with different densities, the equation can be converted to:

$$v_h = v^- \times \frac{8.76}{\rho} \quad (3.22)$$

where ρ is the density of the material.

The corrosion resistance of membrane materials is evaluated by measuring the bending strength or mass change of the sample before and after corrosion [52]. The concentration of sulfuric acid solution used is 3.0 mol/L, the concentration of sodium hydroxide is 6.0 mol/L, and the volume of sulfuric acid or sodium hydroxide used in the corrosion experiment is 0.5 L.

$$C = \frac{|m_f - m_i|}{A} \quad (3.23)$$

where C is the mass change per unit area (g/m^2); A is the initial external surface area of the sample (m^2); m_i is the mass of the sample before corrosion (g); and m_f is the mass of the sample after corrosion (g).

Figure 3.9 shows the changes in various indicators of the silicon carbide ceramic support before and after corrosion [52]. First, corrosion of the membrane by an 1% NaOH solution is examined. After a long period of corrosion, as shown in Fig. 3.9a, the neck connection becomes loose and the surface microstructure changes significantly. The acid resistance of the SiC support is better than its alkali resistance. During continuous corrosion in a 20% H_2SO_4 solution for nearly 100 h, the support underwent a decrease in the bending strength from the initial value of 22.5 MPa to about 17.5 MPa, and the strength remained stable after 40 h, as shown in Fig. 3.9b. After the same period of corrosion in an 1% NaOH solution, the SiC support maintained a bending strength of around 15 MPa. It can be concluded from the mechanical property characterization, that the SiC support has good acid and alkali corrosion resistance. As can be seen in Fig. 3.9c, the SiC support has almost no mass loss in sulfuric acid, and the mass loss in alkali is lower than 1%, indicating that the prepared silicon carbide porous support shows very good acid and alkali corrosion resistance.

3.3 Performance Parameters and Evaluation Methods

3.3.1 Gas Permeability

Gas permeability is the volume of gas that passes through the membrane sample per unit area and per unit time under constant temperature and unit pressure difference [53], which is calculated as follows:

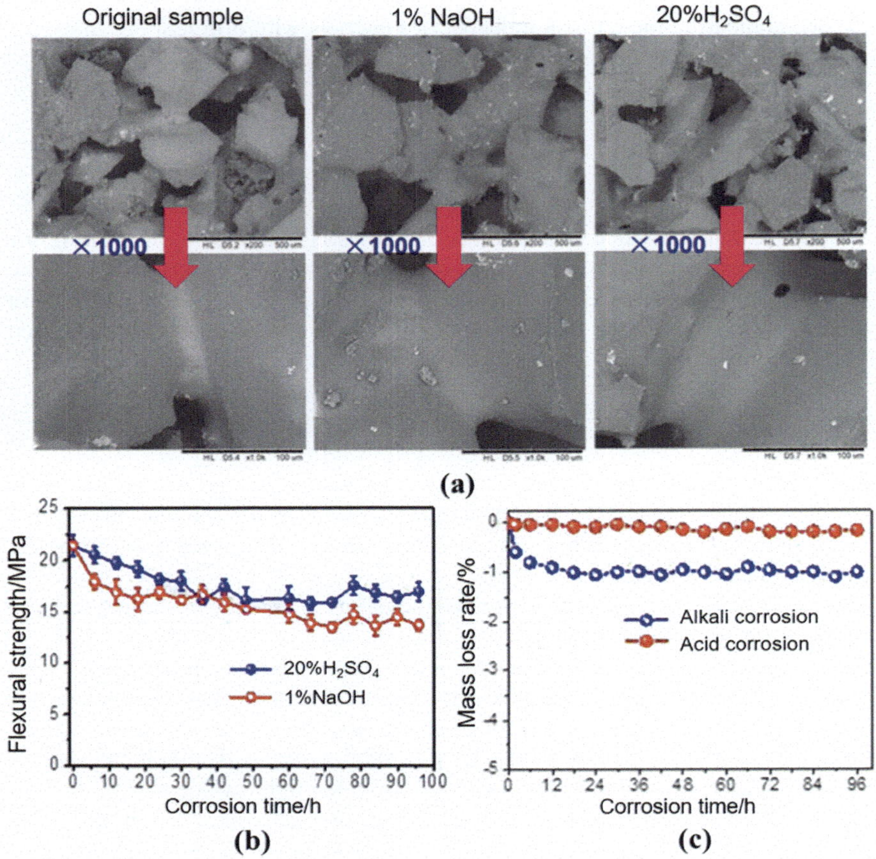

Fig. 3.9 **a** Changes in the microstructure of the silicon carbide support before and after corrosion; **b** changes in the bending strength of the silicon carbide support in acid and alkali solutions; and **c** changes in the mass loss rate of the silicon carbide support in acid and alkali solutions

$$Q_g = \frac{\Delta P}{\Delta t} \times \frac{V}{S} \times \frac{T_0}{p_0 T} \times \frac{24}{p_1 - p_2} \tag{3.24}$$

where Q_g is the amount of gas permeation of the membrane material (cm^3/(m^2 d Pa)); $\frac{\Delta P}{\Delta t}$ is the arithmetic average of the gas pressure change in the low-pressure chamber per unit time during stable permeation (Pa/h); V is the volume of the low-pressure chamber (cm^3); S is the testing area of the sample (m^2); T is the testing temperature (K); $p_1 - p_2$ is the pressure difference on both sides of the sample (Pa); and T_0 and p_0 are the temperature (273.15 K) and pressure (1.0133×10^5 Pa) under standard conditions, respectively.

The gas permeation coefficient can reflect the strength of gas permeability, which is calculated as follows:

$$P_g = \frac{\Delta P}{\Delta t} \times \frac{V}{S} \times \frac{T_0}{p_0 T} \times \frac{D}{p_1 - p_2} = 1.1574 \times 10^{-9} Q_g D \tag{3.25}$$

where P_g is the gas permeation rate of the membrane material (cm^3 cm/(cm^2 s Pa)); $\frac{\Delta P}{\Delta t}$ is the arithmetic average of the gas pressure change in the low-pressure chamber per unit time during stable permeation (Pa/s); T is the testing temperature (K); and D is the thickness of the sample (cm).

3.3.2 *Filtration Efficiency*

Filtration efficiency is calculated based on the size of penetrable particles and the number of filtered dust particles. In most cases, the filtration efficiency depends on factors such as filtration resistance, wind speed, filtration area, etc. Filtration efficiency is defined as the ratio of the difference in fine particle concentration before and after filtration to the fine particle concentration before filtration [54]. The test of filtration efficiency must apply to both non-oily and oily particles. Generally, non-oily particles are sodium chloride (NaCl) particles, and oily particles are dioctyl sebacate (DEHS), dioctyl phthalate (DOP) or other applicable oil (such as paraffin oil) particles [55]. The particle efficiency [56] is calculated as follows:

$$E = \frac{C_0 - C}{C_0} \times 100\% \tag{3.26}$$

where E represents the full particle size dust filtration efficiency (%); C_0 represents the dust concentration in the dust-containing gas (mg/m^3); and C is the dust concentration on the permeation side of the membrane sample (mg/m^3).

3.3.3 *Decontamination Coefficient and Quality Factor*

The purification coefficient, also known as the decontamination factor, is usually an important parameter for measuring the extent of removal of a certain solid particulate impurity during gas cleaning, which is specifically expressed as follows:

$$DF = \frac{Q_0}{Q_i} \tag{3.27}$$

$$DE = \frac{Q_0 - Q_i}{Q_0} \times 100\% \tag{3.28}$$

where DF is the decontamination coefficient; DE is the decontamination rate; Q_0 is the content of the particulate matter in the air before purification; and Q_i is the content of the solid particulate matter in the air after purification.

The comprehensive filtration performance of membrane materials can be evaluated by the quality factor [57, 58], which is specifically calculated as follows:

$$QF = -\frac{\ln(1-R)}{\Delta P} \tag{3.29}$$

where QF is the quality factor, R is the removal efficiency of the particle, and ΔP is the pressure drop.

3.3.4 *Filtration Resistance*

Filtration resistance is a key factor in evaluating the efficiency of gas cleaning membrane materials [59, 60]. Filtration resistance is the static pressure difference before and after the filtration of membrane materials at a certain wind speed or volume [61], the unit is Pa.

The formula is as follows [62]:

$$P = P_2 - P_1 \tag{3.30}$$

where P is the filtration resistance, P_1 is the system resistance before filtration, P_2 is the total resistance of the membrane material and the system after filtration, the units are Pa.

3.3.5 *Service Life*

The service life of a membrane is closely related to the membrane material, membrane structure, operating conditions, etc. For example, the concentration, flow rate temperature, etc., of dust-containing gas can affect the internal structure of the membrane material. Additionally, as the filtration progresses, back blowing is more frequently needed, and the residual pressure drop becomes larger. Therefore, the service life of a membrane is usually defined as the using time during which the permeation rate of the membrane drops to the allowable value. The service life of the membrane can be determined based on experience, usually by measuring the change in the air permeability rate during a certain period of time, and then by extrapolating to the allowable value of air permeability rate. The obtained time is the service life of the membrane. In addition, the service life of the membrane can also be quickly obtained by a temperature rise experiment. That is, under a certain pressure, measuring the change in gas permeation flux and temperature of the membrane as a function of time, and calculating the service life of the membrane at different operating temperatures.

References

1. Pan B, Chen JH, Zhang F, et al. Porous TiO_2 aerogel-modified SiC ceramic membrane supported MnO_x catalyst for simultaneous removal of NO and dust. J Membr Sci. 2020;611: 118366.
2. Hu M, Yin LH, Zhou HX, et al. Manganese dioxide-filled hierarchical porous nanofiber membrane for indoor air cleaning at room temperature. J Membr Sci. 2020;605: 118094.
3. Zhong ZX, Li DY, Zhang BB, et al. Membrane surface roughness characterization and its influence on ultrafine particle adhesion. Sep Purif Technol. 2012;90:140–6.
4. Wei W, Zhang W, Jiang Q, et al. Preparation of non-oxide SiC membrane for gas cleaning by spray coating. J Membr Sci. 2017;540:381–90.
5. Wikol M, Hartmann B, Brendle J, et al. Expanded polytetrafluoroethylene membranes and their applications. New York: Informa Healthcare USA, Inc; 2008. p. 619–40.
6. Makhetha TA, Moutloali RM. Incorporation of a novel Ag-Cu@ZIF-8@GO nanocomposite into polyethersulfone membrane for fouling and bacterial resistance. J Membr Sci. 2021;618: 118733.
7. Huang Y, Huang QL, Liu H, et al. Preparation, characterization, and applications of electrospun ultrafine fibrous PTFE porous membranes. J Membr Sci. 2017;523:317–26.
8. Jo ES, An X, Ingole PG, et al. CO_2/CH_4 separation using inside coated thin film composite hollow fiber membranes prepared by interfacial polymerization. Chin J Chem Eng. 2017;25(3):278–87.
9. Koti RC, Shailaja D. Improving hydrophobicity of polyurethane by PTFE incorporation. J Appl Polym Sci. 2015;132(47):1–5.
10. Yuan K, Feng S, Zhang F, et al. Steric configuration-controllable carbon nanotubes-integrated SiC membrane for ultrafine particles filtration. Ind Eng Chem Res. 2020;59(44):19680–8.
11. Zhu X, Feng S, Zhao S, et al. Perfluorinated superhydrophobic and oleophobic SiO_2@PTFE nanofiber membrane with hierarchical nanostructures for oily fume purification. J Membr Sci. 2020;594: 117473.
12. Qiao H, Feng S, Low ZX, et al. Al-DTPA microfiber assisted formwork construction technology for high-performance SiC membrane preparation. J Membr Sci. 2020;594: 117464.
13. Xu CN, Xu C, Han F, et al. Fabrication of high performance macroporous tubular silicon carbide gas filters by extrusion method. Ceram Int. 2018;44(15):17792–9.
14. Wang XL, Li ZY, Dong ZL, et al. Comparison of three methods for testing porosity of organic flat membranes for water treatment. Ind Water Wastewater. 2019;50(5):57–62.
15. Cai JC, Sun RS, Zhang ZH, et al. Study on hydraulic permeability of electrospun fiber membrane. Membr Sci Technol. 2018;38(3):41–6.
16. Zhang CC, Huang MH, Li BB, et al. Preparation of forward osmosis composite membrane with electrospun polysulfone support layer. Water Treat Technol. 2017;43(8):44–9.
17. Shi J, Yuan Q, Gao CJ. Membrane technology handbook. Beijing: Chemical Industry Press; 2001.
18. General Administration of Quality Supervision, Inspection and Quarantine of the People's Republic of China. Mechanical measurement method for the thickness of plastic films and sheets: GB/T 6672-2001. Beijing: China Standards Publishing House; 2001.
19. Committee NSM. Scanning electron microscope measurement method for the thickness of precious metal composite material coatings: GB/T 38783-2020. Beijing: China Standards Publishing House; 2020.
20. Yi HL, Ding X, Chen SH. Anisotropy and failure criterion of tensile properties of PES/PVC film materials. J Compos Mater. 2005;22(6):98–102.

21. Li WG, Shao JX, Kou HB, et al. Research progress on theoretical characterization methods of high-temperature mechanical properties of materials. J Solid Mech. 2017;2:93–123.
22. Li XH. Research on the hydrophobicity and tensile properties of electrospun fiber membranes. Shanghai: Donghua University; 2010.
23. National Standardization Management Committee of China. Plastics—determination of tensile properties - Part 1: general principles: GB/T 1040.1-2018. Beijing: China Standards Publishing House; 2018.
24. Zhang LB, Wang JQ, Yang SR, et al. Preparation and performance characterization of graphene-polyimide composite films. J Polym. 2014;11:1472–8.
25. Zhang Li, Li PW, Yang ZM, et al. Preparation and performance characterization of high-performance modified polyvinyl alcohol films. Funct Mater. 2020;51:4153–9.
26. National Standardization Management Committee of China. Textiles—tensile properties of fabrics—Part 1: determination of breaking strength and elongation at break (strip method): GB/T 3923.1-2013. Beijing: China Standards Publishing House; 2013.
27. National Standardization Management Committee of China. Test method for bending strength of fine ceramics: GB/T 6569-2006. Beijing: China Standards Publishing House; 2006.
28. Feng SS, Zhong ZX, Zhang F, et al. Amphiphobic polytetrafluoroethylene membranes for efficient organic aerosol removal. ACS Appl Mater Interfaces. 2016;8(13):8773–81.
29. National Standardization Management Committee of China. Measurement of contact angle of plastic film with water: GB/T 30693-2014. Beijing: China Standards Publishing House; 2014.
30. National Standardization Management Committee of China. Measurement method for contact angle of nanofilm: GB/T 30447-2013. Beijing: China Standards Publishing House; 2013.
31. National Standardization Management Committee of China. Test method for super double repellency of nanomaterials: GB/T 26490-2011. Beijing: China Standards Publishing House; 2011.
32. Luo ZY, Han W, Liu KQ, et al. Study on high temperature mechanical properties of in-situ molybdenite combined with silicon carbide support. Refract Mater. 2020;54:201–4.
33. Xiong R, Sun GC, Si KK, et al. Pressure drop prediction of ceramic membrane filters at high temperature. Powder Technol. 2020;364:647–53.
34. Yaoming W. Design, preparation and characteristics of pore gradient ceramic fiber membrane for high temperature flue gas cleaning. Wuhan: Wuhan University of Technology; 2007.
35. Rudnik ERI, Wojciechowski C, Weniarska A. Preliminary studies on polymer membrane structure by the DSC method. Polym Sci. 1999;4(44):292–3.
36. Wu JX, Fang X, Li YC, et al. PTFE/Al/Al_2O_3 reaction material's mechanical properties and reaction characteristics. Eng Plastics Appl. 2018;12:105–9.
37. Ipiña AA, Urrutia ML, Urrutia DL, et al. Thermal oxidative decomposition estimation combining TGA and DSC as optimization targets for PMMA. J Phys Conf Ser. 2018;1107(3): 032011.
38. Wang YL. Preparation and performance study of Aramid 1313/PANOF needle-punched filter felt. Hangzhou: Zhejiang University of Technology; 2019.
39. Ebrahimpour O, Chaouki J, Dubois C. Diffusional effects for the oxidation of SiC powders in thermogravimetric analysis experiments. J Mater Sci. 2013;48:4396–407.
40. Makarian K, Santhanam S, Wing ZN. Coefficient of thermal expansion of particulate composites with ceramic inclusions. Ceram Int. 2016;42(15):17659–65.
41. Hirata Y. Theoretical analyses of thermal shock and thermal expansion coefficients of metals and ceramics. Ceram Int. 2015;41(1):1145–53.
42. Zhang F, Lei Y, Zhang Z, et al. Negative thermal expansion coefficient of Sc-doped indium tungstate ceramics synthesized by co-precipitation. Ceram Int. 2020;46(6):7259–67.
43. Standardization Administration of China. Determination of thermal expansion characteristic parameters of metal materials: GB/T 4339-2008. Beijing: China Standards Publishing House; 2008.
44. Zhang GH, Zhong ZX, Xing WH. Thermal shock resistance of porous ceramic materials. J Mater Heat Treat. 2011;32:6–9.

45. State Administration for Market Regulation. Test methods for engineering ceramics: GB/T 16534-1996. Beijing: China Standards Publishing House; 1996.
46. Standardization Administration of China. Test methods for ceramic tiles Part 9: Determination of thermal shock resistance: GB/T 3810.9-2016. Beijing: China Standards Publishing House; 2016.
47. Standardization Administration of China. Test methods for thermal shock resistance of refractory materials: GB/T 30873-2014. Beijing: China Standards Publishing House; 2014.
48. Ministry of Metallurgical Industry of the People's Republic of China. Test methods for thermal shock resistance of refractory products: YB 4018-1991. Beijing: China Standards Publishing House; 1991.
49. Han F. Research on the preparation of high-performance silicon carbide ceramic membrane supports. Nanjing: Nanjing University of Technology; 2017.
50. Standardization Administration of China. Test methods for ceramic tiles Part 13: determination of resistance to chemical corrosion: GB/T 3810.13-2016. Beijing: China Standards Publishing House; 2016.
51. Standardization Administration of China. Determination of intergranular corrosion resistance of heat-treated aluminum alloys in metal and alloy corrosion: GB/T 36174-2018. Beijing: China Standards Publishing House; 2018.
52. Ministry of Industry and Information Technology of the People's Republic of China. Test method for acid and alkali corrosion resistance of fine ceramics: JC/T 2138-2012. Beijing: China Building Materials Industry Publishing House; 2012.
53. State Administration of Quality Supervision, Inspection and Quarantine. Test method for gas permeability of plastic films and sheets by pressure difference method: GB/T 1038-2000. Beijing: China Standards Publishing House; 2000.
54. Feng S, Li D, Low ZX, et al. ALD-seeded hydrothermally-grown Ag/ZnO nanorod PTFE membrane as efficient indoor air filter. J Membr Sci. 2017;531:86–93.
55. Committee NSM. Test method for fine particle filtration performance of textiles: GB/T 38413-2019. Beijing: China Standards Publishing House; 2019.
56. Standardization Administration of the People's Republic of China. Test method for dust filtration performance of industrial filter cloth: GB/T 38019-2019. Beijing: China Standards Publishing House; 2019.
57. Zhao Y, Low ZX, Feng S, et al. Multifunctional hybrid porous filters with hierarchical structures for simultaneous removal of indoor VOCs, dusts and microorganisms. Nanoscale. 2017;9(17):5433–44.
58. Zhong Z, Xu Z, Sheng T, et al. Unusual air filters with ultrahigh efficiency and antibacterial functionality enabled by ZnO nanorods. ACS Appl Mater Interfaces. 2015;7:21538–44.
59. Liu J, Zhang H, Gong H, et al. Polyethylene/polypropylene bicomponent spunbond air filtration materials containing magnesium stearate for efficient fine particle capture. ACS Appl Mater Interfaces. 2019;11:40592–601.
60. Zhang W, Deng S, Wang Y, et al. Modeling the surface filtration pressure drop of PTFE HEPA filter media for low load applications. Build Environ. 2020;177: 106905.
61. Standardization Administration of the People's Republic of China. Test method for efficiency and resistance of high-efficiency air filters: GB/T6165-2008. Beijing: China Standards Publishing House; 2008.
62. National Standards Bureau. Test method for air flow resistance of filtering particulate respirators: GB/T 6224.4-1986. Beijing: China Standards Publishing House; 1986.

Chapter 4
Membranes for Gas Cleaning at Low and Medium Temperatures

4.1 Biaxially Stretched Nanofiber Membranes

4.1.1 Overview

Biaxial stretching is one of the main methods for porous membrane preparation. In such a process, the polymer cast sheet is stretched in the longitudinal and transverse directions at a certain temperature and set speed, and after heat setting, a thin membrane material with a porous structure is obtained. At present, the biaxial stretching process is widely used to prepare porous polymer membranes. Various polymers, namely polyethylene (PE), polypropylene (PP), polyamide (PA), polyvinylidene fluoride (PVDF), and polytetrafluoroethylene (PTFE), have been utilized in this process [1–4]. Notably, PTFE membrane represents the earliest porous membrane material produced through biaxial stretching. As illustrated in Fig. 4.1, characterized by a highly regular molecular chain without branches and a symmetric arrangement of two fluorine atoms connected to each carbon atom, PTFE qualifies as a crystalline polymer and exhibits exceptional solvent resistance and chemical stability [5, 6]. The PTFE membrane, derived from bidirectional stretching, boasts high porosity, tunable pore sizes, substantial gas flux, alongside superior attributes like temperature resistance, wear resistance, and corrosion resistance. These qualities collectively position it as an exemplary membrane for gas cleaning at low and mediums.

4.1.2 Bidirectional Stretching Technology

4.1.2.1 Introduction to Bidirectional Stretching Technology

Stein [7] was the pioneer in the initial creation of PTFE porous membranes through stretching in 1958. Subsequently, the Gore Company in the United States filed the

Z. Zhong and W. Xing, *Membrane Technology for Air Cleaning*,
https://doi.org/10.1007/978-981-96-1144-7_4

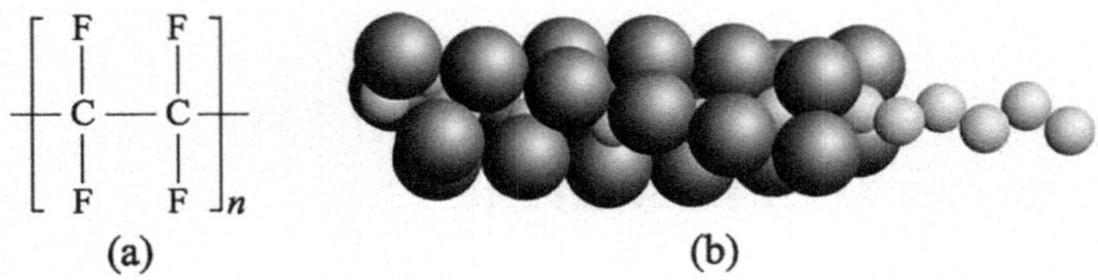

Fig. 4.1 Chemical molecular structure (**a**) and helical structure schematic (**b**) of PTFE

first patent [8] and commenced large-scale production in 1976. Over nearly five decades of relentless evolution, the methodologies and procedures for fabricating PTFE porous membranes have markedly matured, concurrently broadening their spheres of application.

Figure4.2a delineates the procedural outline for fabricating PTFE membranes through stretching [9]. After mixing Initially, a blend of PTFE resin and liquid lubricant undergoes compression into a blank. Subsequently, PTFE thin sheets are crafted via extrusion, calendering, and related processes. Initially, a blend of PTFE resin and liquid lubricant undergoes compression into a blank. Subsequently, PTFE thin sheets are crafted via extrusion, calendering, and related processes. The stretching process primarily bifurcates into two variants: One involves singular longitudinal stretching to achieve a slender, elongated fiber pore configuration, illustrated in Fig. 4.2b (ii, iii); the other is to carry out bidirectional stretching at the same time to obtain a network porous structure constructed by fibers and nodes, depicted in Fig. 4.2b (iv). After stretching, notable alterations manifest wherein the membrane density notably diminishes, stretching strength amplifies, a plethora of pores emerges, and dimensional stability experiences significant enhancement.

4.1.2.2 The Influence of Stretching Process on the Basic Properties of the Membrane

Bidirectional stretching stands as the pivotal process in membrane preparation, allowing for nuanced control over membrane characteristics like thickness, pore size, porosity, and uniformity by modulating parameters such as lateral and longitudinal stretching ratio, stretching rate, and heat setting temperature. According to the molecular weight of the raw material, under typical conditions of a stretching ratio ranging from 6 to 10 times, a stretching rate between 2 and 5 m/min, and a stretching temperature within the range of 180–320 °C, a membrane material with favorable pore distribution and heightened porosity can be obtained [10, 11]. Notably, a lower stretching ratio results in fewer membrane openings and reduced porosity, whereas an excessively high ratio can induce uneven distribution of pore size, fiber thickness, and overall unevenness. Augmenting the stretching rate mitigates stress concentration, amplifies deformation rates, and fosters smaller nodes, yielding a membrane characterized by shorter fibers, a more uniform structure, smaller pore size, and heightened porosity. Lower stretching temperatures facilitate better membrane uniformity but

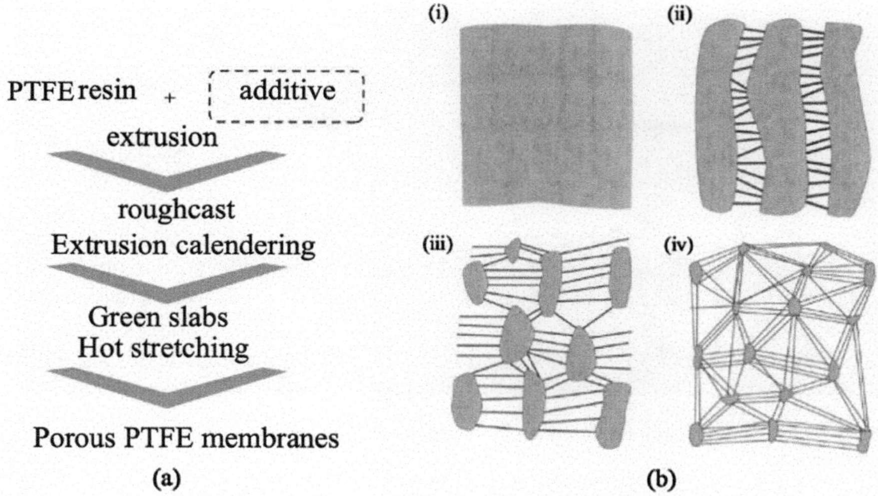

Fig. 4.2 PTFE membrane preparation process flowchart (**a**) and PTFE morphology schematic at different stretching stages (**b**). (i is the PTFE sheet before stretching, ii is the band-shaped crack at the initial stage of stretching, iii is the node formation stage or unidirectional stretching process, iv is the network node formation stage or bidirectional stretching process)

compromise ductility, impeding higher-multiple stretching. Conversely, excessively low temperatures limit molecular chain activity, rendering the membrane prone to breakage; Elevated stretching temperatures enhance membrane ductility, yet excessive heat can result in a thin edge and thick middle, potentially disrupting the folding state of fiber bundles, leading to pore closure and hindered reopening.

4.1.3 Structure and Performance of Polytetrafluoroethylene (PTFE) Membranes

4.1.3.1 Structure of Polytetrafluoroethylene (PTFE) Membranes

The biaxially stretched PTFE porous membrane is a network structure composed of numerous nodes and fibers interconnected (Fig. 4.3). Tailoring the stretching conditions allows for the creation of PTFE porous membranes with varying pore architectures to meet specific requirements. The direction of stretching dictates the orientation of the fibers within the membrane. In uniaxially stretched membranes, fibers typically maintain a parallel alignment, whereas in biaxially stretched ones, fibers diverge, forming a networked pattern. The membrane's porosity is greatly influenced by the proportion of pores formed between fibers, the quantity, and dimensions of the nodes. Consequently, the stretching parameters play a pivotal role in determining both the porosity and the distribution of pore sizes in the resulting porous

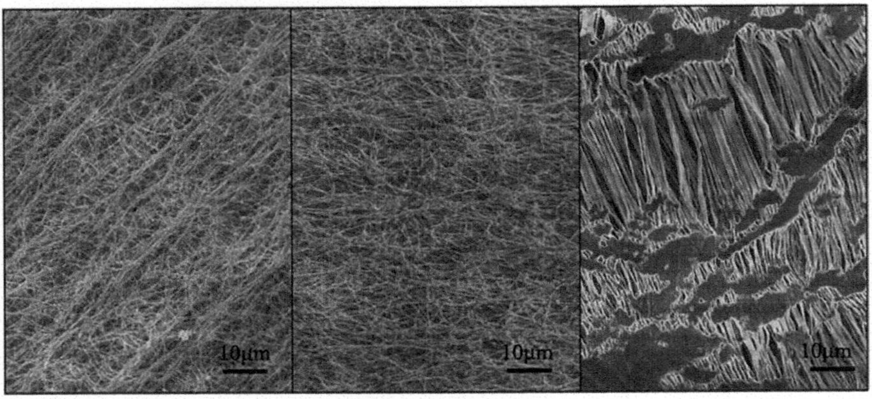

Fig. 4.3 Scanning electron microscope (SEM) image of Polytetrafluoroethylene (PTFE) porous membrane

membrane. Mechanically stretched PTFE porous membranes exhibit a pore size distribution spanning 0.1–10 μm, boasting a porosity that can surpass 50% and even reach levels exceeding 90%. Additionally, their thickness can be controlled within the range of 5–50 μm, featuring a relatively narrow pore size distribution. Moreover, these PTFE porous membranes retain the inherent chemical stability, thermal resilience, resistance to acid and alkali corrosion, lubricity, hydrophobicity, and electrical neutrality characteristic of PTFE materials despite the mechanical stretching process.

4.1.3.2 Performance of Polytetrafluoroethylene (PTFE) Membranes

The biaxially stretched PTFE membrane exhibits specific attributes including a narrow distribution of pores, reduced thickness, minimal gas resistance, favorable surface lubricity, and notable hydrophobic properties. Employed in filtration processes, this membrane operates as a surface filter, effectively intercepting a substantial portion of dust on its surface, thus forestalling dust penetration and the subsequent rise in resistance. The membrane's surface lubricity facilitates the easier dislodgment of dust particles through back blowing or shaking, enhancing the membrane's continuous operational duration and the efficiency of back blowing procedures. In contrast to traditional filter materials, which often possess larger pore sizes resulting in lower retention rates for smaller particles, as depicted in Fig. 4.4a, PTFE membrane materials overcome this limitation by featuring thinner membrane layers and finer fibers, resulting in reduced pore sizes. Consequently, they achieve more efficient retention of small particles while maintaining lower resistance during continuous operations. Furthermore, when examining continuous operation, traditional filter materials experience a rapid increase in resistance, necessitating frequent back blowing. This high resistance coupled with recurrent back blowing significantly

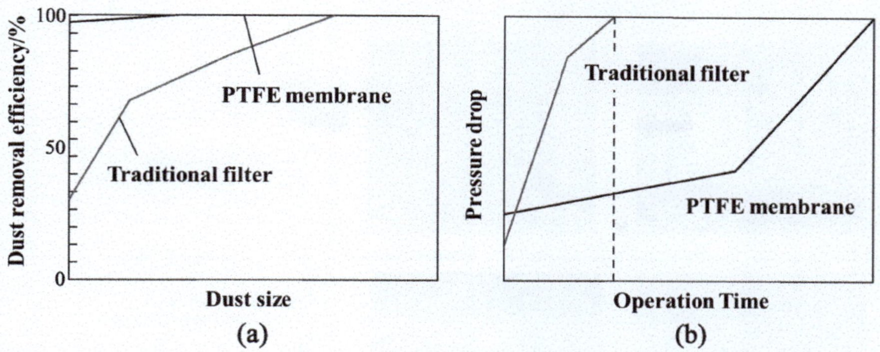

Fig. 4.4 **a** The impact of dust particle size on the retention performance of traditional filter materials and PTFE membranes and **b** the impact of operating time on pressure loss for traditional filter materials and PTFE membrane materials

escalates the energy consumption of the dust collector. Conversely, PTFE membranes demonstrate lower overall operating pressure drops, reducing the frequency of back blowing and subsequently enhancing operational energy efficiency and prolonging service life, as illustrated in Fig. 4.4b.

4.1.4 Application of Polytetrafluoroethylene (PTFE) Membranes

4.1.4.1 Industrial Application of Polytetrafluoroethylene (PTFE) Membranes

As global air pollutant emission standards undergo continual enhancement across diverse nations, the utilization of PTFE membranes in industrial dust control has experienced burgeoning prevalence. Stringent limitations govern exhaust emissions from industrial processes spanning petrochemicals, metallurgy, and power plants, challenging conventional dust removal methodologies to comply with increasingly rigorous emission standards. PTFE membrane materials exhibit elevated efficiency in purifying dust-laden gases, ensuring outlet dust concentrations remain below 5 mg/m^3, thus achieving ultra-low emission standards. Simultaneously, these materials boast high-temperature resistance, enabling prolonged stable operation at 250 °C, with a maximum operational capacity reaching 280 °C, aligning with emission requisites across various pivotal industries. Moreover, the inherent characteristics of PTFE encompass a low friction coefficient, excellent hydrophobicity, and robust chemical stability. Consequently, it finds extensive application in the realm of high-value-added powder recovery, encompassing domains such as traditional Chinese

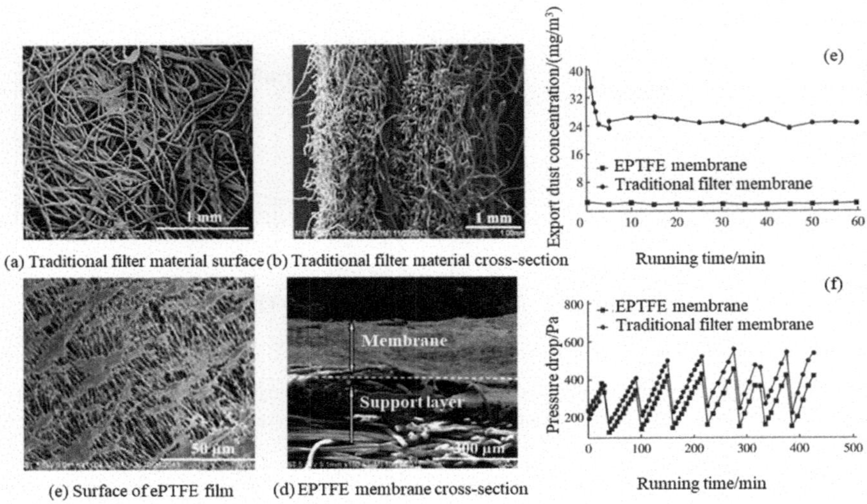

Fig. 4.5 Surface and cross-sectional scanning electron microscope images of traditional filter material and ePTFE membrane (**a**–**d**); Changes in outlet dust concentration and filter pressure drop of traditional filter material and ePTFE membrane with operating time (**e**, **f**)

medicine powder recovery, powder food recovery, and metal powder reclamation, among others.

Wang et al. [12] implemented the dust removal membrane technology devised by the Membrane Science and Technology Research Institute at Nanjing University of Technology in a boiler flue gas dust removal experiment conducted at Zhongyan Jintan Salt Chemical Co., Ltd. Within the experimental setup, both the ePTFE membrane and the conventional filter bag were integrated into the same dust removal apparatus. Comparative analysis of the outlet dust concentration and filter pressure drop between the ePTFE membrane and the traditional filter material dust collector was conducted under identical operational conditions, as depicted in Figs. 4.5. The operational findings indicate that the outlet concentration subsequent to the passage of flue gas through the ePTFE membrane stands at 2 mg/m^3, markedly inferior to the 25 mg/m^3 observed with conventional filter bags. Furthermore, the holistic energy consumption associated with employing ePTFE membrane-based dust removal demonstrates a noteworthy reduction, thereby showcasing pronounced advantages in terms of energy conservation and environmental preservation.

4.1.4.2 Applications of Polytetrafluoroethylene (PTFE) Membranes in the Civilian Field

(1) Air cleaning products

The excellent chemical stability, thermal stability, super hydrophobicity, and antibacterial properties due to the presence of fluorine groups make PTFE suitable for air cleaning. In contrast to the symmetrical composition of the HEPA filter, the structural configuration of the PTFE composite membrane transitions the filtration mechanism from "deep filtration" to "surface filtration". This transition enables the maintenance of elevated filtration efficacy while concurrently diminishing filtration resistance, thereby prolonging the operational lifespan. Regarded as one of the foremost air cleaning materials globally, it stands as the optimal selection for an array of applications such as vacuum cleaners, air filter components, air cleaning apparatus, high-efficiency gas filters, among others. Detailed technical parameters are presented in Table 4.1 for reference.

(2) Personal Protective Equipment

Mask isolation is the most direct and effective way to prevent the spread of respiratory system viruses. Technological advancements have spurred iterative enhancements in mask filter materials, progressing from gauze and non-woven fabric to membrane materials. Among these, two widely employed materials are melt-blown cloth and innovative products. Melt-blown cloth, engendered through high-temperature melt-blown processes, orchestrates controlled fiber accumulation pore sizes, swiftly bonding to create a porous, fluffy structure with commendable and good wrinkle resistance. Characterized by fiber pore diameters spanning 0.5–10 μm, its fine texture facilitates effective dust entrapment owing to its inherent static electricity, rendering it the predominant material for mask filtration across prevailing markets. Nevertheless, the static charge within the melt-blown cloth diminishes progressively throughout production, packaging, factory handling, transportation, and utilization phases, thereby resulting in a decline in its protective efficacy. The PTFE nanomembrane exhibits an inherent fibrous microporous arrangement, boasting a porosity exceeding 85%, while enabling controlled pore sizes ranging from 0.01 to 10 μm. Masks engineered with PTFE as the fundamental constituent demonstrate elevated barrier efficiency alongside prolonged operational longevity. Nano-modified PTFE membrane, endowed with antibacterial and antiviral attributes, manifests a noteworthy capacity to effectively eliminate viruses, bacteria, and other microorganisms, thus presenting a promising trajectory for future mask development. Moreover, from the standpoint of human physiological safeguarding, PTFE membrane material showcases commendable filtration and barrier capabilities, in addition to oil and water repellency, resistance to blood stain penetration, flame retardancy, disinfection resilience, and favorable breathability. Consequently, it emerges as an exemplary protective apparel component. The integration of PTFE membrane with diverse fabric types yields superior clothing materials, excelling in waterproofing, breathability, thermal insulation, quick-drying properties, and enduring wash resistance. Its

Table 4.1 Main technical parameters of PTFE membrane

Pore size/μm	Porosity/%	Filtration efficiency/%	Air resistance (5.33 cm/s)/Pa	Thickness/μm	Maximum usage temperature/°C	Service life/year
0.2–5	80–95	> 99.9	120	10–100	260	> 3

versatility renders it a superior choice for specialized attire, including but not limited to storm jackets, training suits, radiation protection garments, and firefighting [13].

4.1.5 Other Biaxially Stretched Nanofiber Membranes

In addition to PTFE, several polymers, including PE, PP, PA, PET, and PVDF can yield porous membrane materials through the process of biaxial stretching. However, the stretching capacity of these materials is often constrained due to the inherent physical properties of the polymer, resulting in varying fiber structures. Wan et al. [2] devised a continuous biaxial stretching technique to produce a PE nanofiber membrane. This membrane exhibits characteristics typified by fine fibers, uniform pore dispersion, and commendable mechanical attributes. Figure 4.6 illustrates the scanning electron microscope (SEM) depiction of the biaxially stretched PE nanofiber membrane, demonstrating a pore size distribution within the range of 0.01–0.05 μm.

Ding et al. [14] conducted an intricate analysis, examining the three-dimensional progression of β-crystallization and microporous structural alterations within iPP (isotactic polypropylene) throughout continuous biaxial stretching. Through computational derivations and the utilization of scanning electron microscope imagery (depicted in Fig. 4.7), their findings revealed notable phenomena. During the longitudinal stretching phase, a pronounced cavitation occurrence within the lamellae ensued, resulting in diminished three-dimensional directional alignment. This process yielded a proliferation of coarse fibers owing to the three-dimensional asymmetric stacking of lamellar structures. Consequently, these coarse fibers impeded the subsequent transverse stretching phase by impeding the slippage and distortion of the thin lamellar slices, consequently engendering an uneven distribution of pore sizes. Conversely, the radially expanding spherulites experienced considerable slippage during longitudinal stretching. As a consequence of the evenly dispersed lamellar structures, these spherulites demonstrated heightened alignment in the three-dimensional orientation, thereby generating finer fibers. Consequently, the post-biaxial stretching surface exhibited a more uniformly distributed pore size, albeit with larger pore dimensions.

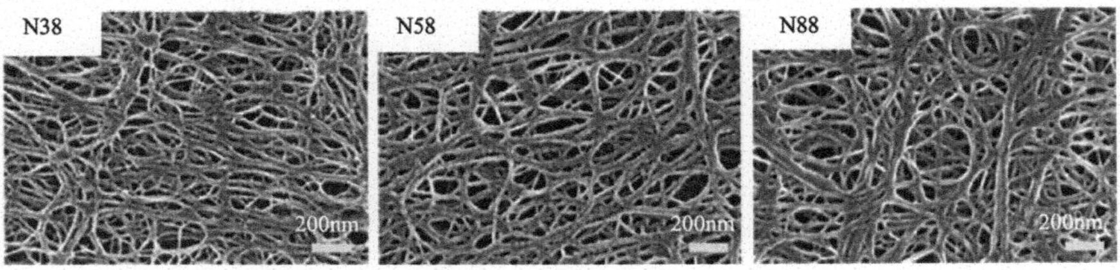

Fig. 4.6 Biaxially stretched PE nanofiber membranes [2]

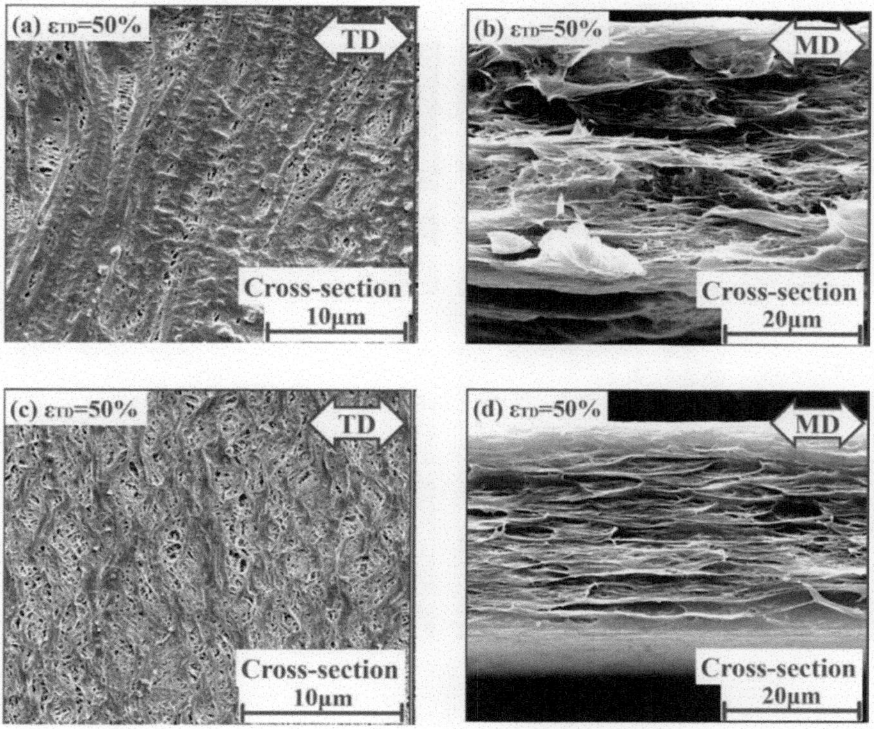

Fig. 4.7 β-iPP membrane cross-section electron microscope image: **a**, **c** transverse (TD); **b**, **d** longitudinal (MD) [14]

Feng et al. [4] conducted co-extrusion of polypropylene (PP) and nylon 6 to produce a microporous membrane featuring a through-hole structure via stretching processes. This membrane demonstrated a pore size of approximately 100 nm, accompanied by a porosity level of 52% (Fig. 4.8). The development of this specific pore architecture proved to be highly contingent on both the blend's composition and the procedural parameters. Notably, a decrease in the concentration of nylon 6 or a reduction in stretching ratio inversely correlated with the likelihood of forming the desired pore structure within the blend. The biaxial stretching of the polymer to generate perforations necessitates specific conditions: (1) The base polymer can crack under stress; (2) the presence of a dispersible rigid polymer whose molecular chains do not align during stretching but align with the base polymer; (3) a requisite dispersion domain size of approximately 1 mm; (4) the utilization of compatibilizers (such as macromolecular coupling agents) or other adhesives to ensure adequate adhesion and interfacial tension between the base and dispersed phases.

Typically, certain materials mentioned above tend to undergo crystallization during the stretching process due to their limited extensibility. This phenomenon results in biaxially stretched membranes with reduced pore sizes and increased

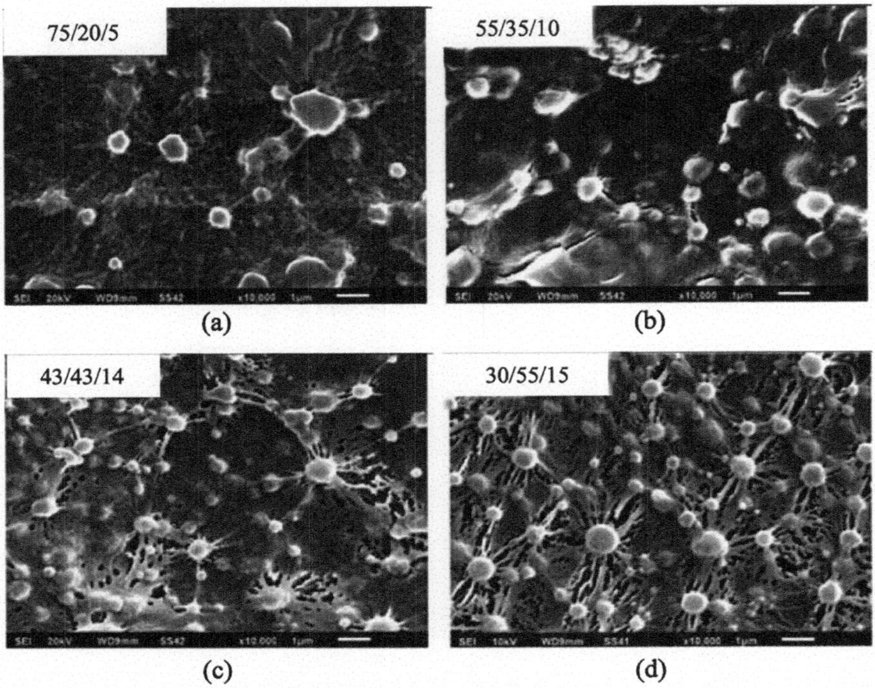

Fig. 4.8 Electron microscopy images of biaxially stretched membranes with various PP/Nylon 6/PP-g-MA blend ratios (PP-*g*-MA is maleic anhydride grafted polypropylene) [4]

membrane thickness. Consequently, their utilization in gas cleaning applications remains relatively limited.

4.2 Electrospun Nanofiber Membranes

4.2.1 Overview

Membrane separation technology represents an innovative and highly efficient approach to the elimination of $PM_{2.5}$ particulate matter from air. Amidst the myriad of gas cleaning membrane materials, significant research endeavors have been directed towards assessing the filtration efficacy of nanofiber membranes for $PM_{2.5}$ removal. The heightened interest stems from dual factors: firstly, the nanofiber membrane's substantial porosity and expansive specific surface area significantly augment the interception capacity of nanofibers towards $PM_{2.5}$ particles. This enhanced interception facilitates heightened contact between $PM_{2.5}$ particles and nanofibers, leading to their adherence by the fibers, thereby achieving elevated filtration efficiencies. Secondly, the scale of nanofibers closely aligns with the average free path of gas molecules, approximately 66 nm. Leveraging the "slip effect," this alignment contributes to a reduction in filtration resistance, thereby optimizing filtration performance.

The evaluation indicators of nanofiber membrane filtration performance typically encompass filtration efficiency and filtration resistance. Typically, a decrease in the pore size of nanofiber membranes leads to heightened filtration efficiency, thereby enhancing gas cleaning capabilities. However, this reduction in pore size concurrently elevates filtration resistance. Thus, the challenge lies in mitigating filtration resistance while preserving high filtration efficiency—a focal point in nanofiber membrane research. Addressing diverse airborne pollutants such as bacteria, microorganisms, volatile organic compounds (VOCs), and ultrafine dust necessitates multifunctional attributes within gas cleaning materials. Consequently, current research trends emphasize strategies to impart multifunctionality to nanofiber membranes.

4.2.2 Electrospinning Technology

Electrospinning technology stands as a versatile and efficient methodology for fabricating nanofiber membranes, offering adjustable fiber diameters ranging from a few nanometers to several micrometers. Its inception traces back to Formhals' pioneering utilization of electrostatic forces in 1934 to produce polymer fibers, marking the seminal advent of electrospinning technology [15]. Notably, in the 1990s, the Reneker research group systematically investigated the electrospinning process for various polymer nanofibers and explored their diverse applications [16, 17], garnering considerable attention and interest from scholars globally. Simultaneously, Reneker and collaborators amalgamated theoretical investigations with practical applications, expanding the scope of nanofiber development towards multifunctionality. This concerted effort significantly contributed to advancing electrospinning technology, fostering its commercial viability.

The applicability spectrum of materials amenable to electrospinning technology is broad and encompasses an array of substances, ranging from synthetic and natural polymers to metals, ceramics, polymer alloys, and solutions infused with chromophores, nanoparticles, or active agents.

The typical laboratory setup for electrospinning comprises essential components: a liquid delivery system, a high-voltage power source, a spinneret, and a collector, as shown in Fig. 4.9a [18]. Initially, the spinning solution, regulated by a micro-injection pump, is dispensed through the spinneret at a controlled rate. The spinneret, linked to a high-voltage power supply, establishes an electrostatic field—typically in the range of tens of kilovolts—between the spinneret and a grounded metal receiving plate. As the solution is extruded, it acquires surface charge due to the electric field, progressively elongating and attenuating until solidification occurs upon deposition onto the collector, resulting in the formation of a three-dimensional porous nanofiber membrane. During this process, the spinning solution initially adopts a charged droplet configuration at the spinneret. As the electric field force and the droplet's surface tension achieve equilibrium, the charged droplet stabilizes at the spinneret's end. Further augmentation of the electric field induces a conical transformation in the charged droplet, commonly recognized as a "Taylor cone" [19], exhibiting a half-angle of 49.3°. With continued escalation of the electric field, the charged jet emanates from the apex of the Taylor cone. During this phase, the jet experiences a brief acceleration period within the electric field, characterized by low velocity and a balanced surface force, assuming a near-linear trajectory with minimal fluctuations, termed as a steady-state jet. After traversing a short distance in this linear path, the steady-state jet succumbs to bending instability, leading to lateral deviation and the formation of a ring structure. Concurrently, the jet undergoes substantial elongation and attenuation. Subsequent to this stage, these attenuated jets undergo further bending instability, engendering secondary rings, exemplified in Fig. 4.9b, delineating a non-steady-state jet progression. This iterative process continues until solvent evaporation reaches completion within the jet or until the fiber diameter diminishes to a size where bending instability ceases to exert influence. Throughout the entirety of the electrospinning process, the resultant fiber morphology is significantly influenced by multiple factors encompassing polymer molecular weight, spinning solution viscosity, surface tension, conductivity, spinning voltage, spinning rate, collector distance, as well as ambient conditions such as temperature, humidity, and pressure. Manipulation and control of these influential variables afford the means to fabricate nanofiber membranes with varied diameters or specialized structural configurations.

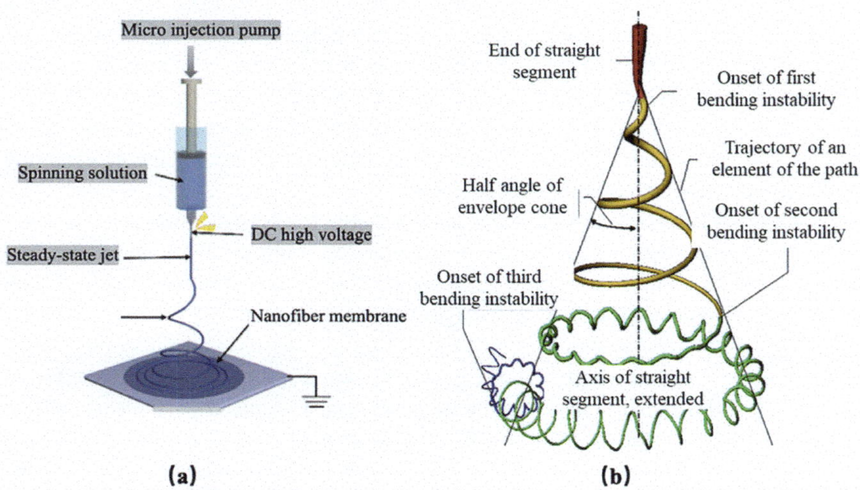

Fig. 4.9 Schematic diagram of a typical electrospinning device (**a**) and the instantaneous motion path of an electrospinning jet (**b**) [18]

4.2.3 *Classification of Electrospun Nanofiber Membranes*

Electrospinning technology stands as a recognized and efficient approach for fabricating nanofiber membranes, yielding products characterized by excellent flexibility, diverse structural patterns, and adjustable wettability. Categorized by their primary constituent materials, electrospun nanofiber membranes are broadly classified into polymer-based, ceramic-based, and carbon-based variants.

4.2.3.1 Polymer-Based Nanofiber Membranes

The superior attributes of polymers, including high flexibility, favorable processability, and cost-effectiveness, establish them as primary materials in the fabrication of electrospun nanofiber membranes. Thus far, researchers have successfully developed a myriad of polymer-based nanofiber membranes, broadly categorized into single-layer and multi-layer structures based on their layer count. Each configuration presents distinct advantages and drawbacks, aligning with specific preparation methodologies.

Single-layer polymer nanofiber membranes are typically fabricated through two primary techniques: direct electrospinning and colloidal electrospinning. Direct electrospinning involves the creation of single-layer nanofiber membranes by electrospinning a high-surface-energy polymer solution. Polyacrylonitrile (PAN) is frequently utilized due to its excellent electrospinnability and widespread availability. Following the direct electrospinning of PAN solutions, the resultant PAN nanofiber membrane's wettability can be adjusted through alkali treatment. This treatment involves alkaline hydrolysis, converting a portion of the nitrile groups within large PAN molecules into carboxyl groups, thereby rendering the PAN membrane hydrophilic and oleophobic when submerged [20, 21]. Colloidal electrospinning refers to the electrospinning of a polymer solution dispersed with high-surface-energy nanoparticles (NPs), suitable for one-step manufacturing of biomimetic superhydrophilic membranes without further treatment. Manipulating the quantity of nanoparticles enables precise control over the morphology and surface energy of the resulting nanofibers. This versatile technique accommodates various nanoparticle types, spanning from zero-dimensional to two-dimensional nanomaterials, successfully integrating them into electrospun nanofibers.

Single-layer nanofiber membranes encounter challenges in achieving concurrent optimization of mechanical strength, permeability, and selectivity. Enhancing fiber deposition within a single-layer membrane can enhance separation efficiency and mechanical performance. However, augmenting membrane thickness tends to diminish permeation flux. Motivated by natural multilayer structures employed in water filtration, such as the stratified compositions observed in vegetation, gravel, sand, and soil layers, there is a growing focus on engineering multilayer nanofiber membranes possessing distinctive wettability profiles for oil/water separation. This approach aims to overcome the inherent limitations associated with single-layer membranes. An elementary yet efficacious approach in fabricating multilayer membranes involves direct electrospinning of a crosslinkable nanofiber ultra-thin layer onto the surface of an electrospun nanofiber membrane, followed by subsequent crosslinking treatment aimed at diminishing the coating's pore size. Multilayer nanofiber membranes with crosslinked networks demonstrate promising efficacy in oil/water separation. The establishment of a bonded structure not only reduces pore size, enhancing separation efficiency, but also upholds structural stability, ensuring consistent filtration performance. However, the bonded structure among nanofibers may impede membrane porosity and interconnectivity, consequently hindering liquid permeation. Furthermore, the limited productivity associated with this technology constrains its practical application.

4.2.3.2 Ceramic-Based Nanofiber Membranes

The potential of polymer-based nanofiber membranes in practical applications like oil/water separation is substantial. However, inherent limitations persist, including susceptibility to high temperatures, organic solvents, and oxidants. In addressing these drawbacks, ceramic nanofiber membranes are considered good candidates to overcome these shortcomings due to their good thermal stability, corrosion resistance, and relative inertness. Typically, ceramic nanofibers can be prepared through a two-step process: (1) the creation of composite nanofibers composed of a polymer template and a sol–gel inorganic precursor by electrospinning; (2) the calcination of these composite nanofibers in air to eliminate the polymer phase. It is crucial to note that calcination may cause ceramic nanofibers to shrink due to the decomposition of the polymer template and the densification of nanocrystals. This phenomenon often results in fractures within of ceramic nanofibers, rendering the ceramic membranes fragile and compromising their poor mechanical performance consequently affects their practical application. Researchers have explored various strategies to address this issue, such as adjusting calcination conditions, introducing doping metal ions, or employing various compounds. These approaches have led to the development of various flexible ceramic nanofiber membranes, including SiO_2, SiO_2/TiO_2, etc. In recent years, researchers have prepared SiO_2/TiO_2 nanofiber membranes by electrospinning SiO_2/TiO_2 mixed sol–gel solution, resulting in fiber membrane characterized by worm-like pores. The worm-like porous structure can give the ceramic membrane excellent flexibility, facilitating easy recovery in the oil/water separation process. Moreover, compared to solid-structured nanofibers, the worm-like structure provides more active sites, such as Si–OH and Ti–OH groups. These additional active sites enable the membrane to selectively capture molecules of higher polarity surface energy.

4.2.3.3 Carbon-Based Nanofiber Membranes

Carbon-based nanofiber membranes represent an alternative category of nanofiber materials with anti-swelling and high chemical stability. Differing from ceramic-based nanofibers, carbon-based nanofibers have inherent hydrophobic and oleophilic properties, enabling easy regeneration after use. Notably, oil stains in carbon membranes can be easily eliminated through combustion [22]. However, due to the inherent brittleness of the carbon matrix, conventional carbon nanofiber membranes often fail to withstand physical impacts encountered during the separation process. In recent years, through carefully designed fiber microstructures, carbon nanofibers have excellent structural stability under mechanical deformation. For example, Liu et al. [23] manufactured flexible large-pore carbon nanofiber membranes through a sublimation etching strategy. They used terephthalic acid (PTA) as a porogen directly incorporated in to the PAN solution for the electrospinning process. During the subsequent carbonization process, the PTA gradually sublimated, leaving large pores in/on the carbon nanofibers. The resulting large-pore carbon membranes can be folded into several layers and quickly restored to its original state after removing the load. This behavior stems from the efficacy of the large pores in alleviating stress concentration during fiber bending deformation.

4.2.4 Structure and Performance of Functional Electrospun Nanofiber Membranes

The polymers employed in electrospinning for the production of nanofiber membranes encompass a range of materials, including polylactic acid (PLA), polyamide (PA), polyacrylonitrile (PAN), polyester (PET), polymethyl methacrylate (PMMA), polyetherimide (PEI), polyethylene oxide (PEO), polyvinyl alcohol (PVA), polyvinylpyrrolidone (PVP), cellulose acetate (CA), polysulfone (PSU), etc. [24]. The prepared nanofibers with a narrow diameter distribution and smooth surface tend to stack tightly, as shown in Fig. 4.10a, culminating in nanofiber membranes with a high surface density, which is not conducive to air permeation, making the filtration resistance high [25]. For example, when filtering aerosol particles with a diameter of less than 300 nm [e.g., NaCl, dioctyl phthalate (DOP), etc.], under the test conditions of an airflow speed of 5.3 cm/s, the filtration efficiency of these ordinary polymer nanofiber membranes with a smooth surface structure reaches more than 99%. However, the filtration resistance is generally above 300 Pa [26, 27]. Existing research shows that the fiber diameter, fiber surface structure, and the volume fraction of fibers in the fiber membrane have a significant impact on the gas cleaning performance of the nanofiber membrane [28, 29]. In order to achieve high filtration efficiency while reducing filtration resistance, researchers have devised diverse nanofiber membrane structures such as bead, spider web, branch structures, and

Fig. 4.10 FESEM images of electrospun nanofiber membranes with different structures **a** smooth structure; **b** bead structure; **c** spider web structure; and **d** branch structure

composite structures, as shown in Figs. 4.10b–d and 4.11, aiming to ameliorate their filtration resistance [30].

4.2.4.1 Bead Structure

Electrospinning technology can be used to prepare PLA nanofiber membranes with a porous bead-on-string structure, as shown in Fig. 4.10b. A certain number of large-sized beads are used to reduce the volume fraction of fibers in the nanofiber membrane, increase the gas permeability of the nanofiber membrane, and thus reduce filtration resistance. Wang et al. [25] demonstrated that operating at an airflow speed of 5.8 cm/s, the porous bead-on-string nanofiber membrane achieved a filtration efficiency of 99.997% for NaCl particles with an average diameter of 260 nm, while sustaining a filtration resistance of merely 165.3 Pa. Ogi et al. [31] conducted gas cleaning tests on PAN nanofiber membranes with smooth surfaces and bead-on-string structures, comparing their filtration performance concerning aerosols. The findings notably favored the bead-on-string structured PAN nanofiber membrane, displaying superior filtration efficacy. Furthermore, Yun et al. [28] compared the filtration performance of PAN nanofiber membranes and PMMA bead-on-string nanofiber membranes. Their investigations revealed that while both membranes achieved equivalent filtration efficiency for NaCl particles under 300 nm in diameter, the filtration resistance of the PMMA bead-on-string nanofiber membrane was about one-third of that of the PAN nanofiber membrane.

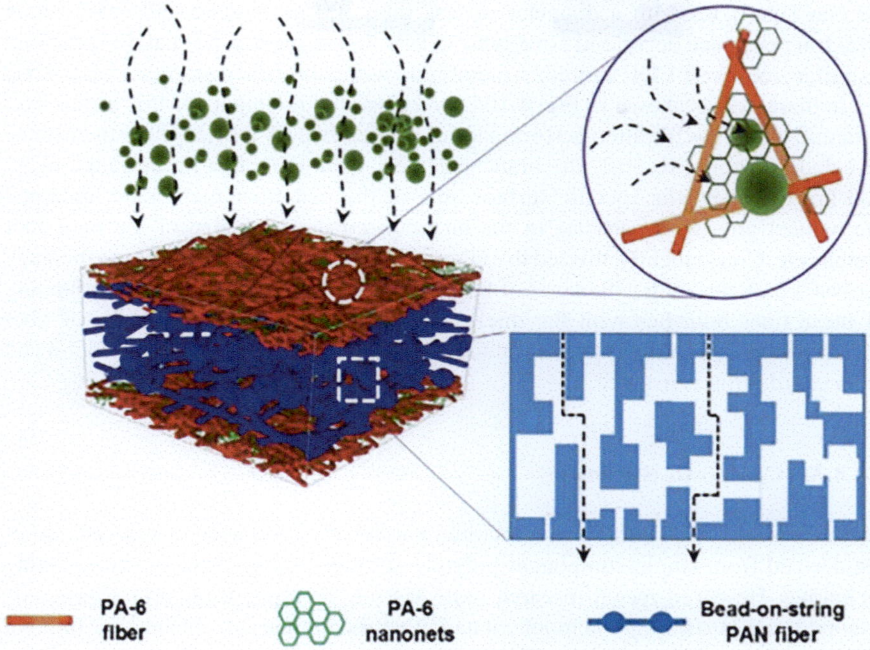

Fig. 4.11 Schematic diagram of the sandwich-structured PA-6/PAN/PA-6 composite nanofiber membrane [37]

4.2.4.2 Spider Web and Branch Structures

The team led by Prof. Ding at Donghua University pioneered the development of a 2D spider web structure nanofiber membrane (nanofiber/nets) with a diameter of less than 100 nm for gas filtration through a one-step electrospinning/netting technique. For example, a layer of PA-66 nanospider web fiber membrane is deposited on a PP microfiber nonwoven material with a surface density of 9.8 g/m^2, as shown in Fig. 4.10c [20]. The fiber diameter of this PA-66 nanospider web fiber membrane is within 360 nm. The conventional PA-66 nanofibers, which are thicker in diameter and randomly distributed, are interconnected by a thin layer of spider web-like structure with a diameter of about 30 nm. When the thickness of the PA-66 nanospider web fiber membrane is 3.96 μm, its filtration efficiency for NaCl particles with a diameter of less than 300 nm is close to 95%, and the resistance is less than 100 Pa. This fiber structure with a diameter of about 30 nm can promote the slip effect of gas molecules, thereby reducing filtration resistance, while concurrently decreasing pore size and augmenting filtration efficiency. Liu et al. [32] prepared a bio-based PA-56 nanospider web fiber membrane for gas filtration, using the small size effect and excellent porosity of the nanospider web structure with a diameter of about 20 nm, achieving high-efficiency and low-resistance gas filtration. The filtration efficiency for NaCl particles with a diameter of less than 300 nm reached 99.995% while filtration resistance decreased notably to 111 Pa. Subsequently, Li et al. [33] devised a similar tree-like PVDF nanofiber membrane with a branched structure of 5–1000 nm in diameter, as shown in Fig. 4.10d. Investigating the impact of this multi-level structure on the gas filtration performance of the tree-like PVDF nanofiber membrane revealed intriguing results. The branched fibers with diameters in the range of 5–100 nm increased the specific surface area of the nanofiber membrane, enlarged the contact area with particles in the air, reduced the pore size of the nanofiber membrane. Consequently, this led to a significant enhancement in filtration efficiency for NaCl particles with a diameter of 260 nm (99.999%). Moreover, the overlapping of these finer branches with the thicker main fibers reduced the density of fiber accumulation in the nanofiber membrane, which is beneficial for gas permeation and reduces filtration resistance.

4.2.4.3 Composite Structure

The bead-string and spider-web structured nanofiber membranes mentioned earlier are derived from single-component polymer solutions through the electrospinning technique. However, recent research indicates that incorporating certain inorganic nanoparticles into a single-component nanofiber membrane can enhance the internal porosity of organic/inorganic composite nanofiber membranes, reducing fiber accumulation density and consequently lowering gas filtration resistance. Wang et al. [21] utilized a dual-needle electrospinning system to fabricate a SiO_2/PAN composite nanofiber membrane with a dual-diameter distribution. Gas filtration test results show that the SiO_2 /PAN composite nanofiber membrane has a filtration efficiency of up to 99.989% for solid pollutant particles below 300 nm in diameter, accompanied by a filtration resistance of 117 Pa. In a similar vein, Li et al. [34] added SiO_2 nanoparticles into the PEI solution and directly electrospun the mixture to generate a SiO_2/PEI composite nanofiber membrane. This composite membrane demonstrated remarkable filtration performance for solid pollutant particles under 300 nm in diameter, achieving a filtration efficiency of 99.992% and a notably low filtration resistance of 61 Pa. Furthermore, Wang et al. [35] a TiO_2/PLA composite nanofiber membrane designed for air cleaning. Notably, this membrane exhibited a filtration efficiency of 99.996% for NaCl particles with a diameter of 260 nm, accompanied by a filtration resistance of 128.7 Pa. Additionally, owing to the exceptional photocatalytic activity of TiO_2 nanoparticles, exposure to ultraviolet light facilitated the generation of hydroxyl radicals (·OH). These radicals induced peroxidation of polyunsaturated phospholipids on bacterial cell membranes, leading to their inactivation. Consequently, the TiO_2/PLA composite nanofiber membrane showcased notable antibacterial properties, achieving a 99.5% inhibition rate against Staphylococcus aureus.

In addition, another type of composite structure refers to a composite nanofiber membrane with a double-layer or multi-layer structure obtained by stacking two or more layers of nanofiber membranes. For example, Wang et al. [36] conducted an investigation wherein they stacked a PLA nanofiber membrane (PLA-N) with a diameter of around 150 nm atop a porous PLA nanofiber membrane (PLA-P) measuring approximately 1.71 μm in diameter, creating a double-layer composite nanofiber membrane. This composite membrane's filtration performance was evaluated concerning NaCl aerosol particles (size: 260 nm). Remarkably, the double-layer composite nanofiber membrane, featuring a PLA-N to PLA-P mass ratio of 1/5, achieved an outstanding filtration efficiency of 99.999% alongside a filtration resistance of 93.3 Pa under an airflow velocity of 5.3 cm/s. In a different approach, Yang et al. [37] engineered a sandwich-structured composite nanofiber membrane, as delineated in Fig. 4.11. This composite membrane comprised PA-6 nanoweb layers at the top and bottom, while the central layer consisted of a PAN bead nanofiber membrane. Employed for air filtration purposes, the PA-6/PAN/PA-6 composite nanofiber membrane demonstrated impressive filtration efficacy, reaching 99.9998% for pollutant particles spanning diameters between 300 and 500 nm. This is because the interconnected PA-6 nanoweb structure with a diameter of about 20 nm can promote the slip effect of gas molecules. Concurrently, the PAN bead nanofiber contributed to aperture creation, reducing fiber packing density and consequently diminishing filtration resistance.

4.2.5 Applications of Functional Electrospun Nanofiber Membranes

In the utilization of electrospun nanofiber membranes for air cleaning, susceptibility to attack by bacteria, fungi, molds, and various airborne microorganisms is a prevalent concern. Upon adherence of microorganisms to dust and other pollutant particles deposited on the nanofiber membrane, rapid proliferation occurs. This phenomenon not only severely compromises the efficacy of air cleaning but also curtails the service life of the nanofiber membrane. Moreover, the introduction of formaldehyde and other Volatile Organic Compounds (VOCs) stemming from indoor decoration poses a significant health risk. These gases, when absorbed through the respiratory and digestive tracts, can lead to multi-organ poisoning in the human body. This section mainly introduces the applications of functional electrospun nanofiber membranes in combating bacterial growth, removing VOCs and multifunctional synergy.

4.2.5.1 Antibacterial Nanofiber Membranes

Nanosilver, recognized for its broad-spectrum and highly potent antibacterial properties coupled with low toxicity to human physiology, finds extensive application in gas filtration nanofiber membranes with antibacterial functionalities. Jeong et al. [38] undertook the preparation of a transparent silver nanowire infiltration network specifically designed for air cleaning systems. Their study demonstrated that this transparent silver nanowire infiltration network achieved a remarkable filtration efficiency of 99.99% for $PM_{2.5}$ particles while exhibiting commendable antibacterial performance. In a different approach, Ma et al. [39] fabricated a polyethylene terephthalate/chitosan (PET/CTS) composite nanofiber membrane through electrospinning. They subsequently deposited a layer of nanosilver onto the fiber membrane's surface using a reduction method. Bacteriostatic experiments revealed the efficacy of the silver-loaded PET/CTS composite nanofiber membrane, displaying sterilization rates of 99.97% and 99.99% against Staphylococcus aureus and Escherichia coli, respectively. Furthermore, Neeta et al. [40] introduced $AgNO_3$ into cellulose acetate (CA), polyacrylonitrile (PAN), and polyvinyl chloride (PVC) polymer solutions. Employing N, N-dimethylformamide as a solvent and reducing agent, they utilized an in-situ reduction method to directly fabricate silver-loaded nanofiber membranes. Antibacterial assessments revealed the commendable antibacterial efficacy of the silver-loaded nanofiber membrane. Montazer et al. [41] employed an in-situ synthesis method to uniformly embed a layer of Ag nanoparticles onto the surface of nylon nanofibers. This method facilitated the creation of a silver/nylon composite nanofiber membrane showcasing substantial antibacterial properties against Escherichia coli and Staphylococcus aureus.

In addition, biomass-derived nanofiber membranes serve as viable alternatives in antibacterial air cleaning applications. For instance, Choi et al. [42] introduced Sophora flavescens extract into a polyvinylpyrrolidone (PVP) solution and utilized electrospinning to produce a herbal extract-doped (HEI) nanofiber membrane for air cleaning purposes. Their findings showcased the HEI nanofiber membrane's significant inhibition rate of 99.98% against Staphylococcus aureus, coupled with an impressive filtration efficiency of 99.99%. Furthermore, Souzandeh et al. [43] reported on a cross-linked protein nanofiber gas cleaning membrane renowned for its potent inhibitory effects against Shigella and Staphylococcus aureus.

4.2.5.2 VOCs Removal Nanofiber Membranes

Data from the U.S. Environmental Protection Agency (EPA) indicates indoor concentrations of volatile organic compounds (VOCs) are ten times higher than outdoor levels [44]. Materials used in indoor decor, such as synthetic building materials, furniture, and household products, frequently emit carcinogenic gases like formaldehyde and toluene. These compounds can adversely affect respiratory and central nervous systems, causing irreversible harm to human health. Thus, the development of functional gas cleaning materials capable of removing VOCs holds immense significance. Presently, VOC removal primarily relies on catalytic degradation technology, commonly utilizing semiconductor metal oxides (e.g., TiO_2, V_2O_5, ZrO_2, CeO_2, MnO_x etc.) as photocatalytic materials. Among these, TiO_2 stands out as one of the most extensively employed photocatalysts. It efficiently catalyzes the degradation of various VOCs, potentially breaking them down into harmless byproducts like CO_2 and H_2O [45].

In a study conducted by Modesti et al. [44], the degradation performance of methanol gas by three forms of TiO_2 photocatalytic membrane—TiO_2 nanofiber membrane, TiO_2 nanoparticle-modified methyl methacrylate-methyl acrylate copolymer hybrid nanofiber membrane (hybrid membrane), and TiO_2 nanoparticle electrostatically sprayed PAN nanofiber multilayer membrane (multilayer membrane, Fig. 4.12a)—was examined. Notably, the multilayer membrane exhibited a near 100% degradation rate of methanol due to the even electrostatic spraying of TiO_2 nanoparticles onto the PAN nanofiber membrane's surface. Under ultraviolet light, this configuration excited more TiO_2 nanoparticles, generating oxidative ·OH radicals. These radicals reacted with methanol gas molecules, oxidizing them to ultimately yield CO_2 and H_2O. Subsequent research observed enhanced photocatalytic degradation efficiency of composite nanofiber membranes for VOCs by combining other semiconductor oxides with TiO_2. Zhu et al. [46], for instance, employed hydrothermal synthesis to coat V_2O_5 nanoparticles onto electrospun TiO_2 nanofiber membrane, resulting in a V_2O_5/TiO_2 composite nanofiber membrane (Fig. 4.12b). Their investigation into the effect of V_2O_5/TiO_2 ratio on acetone gas catalytic degradation performance unveiled an optimal 5% mass fraction of V_2O_5 achieving a 100% conversion rate at 320 °C, nearly tenfold higher than that of pure TiO_2 nanofiber membrane. This improvement stemmed from the electronic interaction between V_2O_5 and TiO_2, enhancing their catalytic efficiency. Further, Zhu et al. [47] employed a hydrothermal method to uniformly grow MnO_x nanoneedles on the surface of electrospun TiO_2 nanofiber membrane, resulting in a hierarchically structured MnO_x/TiO_2 composite nanofiber membrane (Fig. 4.12c). This composite exhibited exceptional catalytic degradation efficiency for lower concentrations of acetone gas, achieving a 90% conversion rate at 290 °C. Furthermore, Chen et al. [48] utilized electrospinning technology to fabricate a mesoporous-macroporous $Ce_{1-x}Zr_xO_2$ nanofiber membrane (Fig. 4.12d). Their catalytic degradation experiments revealed high conversion rates for both benzene and toluene.

The TiO_2 nanofiber membrane mentioned previously, used for the photocatalytic degradation of VOCs, typically demands higher temperature conditions to eliminate over 90% of the pollutant gases. Additionally, adsorption serves as a common method for VOC removal, often achieving high rates of elimination at room temperature. The activated carbon nanofiber membrane (ACNF) emerges as a prevalent adsorption-based material for VOC gas cleaning due to its ultra-high specific surface area, usually around 1000 m^2/g, facilitating enhanced pollutant adsorption. Consequently, extensive research endeavors aim to boost ACNF's adsorption capacity for VOCs by augmenting its specific surface area. Oh et al. [49] introduced manganese acetate precursor into the PAN solution, generating an Mn-PAN precursor nanofiber membrane via electrospinning, and subsequently activated it, yielding Mn-ACNF. This enhancement elevated the specific surface area by nearly half compared to ACNF, increasing the toluene adsorption capacity to 68 g/100 g. Similarly, Guo et al. [50] fabricated an electrospun graphene oxide/activated carbon nanofiber membrane (GO/ACNF), leveraging GO's high-porosity structure to augment ACNF's specific surface area. Their findings demonstrated that GO/ACNF achieved notable adsorption capacities for benzene and butanone gas at 83.2 cm^3/g and 130.5 cm^3/g, respectively. Moreover, Tian et al. [45] combined a TiO_2 nanofiber membrane with an activated carbon fiber membrane to craft a TiNF/ACF porous composite fiber membrane.

This amalgamation harnessed the synergistic effects of TiO_2 nanofibers' photocatalytic properties and ACF's adsorption to eliminate toluene. At higher toluene concentrations (4600 μL/L), the predominant action involved adsorption, achieving an adsorption amount of up to 98.9%. Conversely, at lower toluene concentrations (115 μL/L), catalysis became the primary mechanism, resulting in a degradation rate of 100%. Additionally, Hu et al. [51] successively utilized ZIF-67 and manganese oxide as fillers to craft ZIF-67/PS and MnO_2/PS nanofiber membranes (depicted in Fig. 4.13). By controlling the layered porous structure, these membranes effectively removed SO_2 or formaldehyde gas while ensuring efficient solid particle filtration. The multi-level structured pores offered an expansive specific surface area, enhancing nanofiber contact with gaseous pollutants, while maintaining high gas permeability to minimize filtration resistance.

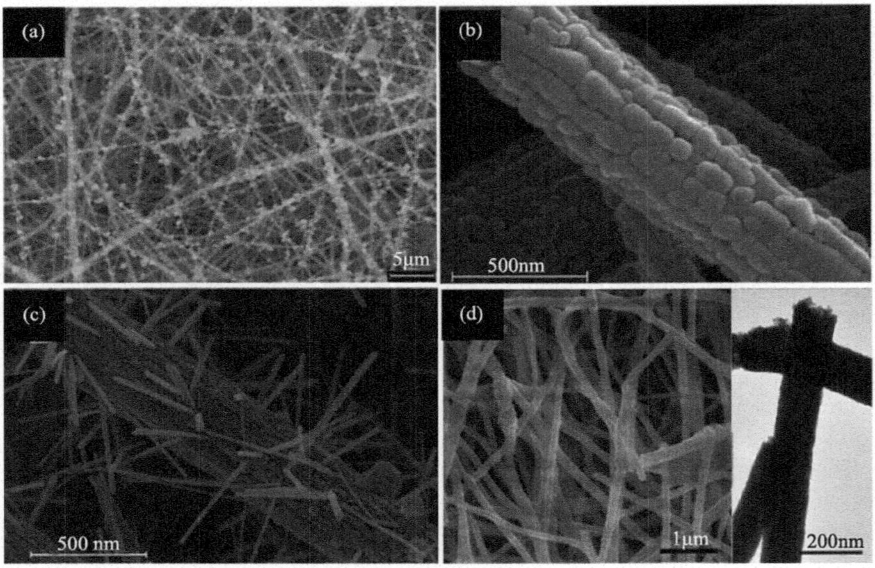

Fig. 4.12 FE-SEM images of TiO_2 based nanofiber membranes. **a** TiO_2/PAN multilayer nanofiber membrane [44]; **b** V_2O_5/TiO_2 nanofiber membrane [46]; **c** MnO_x/TiO_2 composite nanofiber membrane [47]; **d** mesoporous-macroporous $Ce_{1-x}Zr_xO_2$ nanofiber membrane FESEM and TEM image [48]

4.3 Special Wettability Nanofiber Membranes

4.3.1 Overview

In gas cleaning systems, aside from common solid particles, oily pollutants are often present [38, 52–54]. However, conventional membrane materials encounter several challenges in various applications, particularly in high humidity environments, systems containing microorganisms, and in powder recovery systems prone to dampness. For instance, in the cleaning of oily flue gas, the membrane surfaces tend to accumulate oily components, leading to a rapid decline in filtration efficiency and a substantial reduction in lifespan [55–57]. In environments with elevated humidity, condensation occurs due to external gas cooling, resulting in the accumulation of captured particles on the membrane surface. Moreover, in room temperature cleaning settings, microbial adsorption and growth on the membrane surface can diminish filtration efficiency, leading to secondary pollution concerns [58–60]. The structure and function of conventional. Conventional nanofiber membranes exhibit relatively simplistic structures and functions, limiting their optimal performance. By integrating structural design and surface modifications in nanofiber membranes, functionalization can be achieved, significantly enhancing the efficiency.

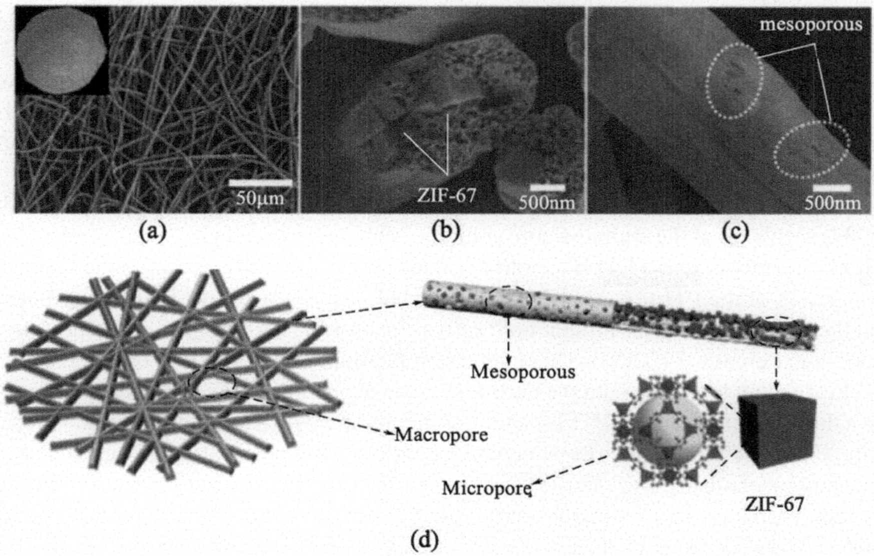

Fig. 4.13 Scanning electron microscope image of ZIF-67/PS nanofiber membrane (**a**–**c**) and schematic diagram of the structure (**d**) [51]

4.3.2 *Omniphobic Nanofiber Membranes*

4.3.2.1 Preparation of Omniphobic Nanofiber Membranes

The introduction of low surface energy media is the main method for preparing omniphobic nanofiber membranes. This technique combines nanoparticle loading with other methodologies to create secondary microstructures, thereby diminishing the solid–liquid contact surface and achieving hydrophobic and oleophobic properties on the membrane surface [61–63]. The production of omniphobic membranes primarily relies on surface modification methods, such as surface grafting, impregnation, and vapor deposition. However, due to constraints posed by the microporous structure of the membrane material and the demanding gas environment, omniphobic modification necessitates that it doesn't obstruct membrane pores or compromise the material's mechanical and thermochemical performance. Hence, when incorporating low surface energy media, it is crucial to comprehensively consider the structural stability of the membrane material and the bonding strength and method between the media and the membrane material [64, 65].

(1) Surface Grafting Method

Surface grafting refers to initiating low surface energy monomers on the membrane surface. Typically, with chemically inert membrane materials, the process involves initially introducing active components onto the membrane surface as intermediates. Subsequently, the dual-repellent modification of the membrane surface is completed by grafting the monomer onto these intermediates. The distinctive feature of the surface grafting method lies in the chemical bonding of the low surface energy monomer to the microporous membrane surface, thereby enhancing the bonding performance between the monomer and the membrane.

Zhu et al. [66] utilized a nano SiO_2 dispersion as the shell spinning solution and PTFE emulsion as the spinning core solution, effectively enveloping SiO_2 nanoparticles uniformly onto the PTFE nanofibers. This process not only enhances the surface roughness of the PTFE nanofiber membrane but also leverages the hydroxyl groups on the SiO_2 surface, providing ample active sites for the surface grafting of perfluoropolymer chains (PFTMS). This interaction facilitates the formation of a stable Si–O–Si covalent bond between the perfluoro chain segment and the substrate, resulting in a dual-repellent SiO_2@PTFE nanofiber membrane with a multi-level rough structure. Figure 4.14 illustrates the schematic diagram of the preparation process and the grafting mechanism.

The prepared dual-repellent membrane material has contact angles with water and oil (hexadecane) of 173° and 134°, respectively. Following ultrasonic oscillation in an ethanol solution for 3 h, the contact angles for oil and water on the dual-repellent membrane showed minimal decrease, indicating the membrane's remarkable stability.

Utilizing low-temperature plasma to bombard the membrane material surface opens the chemical bonds of the surface molecules. Within this plasma environment, certain chemical bonds of the low surface energy monomer are activated, allowing them to graft onto the base group present on the membrane surface, resulting in the acquisition of a dual-repellent surface [67–69]. Low-temperature plasma exhibits controllable energy levels, strong activation capability, and minimal impact on the substrate [70, 71]. In a study by Feng et al. [72], atomic layer deposition was initially employed to uniformly deposit nano-scale ZnO particles onto the surface of PTFE membrane fibers. This process effectively increased the surface roughness of the fiber membrane, enhanced nanofiber rigidity, and maintained fiber structure stability. Subsequently, through plasma grafting of a low surface energy perfluoromonomer (perfluorodecyl acrylate), the surface energy was reduced, rendering it hydrophobic and oleophobic. Figure 4.15 illustrates the mechanism of the perfluoromonomer's grafting reaction with the PTFE membrane under plasma influence.

Through plasma activation, the bond energy of the CC double bond (620 kJ/mol) is lower than that of the CO double bond (798 kJ/mol), making the CC double bond more active and prone to opening and subsequent reaction with the active unit between the ZnO@PTFE membrane. Consequently, the oil (hexadecane) and water contact angles of this dual-repellent membrane reach 125° and 150°, respectively.

Surface grafting methods can directly graft low surface energy monomers onto the surface of membrane materials, or first introduce active groups onto the membrane

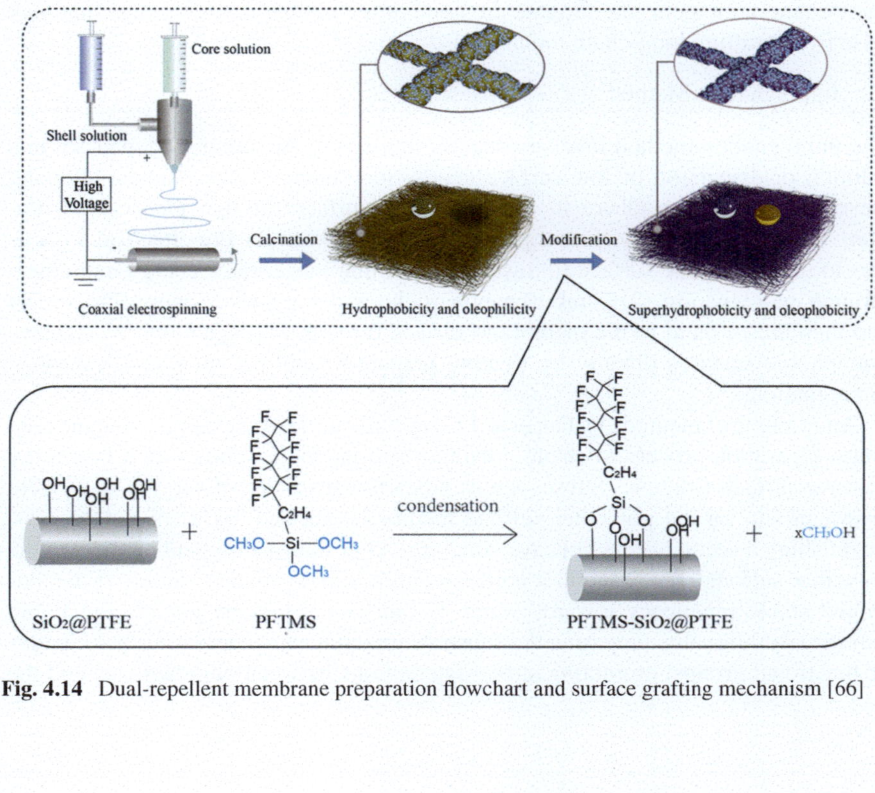

Fig. 4.14 Dual-repellent membrane preparation flowchart and surface grafting mechanism [66]

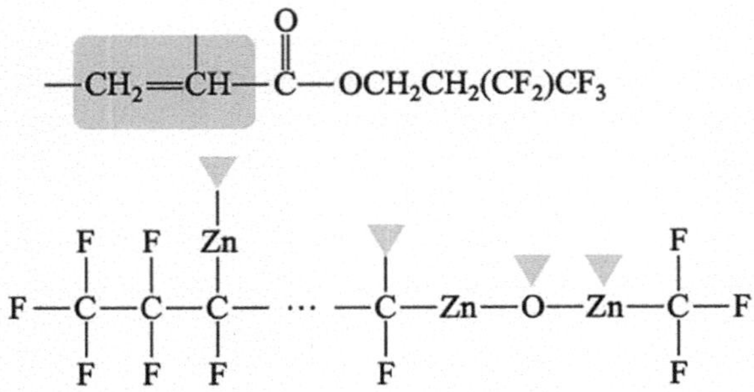

Fig. 4.15 Grafting mechanism of perfluoromonomer on PTFE membrane surface [72]

surface as intermediates. These active groups serve as reaction sites for graft copolymerization, facilitating stable chemical bonding between the low surface energy monomers and the membrane surface. Additionally, modifying the method of monomer introduction, such as introducing low surface energy monomers through atmospheric pressure or spray techniques, can help prevent or minimize the agglomeration effect of liquid surface tension on fibers, preserving the stability of the membrane structure. As plasma technology advances, atmospheric pressure treatment or spray-based monomer introduction could become crucial for large-scale plasma modifications to create omniphobic membranes. Furthermore, opting for short-chain monomers can mitigate the issue of membrane pore blockage that may arise from grafting long-chain monomers.

(2) Impregnation Method

The impregnation method involves directly immersing the membrane material in a solution or dispersion of low surface energy monomers or compounds, allowing these components to adhere to the membrane surface through physical adsorption. There are primarily three operational approaches: (1) Pre-attach nanoparticles to the membrane surface followed by attaching low surface energy monomers through impregnation. (2) Combine nanoparticles with low surface energy monomers and then attach them to the membrane surface through impregnation. (3) Achieve omniphobic surfaces directly by treating prepared nanofiber membranes through impregnation.

An example of method (1). Fan et al. [73] initially utilized the non-solvent induced phase separation process to create polyether sulfone membranes with a first-order concave surface structure. Subsequently, silica nanoparticles of varying sizes were grown in situ on the polyether sulfone membrane surface via a sol–gel process, establishing a second-order concave structure. After fluorinating and modifying the polyether sulfone membrane in a perfluorooctyltriethoxysilane solution, it demonstrated stable repellency towards water, hexadecane, ethylene glycol, and liquid paraffin. Additionally, preparing the substrate material can generate surface charges or functional groups, promoting strong interactions between nanoparticles and the substrate. In a similar vein, Lin et al. [74] initiated their process by creating alkali-treated PVDF membranes through aminosilane functionalization. They then electrostatically grafted negatively charged silica nanoparticles onto the membrane surface, followed by fluorosilane modification to reduce surface energy. The resulting PVDF membrane exhibited superhydrophobic and oleophobic properties, with a water contact angle exceeding 150°, while the contact angles of sodium dodecyl sulfate (SDS) solution and mineral oil were both greater than 130°. Moreover, Khan et al. [75] submerged PES membranes in an ethanol solution of 3-aminopropyltriethoxysilane to induce a positive charge on the membrane surface. This process enhanced the loading amount and binding force of SiO_2 nanoparticles on the membrane surface. Subsequently, a vacuum filtration method was employed to apply a low surface energy polydimethylsiloxane-fluoroalkylsilane (PDMS-FAS) coating onto the PES membrane surface, forming a stable Si–O–Si covalent bond structure between the perfluoromonomer and the membrane substrate.

An example of method method (2). Initially, the nanoparticles underwent modification to achieve omniphobic properties before being affixed to the membrane surface, thereby effectively enhancing the membrane's overall omniphobic performance. In their work, Xu et al. [76] utilized various perfluoromonomers [1H, 1H, 2H, 2H-Perfluorooctyltrichlorosilane (PFOTS), 1H, 1H, 2H, 2H-Perfluorodecyl acrylate (PFDAE), 1H, 1H, 2H, 2H-Perfluorooctyltrimethoxysilane (PFOTMS)] to fluorinate and modify SiO_2 nanoparticles. This process involved grafting low surface energy perfluoroalkyl chains onto the SiO_2 nanoparticle surface, resulting in hydrophobic and oleophobic SiO_2 nanoparticles. Comparatively, PFTMS-grafted SiO_2 nanoparticles achieved superamphiphobic effects, with contact angles of 154° and 171° for oil (hexadecane) and water, respectively. These superamphiphobic nanoparticles were then applied onto porous PTFE fibers using both impregnation and filtration methods. The impregnation method faced challenges in depositing nanoparticles into the pores, leading to poor membrane adhesion and eventual detachment of amphiphobic nanoparticles from the membrane. In contrast, the filtration method yielded better amphiphobic performance in the resulting PTFE fiber membrane. Here, the coating effectively penetrated the pores aided by airflow. Experimental outcomes demonstrated contact angles of 123° and 146° for oil (hexadecane) and water, respectively, highlighting improved performance compared to the impregnation method.

An example using method (3). The nanofiber membrane, prepared initially, underwent direct impregnation to lower its surface energy, resulting in the acquisition of amphiphobic functionality. In contrast to grafting impregnation, this process primarily relies on physical binding for integrating the low surface energy material. Lin et al. [65] fabricated a poly(m-phenylene isophthalamide) (PMIA) nanofiber membrane via electrospinning and subsequently subjected it to fluorination by immersion in a perfluorosilane solution, reducing its surface energy and rendering the PMIA nanofiber amphiphobic. Research findings have demonstrated the excellent thermal stability of the amphiphobic PMIA nanofiber. In a similar vein, Mao et al. [77] initiated their study by producing a silica nanofiber (SNF) membrane through electrospinning. They proceeded to impregnate the newly synthesized fluorinated polyurethane (FPU) onto the fiber surface, resulting in the formation of an SNF/

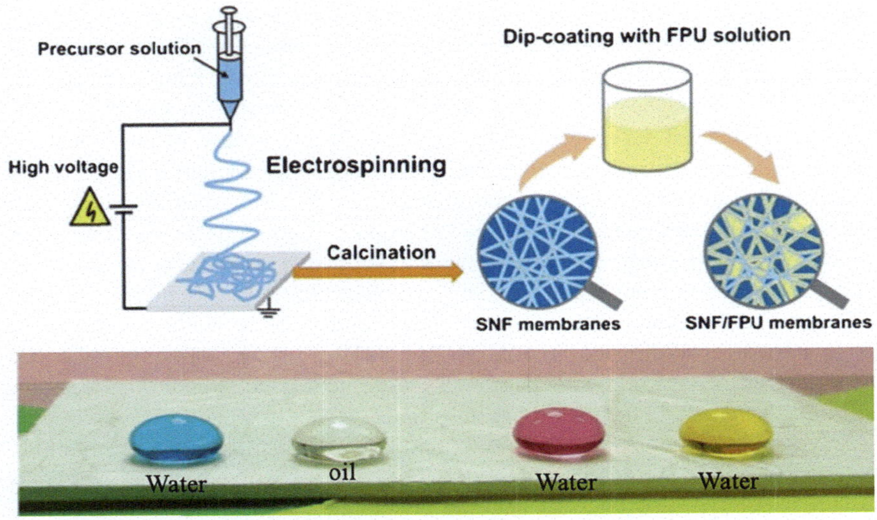

Fig. 4.16 Schematic diagram of the preparation of SNF/FPU fiber membrane (**a**); Optical photos of oil droplets and water droplets on the surface of the amphiphobic membrane (**b**) [77]

FPU nanofiber membrane with amphiphobic characteristics [refer to Fig. 4.16a]. This membrane exhibited commendable waterproof and breathable qualities. Investigations revealed that when the FPU content in the impregnating solution was 1%, the SNF/FPU fiber membrane displayed optimal amphiphobic performance, registering oil and water contact angles reaching 147° and 121°, respectively [refer to Fig. 4.16b].

The impregnation method boasts mild process conditions and simple operation, rendering it suitable for various substrate types with a broad spectrum of applications. However, the influence of the impregnation order on the microstructure of the membrane material must be considered during this process. Achieving this involves matching the nanoparticle size with the monomer type, thereby minimizing structural changes in the membrane caused by impregnation. When combining nanoparticles with membranes, prerequisites such as electrostatic and capillary actions are vital to ensuring sufficient contact and adhesion between the monomer and the membrane's microstructure. Simple physical action alone cannot guarantee the secure integration of nanoparticles with the base membrane, potentially leading to uneven surface performance due to nanoparticle detachment [78]. As a result, superhydrophobic membranes prepared through this method may not be suitable for use in harsh gas cleaning environments.

The mixing of nanoparticles with a low surface energy solution and spraying it onto the membrane material offers repairable coating options, holding significant potential in creating superhydrophobic membrane surfaces [79–81]. Zhou et al. [81] developed a water-based coating system consisting of superhydrophobic nanoparticles (Zonyl@321) and perfluorosilane (FAS) applied through spraying onto

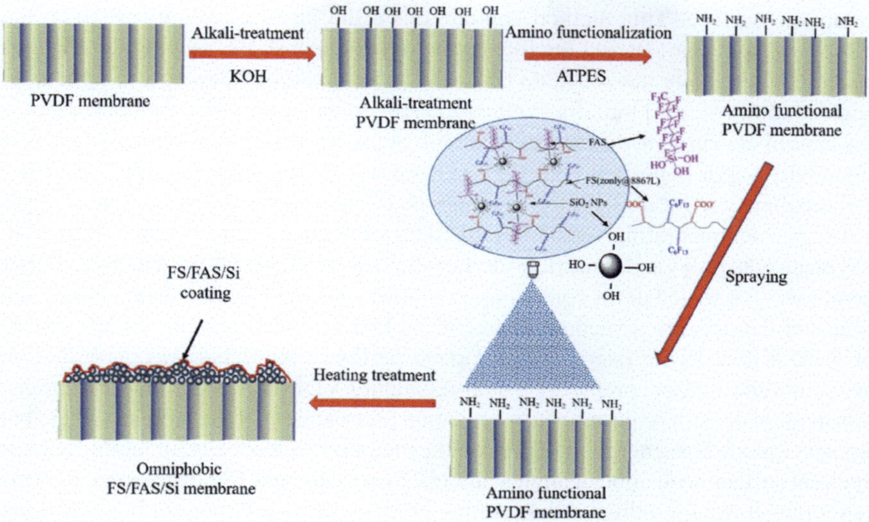

Fig. 4.17 Preparation flow chart of FS/FAS/Si membrane [82]

fabrics, organic or inorganic fiber membranes, and various substrates, all exhibiting commendable superhydrophobic performance. To enhance the bonding strength and coating amount, the membrane material's surface is pre-functionalized to augment surface charge. Li et al. [82] treated PVDF membrane with surface amination, enriching the membrane surface with a substantial positive charge. Employing electrostatic adsorption, a mixed aqueous solution of fluorocarbon surfactant (FS), fluorinated alkyl silane (FAS), and silica nanoparticles were absorbed onto the membrane surface, rendering the PVDF membrane superhydrophobic, as illustrated in Fig. 4.17. The results demonstrate that the resulting superhydrophobic membrane showcases excellent repellency to liquids, with contact angles surpassing 150° for water, diiodomethane, mineral oil, and SDS solution. Compared with the impregnation method, the spraying approach necessitates considering the relationship between nanoparticle size and the membrane material's basic structure, typically favoring smaller nanoparticles. The distinction lies in how spraying mitigates changes in the membrane material's microstructure caused by the impregnation process, preserving the material's fundamental structure and controlling the coating volume by regulating the number of sprays. However, it's important to note that the spraying process primarily focuses on surface modification, lacking efficacy in modifying the membrane material's interior. Consequently, the spraying method excels in creating asymmetric interface membranes.

(3) Vapor Deposition Method

Chemical vapor deposition is a method of using one or several gaseous compounds or monomers containing low surface energy substances to carry out chemical reactions on the membrane surface to generate thin membranes. Physical vapor deposition directly vaporizes low surface energy chemicals and deposits them on the membrane surface. This method capitalizes on the physical affinity between the perfluoro chain segment and the membrane material substrate, making the requirements for the membrane substrate relatively lenient and broadening its application range. Nadir et al. [83] initially synthesized silicon nanoparticles on prepared cellulose nanofibers via an in-situ sol–gel method. Subsequently, perfluoroalkylsilane was chemically vapor-deposited onto their surface to reduce surface energy, imparting dual repellency to SiNPs cellulose nanofibers. Results showed a water contact angle of 150.6°, demonstrating repellency to SDS, mineral oil, and ethanol. Wu et al. [84] placed the polyvinylidene fluoride-hexafluoropropylene (PVDF-HFP) nanofiber membrane in a sealed dryer containing perfluorooctyltrichlorosilane monomer, then subjected the dryer to a vacuum degree of 20 kPa and a temperature of 50–120 °C for 5–40 h in a drying oven. The vaporized perfluoro monomer was deposited on the membrane surface, preparing a dual-repellent PVDF-HFP nanofiber membrane. This method doesn't require surface activation pretreatment of the prepared fibers but deposits low surface energy materials on the membrane physically, obtaining a dual-repellent surface with contact angles of 154.1° for water and 122.6° for ethanol. This technique allows for a thinner fluorination layer on the membrane surface, resulting in minimal impact on the membrane's pore structure, and ensures more uniform monomer adherence to the membrane surface. Similar to the spraying method, vapor deposition primarily acts on the membrane surface, exerting minimal effect on the membrane's interior. Table 4.2 shows a comparison and summary of different dual-repellent membrane modification methods, detailing modification mechanisms, advantages, and disadvantages.

4.3.2.2 Dual-Phobic Membrane Applications and Mechanism in Gas Cleaning

The dual-phobic membrane, characterized by its surface hydrophobic and oleophobic properties, presents notable advantages in addressing pollution and facilitating self-cleaning mechanisms within gas cleaning processes. Across diverse gas cleaning frameworks characterized by high humidity, elevated viscosity, and the presence of microorganisms, the dual-phobic membrane outperforms conventional membrane materials by efficiently capturing particulate matter. Simultaneously, it mitigates condensation and the accumulation of filter cake resulting from microbial adhesion. This remarkable capability significantly enhances both gas cleaning efficiency and the longevity of the membrane material. In the context of oil aerosol systems, standard membrane materials encounter significant limitations. In contrast, the dual-phobic membrane exhibits resilience by maintaining a consistent pressure drop, underscoring its promising potential for practical implementation.

Xu et al. [76] utilized a PTFE dual-phobic membrane prepared through nanoparticle impregnation and filtration for the cleaning of corn oil aerosols. The controlled filtration flow rate stood at 6.67 L/min, covering an area of 7.07 cm^2, with an aerosol concentration of was 1320 mg/m^3, as depicted in Fig. 4.18a. The concentrations of corn oil aerosols post-filtration through the dual-phobic PTFE membrane and

Table 4.2 Comparison of different dual-repellent membrane material modification methods

Modification method	Mechanism	Advantages and disadvantages	References
Wet grafting	Low surface energy monomer chemically bonds with the surface of the microporous membrane	The prepared dual-repellent membrane has good stability, but it needs to introduce active sites on the membrane surface	[66, 85]
Thermal grafting	Crosslinking reaction between membrane material and low surface energy monomer	Requires high-temperature heat treatment, has a significant impact on the structure of the membrane material	[86]
Plasma grafting	Plasma activates the surface, grafting the selected monomer	Energy controllable, strong activation ability, little impact on the substrate itself, but high vacuum requirements for the environment, large-scale application is difficult	[87]
Impregnation method	Low surface energy monomer or dual-repellent nanoparticles are physically adsorbed on the membrane surface	Mild process conditions, simple operation, wide application range, but the impregnation process has a significant impact on the membrane structure, nanoparticles are easy to fall off	[73, 76, 88, 89]
Spray method	Dual-phobic coating atomized and sprayed on the membrane surface	Damage to the surface coating can be repaired by re-spraying, but the ability to modify the internal properties of the membrane material is poor	[81, 82]
Gas phase deposition method	Low surface energy gas molecules deposited on the membrane surface	Forms a thinner fluorination layer on the membrane surface, with little impact on the pore structure of the membrane	[83, 84]

the ordinary PTFE membrane were recorded at 6 mg/m^3 and 21 mg/m^3, respectively. Notably, the dual-phobic membrane exhibited superior corn oil aerosol filtration efficiency, exceeding 99.5%. Upon analysis of filtration resistance, oil droplets adhering to the surface of the ordinary PTFE membrane caused a rapid surge in filtration pressure reaching 35 kPa within 6 min (Fig. 4.18b). In contrast, the filtration pressure of the dual-phobic membrane showed a slight increase under prolonged filtration conditions, ultimately stabilizing at 8.25 kPa. Subsequently, the stability of the dual-phobic membrane underwent evaluation post-ultrasonic cleaning with deionized water and drying. With an increase in the number of washes applied to the dual-phobic membrane, a marginal decline in its retention rate was observed, concomitant with a slight rise in filtration pressure (Fig. 4.18c). This phenomenon may arise from the loss of nanoparticles during the cleaning process, resulting in diminished dual-phobic properties on the membrane surface.

Feng et al. [72] conducted a comparative analysis between ordinary PTFE membranes and ZnO@PTFE dual-phobic membranes, the latter being grafted with perfluorodecyl acrylate, in the filtration performance of incense aerosols. The membrane's pore size measured 5 μm, while gas flow rate was controlled at 1 L/min through valve adjustment, covering a test area of 7.07 cm^2 (with a diameter of $\varphi = 3$ cm), as depicted in Fig. 4.19. The initial filtration pressure drop for the ordinary PTFE membrane was approximately 0.5 kPa, demonstrating an exponential increase with prolonged filtration. In contrast, the double-repellent membrane showcased an initially lower pressure drop, exhibiting a gradual rise over time, indicative of commendable anti-pollution capabilities. Moreover, the double-repellent membrane displayed an approximate 99.5% retention rate for incense aerosols, resulting in an outlet concentration of less than 0.50 mg/m^3. These findings highlight the significant reduction in filtration resistance for oily aerosols with the double-repellent PTFE fiber membrane compared to the original membrane, while consistently maintaining high retention performance.

Figure 4.20a–c illustrates the filtration process of oil and gas aerosols using oleophilic and oleophobic fibers, starting from the fundamental filtration unit. Oleophilic fibers allow oil and gas aerosol particles to traverse through them with the airflow, adhering to these fibers through mechanisms such as diffusion, inertia, and interception, among others. Due to the robust adhesion between oil droplets and the oleophilic fiber surface, the filtered oil droplets firmly attach to the fiber surface. Over time, as filtration continues, an accumulation of oil droplets occurs, resulting in the formation of an oil layer on the oleophilic membrane surface (Fig. 4.20b). This layer, challenging to dislodge due to gravity, obstructs the filtration channel, consequently elevating gas resistance across the filtration membrane (Fig. 4.20d). In contrast, double-repellent fibers, owing to their oleophobic properties, lead to filtered oil droplets appearing as distinct entities on the fiber surface (Fig. 4.21c). As filtration proceeds, the small accumulated oil droplets on the membrane surface gradually coalesce until their gravity surpasses their adhesion to the double-repellent membrane

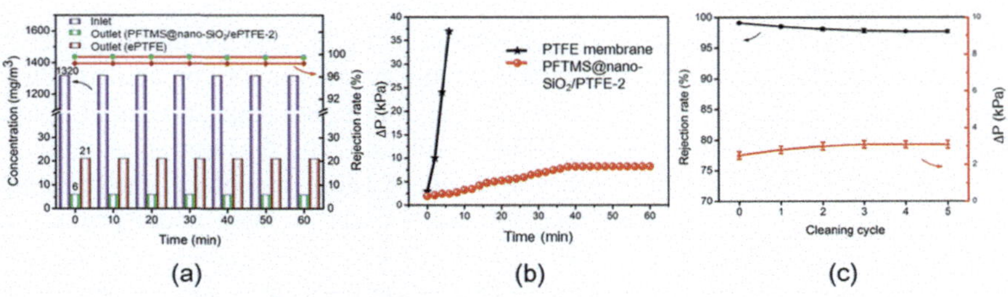

Fig. 4.18 Double-repellent membrane's oil smoke gas retention performance (**a**); Pressure drop changes of PTFE membrane and PTFE double-repellent membrane in the process of filtering oil smoke gas (**b**); and recirculation filtration performance of double-repellent membrane (**c**) [76]

surface. Consequently, these oil droplets smoothly slide off the double-repellent membrane under the force of gravity, endowing the membrane with a regenerative function (Fig. 4.20e). Simultaneously, the reduced adhesion facilitates the cleaning of the double-repellent membrane, enhancing its reusability [90].

Zhu et al. [66] fabricatedSiO_2@PTFE nanofiber membrane through a one-step electrospinning process, subsequently treating it with fluorination to explore the dual-repellent properties of the F-SiO_2@PTFE Nanofiber Membrane in filtering Corn Oil Aerosol. A consistent source of corn oil aerosol was generated via an aerosol generator, possessing a particle size of approximately 0.26 μm and an inlet concentration of 200 mg/m^3. Following 60 min of continuous filtration, the outlet concentrations for the PTFE, SiO_2@PTFE, and F-SiO_2@PTFE nanofiber membranes measured 5.1 mg/m^3, 11.5 mg/m^3, and 1.84 mg/m^3, respectively. The smaller pore size of the PTFE nanofiber membrane, in comparison to the modified membrane, contributed to a lower outlet concentration than that of the SiO_2@PTFE nanofiber membrane. Despite the omniphobic membrane's larger pore size relative to the PTFE fiber membrane, it demonstrated the lowest outlet concentration. These experimental findings underscore the F-SiO_2@PTFE nanofiber membrane's superior filtration efficiency for corn oil aerosol (99.08%), while maintaining a stabilized filtration pressure drop at approximately 2.5 kPa.

Several studies propose that when filtering oily aerosols using an omniphobic membrane, the oil-repellent nature of the fiber surface causes small oil droplets to exhibit a bouncing phenomenon upon reaching the fiber surface at specific speeds [91]. During this bouncing process, these small oil droplets might collide with one another or with subsequent droplets, gradually increasing in size until they become too substantial to bounce, dispersing in the pore channels as droplets. Consequently, larger oil droplets occupy a certain pore volume, thereby elevating gas resistance [92, 93]. Maintaining a constant filtration flow rate escalates the air flow speed within the pore channels, generating a greater driving force for these droplets. When this force surpasses the adhesion force between the droplet and the fiber surface, the

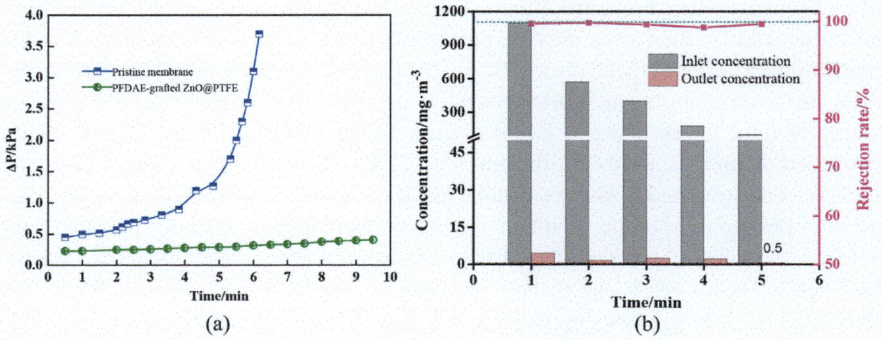

Fig. 4.19 Filtration pressure drop of ordinary PTFE membrane and double-repellent PTFE membrane (**a**); and filtration performance of double-repellent PTFE membrane [72] (**b**)

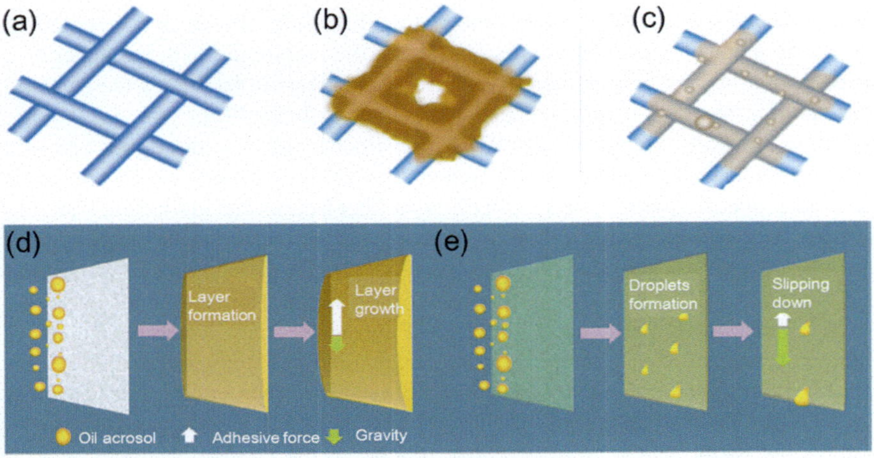

Fig. 4.20 Analysis of the filtration mechanism of oil smoke on fibers and membrane surfaces [90] **a** schematic diagram of the membrane fiber structure before filtration; **b** schematic diagram of oil adhesion on oleophilic membrane fibers; **c** schematic diagram of oil adhesion on oleophobic membrane fibers; and **d, e** analysis of the oil smoke filtration mechanism of ordinary PTFE membrane and double-repellent PTFE membrane

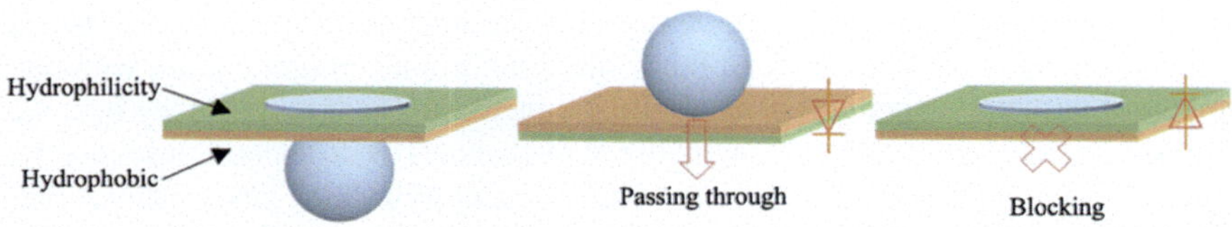

Fig. 4.21 Schematic diagram of the "diode" phenomenon of liquid in asymmetric infiltration membrane

droplet moves in the direction of the gas flow [94]. Additionally, larger oil droplets are influenced by their own gravity, prompting movement. Consequently, through the combined effect of these forces, the oil droplets depart from the fiber membrane, allowing the omniphobic membrane to restore its effective filtration area. This process continues until the discharged liquid equates to the newly collected liquid, establishing a dynamic filtration equilibrium [95–97]. Consistently maintaining a constant inlet concentration and flow rate sustains this dynamic equilibrium. Thus, over time, the omniphobic membrane retains excellent performance in filtering oily aerosols [95–97]. Conversely, for oleophilic fibers, oil droplets tend to rapidly spread across the fiber surface, coalescing at fiber intersections to form a liquid film. This film obstructs membrane pores, diminishing both the effective pore size and the filtration area of the filter [98].

4.3.3 *Asymmetric Wetting Nanofiber Membranes*

In recent years, Janus membranes have garnered increased attention and research interest. Derived from the 'two-faced god' in ancient Roman mythology, Janus membranes earn their name from exhibiting different properties on their front and back surfaces [99]. Among these, Janus membranes showcasing distinct infiltration properties on each side are categorized as asymmetric infiltration membranes. These materials demonstrate structures such as hydrophilic-hydrophobic or oleophilic-oleophobic attributes within their thickness. Driven by these unique surface physicochemical properties, these three-dimensional porous membrane materials facilitate anisotropic transport between their cross-sectional layers, leading to a 'diode' phenomenon in liquid transport [100, 101]. For instance, considering the hydrophilic/hydrophobic membrane illustrated in Fig. 4.21, water droplets readily permeate from the hydrophobic side to the hydrophilic side. However, when the membrane is reversed, it partially impedes the passage of droplets.

4.3.3.1 Structure of Asymmetric Wetting Membranes

Based on their membrane structure, Janus membranes are categorized into three types: A-on-B, A-and-B, and A-to-B, with A and B representing two components exhibiting significant differences in infiltration [99], as depicted in Fig. 4.22. The A-on-B type features an A layer considerably thinner than the B layer, yet it significantly influences the surface properties of the membrane. Conversely, the A-and-B type showcases similar thicknesses for both A and B layers. The A-to-B type demonstrates a gradual transition from A to B across the membrane's cross-section. Similarly, asymmetric wetting membranes are classified based on the cross-sectional wettability changes, divided into abrupt wetting type (opposite wettability) and gradient wetting type (wettability gradient) [101]. Interestingly, the abrupt wetting type encompasses both A-on-B and A-and-B types, while the definition of the gradient wetting type aligns with the A-to-B type. Both abrupt and gradient wetting membrane materials find diverse applications in humidity regulation, fog collection, oil mist filtration, oil–water separation, and emulsification. However, the former exhibits a distinct interface connecting two entirely different wetting areas, whereas the latter necessitates ensuring progressive wetting changes through specialized preparation methods.

(1) Abrupt Wetting Type

The predominant method to fabricate abrupt wetting type asymmetric infiltration membranes involves the separate preparation of two materials possessing different wetting properties, subsequently combining them. For instance, Wu et al. [100] manufactured hydrophobic polyurethane (PU) and hydrophilic polyvinyl alcohol (c-PVA) using electrospinning, followed by crosslinking to produce a nanocomposite fiber membrane. Similarly, Ding et al. employed a combination of materials, such as hydrophobic non-woven fabric/PAN-SiO_2 [102], hydrophobic cellulose acetate/hydrolyzed hydrophilic cellulose acetate [103], and hydrophobic PU/

hydrophilic porous PAN [104], crafting asymmetrically infiltrating membranes through electrospinning. Other researchers have utilized comparable methodologies to create double-layer asymmetrically infiltrating fiber membranes like ZnO-PVDF/PAN [105], PU/CA [106], and TPU/TPU-TBAC [107]. Alternatively, asymmetrically infiltrating membranes can be formed via vacuum filtration by sequentially coating hydrophilic and hydrophobic nanotubes [108] or nanowires [109] onto a porous substrate, followed by peeling them off [110]. While electrospinning and vacuum filtration methods offer precise and convenient control over membrane layer thickness, ensuring interface compatibility between the layers is crucial for membrane stability [99]. Moreover, an asymmetrically infiltrating structure can also be achieved during membrane formation by leveraging the migration or phase separation of immiscible components within the membrane matrix [111, 112] (as depicted in Fig. 4.23).

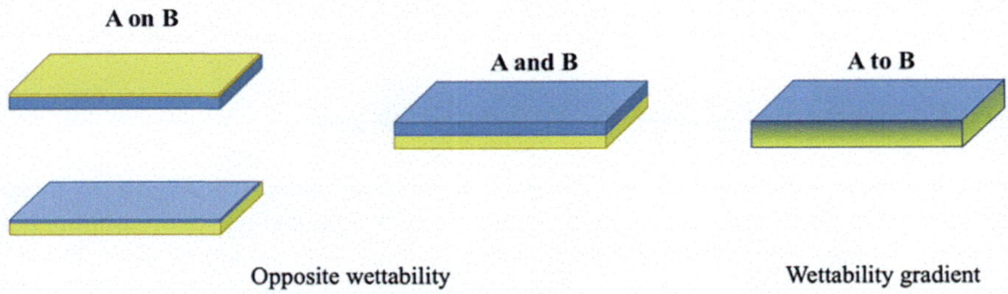

Fig. 4.22 Schematic diagram of asymmetric wetting membrane A-on-B, A-and-B and A-to-B structures

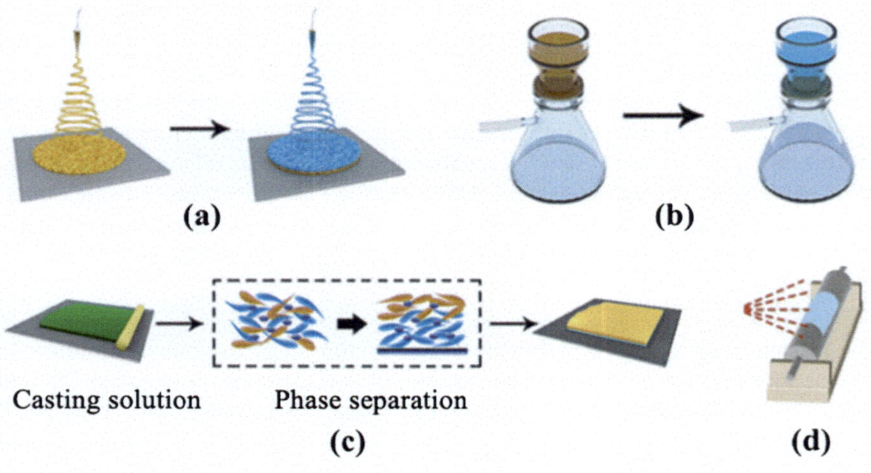

Fig. 4.23 Preparation method of abrupt wetting asymmetrically infiltrating membrane [111, 112] **a** electrospinning method; **b** vacuum filtration method; **c** phase separation method; **d** spray coating method

Similarly, the fabricated fiber membrane can be altered or modified on one side to create an abrupt wetting structure. Notably, single-sided spray coating and electrostatic spray coating represent typical examples, as illustrated in Fig. 4.23 (Mates et al. [113]) conducted single-sided spray coating of nanoclay/aqueous fluoropropyl acrylate copolymer on an ordinary cellulose fabric, as confirmed by Energy Dispersive Spectrum (EDS) analysis, verifying the coating's presence solely on the sprayed side of the membrane. Zeng et al. [114] achieved an asymmetrically infiltrating fabric by electrostatically spraying a thin hydrophobic commercial photoresist (SU-8) coating onto a hydrophilic fabric. The electrostatic spray coating process involves liquid atomization induced by a strong electric field. Modulating the liquid flow rate allows for control over the coating thickness, thus enhancing the manageability of the hydrophobic layer thickness. Beyond the hydrophobic-hydrophilic structure, Wei et al. [115] treated the fiber membrane with immersion coating and single-sided electrostatic spray coating, yielding a fiber membrane possessing a super-hydrophobic-super-oleophilic structure throughout its thickness.

As mentioned above, the creation of abrupt asymmetrically infiltrating membranes involves composite methodologies such as electrospinning, vacuum filtration, phase separation, and single-sided modification techniques like spray coating and electrostatic spray coating. Practically all the methods involved in preparation offer control over the thickness of each layer. However, only a limited number of electrospinning techniques involve adjusting the wettability of a specific layer by introducing substances influencing membrane wettability into the spinning solution. Building upon this observation, a conjecture emerges regarding the developmental trajectory of abrupt wetting asymmetrically infiltrating membrane preparation methods: initially combining two layers of membranes with differing wettability and subsequently fine-tuning the wettability of a specific layer as required through single-sided modification.

(2) Gradient wetting type

Gradient wetting asymmetrically infiltrating membranes are primarily obtained through single-sided modification techniques. Among these, single-sided ultraviolet irradiation was one of the pioneering methods introduced for creating gradient wetting asymmetrically infiltrating membrane structures. In 2010, Wang et al. [116] reported the first instance of a gradient wetting membrane material. Their study employed a two-step approach to construct a wetting gradient from superhydrophobicity to hydrophilicity across the cross-sectional direction. Initially, commercial polyester fabric was impregnated with alkyl-substituted silica containing titanium dioxide. Subsequently, the impregnated fabric underwent treatment with single-sided ultraviolet irradiation (Fig. 4.24a). The silica/titanium dioxide-coated fabric exhibited a superhydrophobic surface. Upon ultraviolet irradiation, the photocatalytic degradation of alkyl by titanium dioxide introduced oxygen-containing hydrophilic groups to the irradiated surface, rendering it hydrophilic, while the opposite side retained its superhydrophobicity. The photocatalytic reaction's intensity relied on light intensity, leading to gradual photodegradation along the fabric's thickness direction. This generated a wetting gradient across the fabric's thickness. Similar

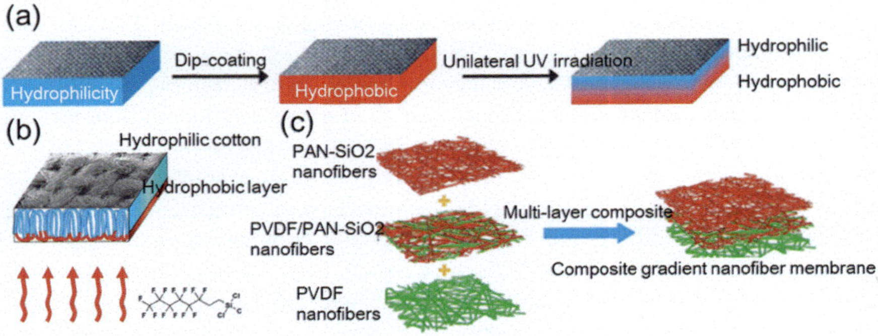

Fig. 4.24 Preparation methods of gradient wetting asymmetric infiltration membranes [116, 117]. **a** Single-sided ultraviolet irradiation method; **b** vapor diffusion method; **c** three-layer composite electrospinning method

to ultraviolet irradiation, plasma treatment can induce the formation of polar groups (such as hydroxyl groups) on solid surfaces. For instance, hydrophobic polytetrafluoroethylene fabric was treated with O_2/H_2 plasma on one side to create a hydrophilic surface, resulting in a hydrophobic-hydrophilic gradient wetting membrane [117]. Another method to obtain a gradient wetting structure involves one-sided deposition through solute diffusion. Tian et al. [117] utilized perfluorooctyltrichlorosilane (PFOTS), known for its reactivity with hydroxyl groups, as a hydrophobic agent. They employed vapor diffusion technology to selectively hydrophobize one side of a hydrophilic cotton fabric (Fig. 4.24b).

Gradient wetting structures can also be created through electrospinning techniques. Ding Bin's team utilized electrospinning to fabricate a gradient wetting structure by incorporating a two-component mixed spinning intermediate layer between the hydrophilic and hydrophobic layers. For instance, Zhao et al. [118] developed a gradient fiber membrane structure composed of polyacrylonitrile-silicon dioxide nanoparticles (PAN-SiO_2) as the outer layer, polyvinylidene fluoride/polyacrylonitrile-silicon dioxide nanoparticles (PVDF/PAN-SiO_2) fibers as the middle layer, and polyvinylidene fluoride (PVDF) as the inner layer (Fig. 4.24c). Similarly, Miao et al. [119] created a three-layer fiber membrane using polyurethane/(polyurethane-hydrolyzed polyacrylonitrile)/hydrolyzed polyacrylonitrile [PU/(CPU-HPAN)/HPAN].

To summarize, the prevalent methods used to achieve gradient wetting structures involve ultraviolet irradiation, plasma treatment, and one-sided deposition, primarily reliant on the modification effect's attenuation along the thickness direction. However, confirming the successful preparation of gradient wetting materials requires advanced equipment and viable methodologies. Notably, unlike the continuous wetting gradient obtained through the aforementioned techniques, the structure achieved by introducing an electrospun mixed spinning transition layer presents a discontinuous wetting gradient.

4.3.3.2 Applications and Mechanisms of Asymmetrically Wetting Membranes

Due to distinct wetting properties across various surfaces, a unique 'liquid diode' mass transfer phenomenon manifests within the three-dimensional structure of most asymmetrically wetting materials. Much like electrical current flowing through a diode in a single direction, liquid exhibits a facile penetration from the hydrophobic side to the hydrophilic side. However, upon reversing the direction, the liquid's penetration is partially impeded. Consequently, asymmetrically wetting membranes hold vast potential applications in moisture (humidity) regulation, fog collection, filtration of oily aerosols, oil–water separation, and emulsification, among others.

In 2010, Wang et al. [116] proposed two water transport mechanisms in their study on directional water transport through the asymmetrically wetting fabric: (1) Water droplets are pulled from the hydrophobic layer to the hydrophilic layer under the influence of the hydrophobic-hydrophilic gradient (Fig. 4.25a); (2) Water evaporates from the hydrophobic side and condenses in the hydrophilic layer. The hydrophobic-hydrophilic structure accelerates the process of water evaporation and condensation. When high-concentration humid air connects the water droplets in the hydrophobic and wet hydrophilic layers, it triggers directional water transport (Fig. 4.25b). Based on this directional water transport mechanism, Zhao et al. [118] developed a three-layer asymmetrically wetting fiber membrane. It consisted of polyacrylonitrile-silicon dioxide (PAN-SiO_2SiO_2) as the superhydrophilic layer, polyvinylidene fluoride (PVDF) as the hydrophobic layer, and PVDF/PAN-SiO_2 as the intermediate layer. This membrane exhibited a filtration efficiency of up to 99.99% for $PM_{2.5}$, a filtration resistance of only 86 Pa, and a water vapor transmission rate of up to 13,612 g/(m^2·d). It proved beneficial for alleviating humidity and resistance issues inside masks caused by breathing during wear. Similarly, Lao et al. [120] developed a 'skin-like' liquid directional transport fabric. They modified hydrophilic cotton fabric with titanium dioxide (TiO_2) nanoparticles coated with 1H, 1H, 2H, 2H-perfluorooctyltriethoxysilane (PFOTES) to achieve superhydrophobicity. A patterned net cover was selectively plasma-treated on the hydrophobic fabric to form porous gradient wetting channels, resembling local sweat glands. This 'skin-like' gradient wetting membrane exhibited water permeability 15 times higher than the best commercial breathable fabric. The unidirectional water transport mechanism of asymmetrically wetting membrane materials also finds application in fog collection. For instance, Cao et al. [121] designed a highly efficient fog collection system by combining a hydrophobic copper mesh onto a hydrophilic cotton fabric. The hydrophobic copper mesh prevents water accumulation on the surface, while the hydrophilic cotton fabric rapidly absorbs water droplets, facilitating continuous fog condensation. This asymmetric wetting fog collection system showed 1.3 times and 2.2 times higher performance compared to the individual hydrophobic copper mesh and hydrophilic cotton, respectively, offering a promising method for freshwater production in desert areas.

In terms of oil aerosol filtration, asymmetric wetting membrane materials featuring oleophilic/oleophobic structures have emerged with significant application potential. Typically, oil aerosols tend to permeate the pores of oleophilic membranes, resulting in oil droplets being carried downstream within the airflow. Consequently, traditional oleophilic membranes exhibit relatively lower filtration efficiency. The introduction of asymmetric wetting membrane materials featuring oleophilic/oleophobic structures aims to enhance oil mist filtration efficiency while mitigating filtration resistance. For instance, as depicted in Fig. 4.26a, a composite membrane comprising a super oleophobic membrane and an oleophilic membrane [91] demonstrates intriguing filtration characteristics. When the super oleophobic membrane is positioned at the forefront and the oleophilic membrane at the rear, the filtration membrane achieves a 98.7% removal efficiency for DEHS aerosol small particles, with a pressure drop of 11.9 kPa. Interestingly, the super oleophobic filtration membrane almost entirely removes DEHS aerosol large particles, but when combined with an oleophilic membrane at the rear, the removal efficiency for these particles drops to 99.74%. This change may be attributed to the alteration in liquid transfer pathways. The addition of the oleophilic membrane appears to facilitate droplet movement from the front to the oleophilic layer, potentially increasing the likelihood of liquid blow-off from the rear oleophilic layer, resulting in a secondary entrainment phenomenon. Consequently, this leads to a reduction in the filtration performance of the membrane. Conversely, reversing the arrangement yields improved performance. In this scenario, the removal efficiency for DEHS aerosol small particles elevates to 99.3%, accompanied by a reduced pressure drop of only 9.0 kPa. The oleophilic membrane at the forefront captures oil droplets, which are then obstructed by the oleophobic membrane at the rear. The rear oleophobic

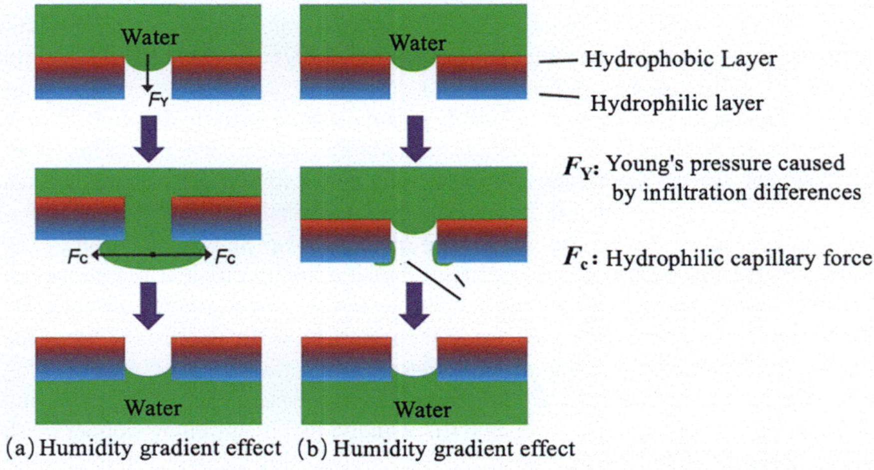

Fig. 4.25 Two possible water transport mechanisms of asymmetrically wetting fabric [116]

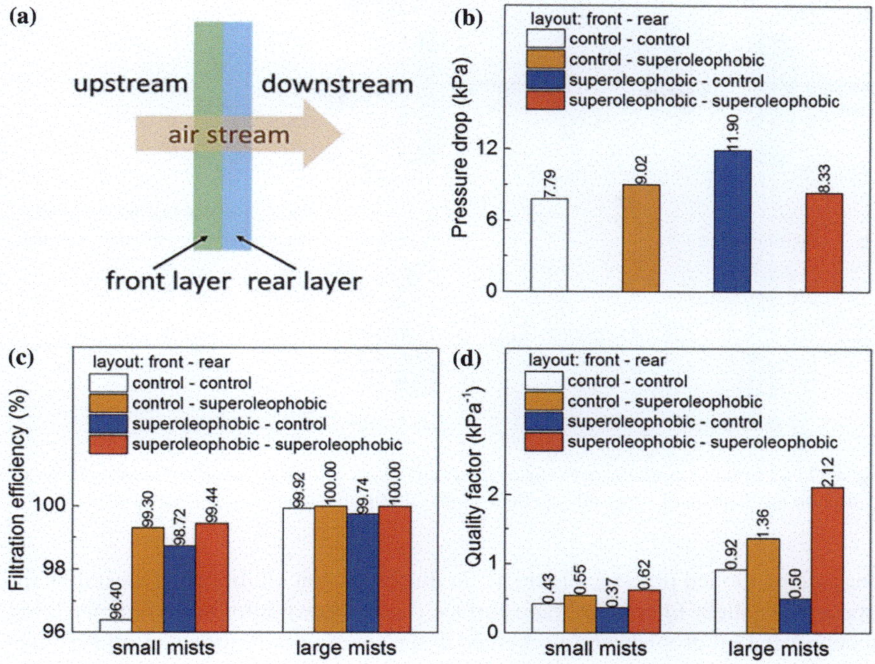

Fig. 4.26 The effect of the arrangement of the double-layer filter (**a**) and the layout of the double-layer filtration membrane (**b–d**) on the oil mist filtration performance [91]

membrane serves as a guide for liquid discharge, curtailing secondary entrainment and ultimately enhancing filtration performance.

Liu et al. [122] achieved a 20% reduction in the steady-state filtration pressure drop of a filter membrane by introducing a thin oleophilic oil-guiding fiber layer ahead of the super oleophobic filtration membrane. When using only the super oleophobic filter membrane for filtration, oil droplets tend to adsorb and aggregate on its surface, eventually clogging some pores. However, incorporating the oleophilic oil-guiding fiber layer encourages oil droplets previously obstructing the super oleophobic membrane to migrate into the oleophilic layer. This action clears the blocked pores, effectively lowering the filtration resistance. Nevertheless, the study involved using the oleophilic membrane and super oleophobic membrane in conjunction rather than creating truly asymmetric wetting membrane materials. To fabricate a monolithic membrane exhibiting varying oil wettability across its thickness, Wei et al. [115] modified a fiber membrane for super oil-affinity by immersing it in a glycerol propoxy triglycerol ether (GPTE) solution. Subsequently, a perfluoroalkyl acrylate copolymer (PFAP) was electrospun onto one side of the super oil-affinity membrane. Testing the prepared membrane for oily aerosol filtration in different directions revealed better filtration performance when the super oil-affinity surface served as the filtration surface compared to the super oil-repellent surface. The mechanism resembled

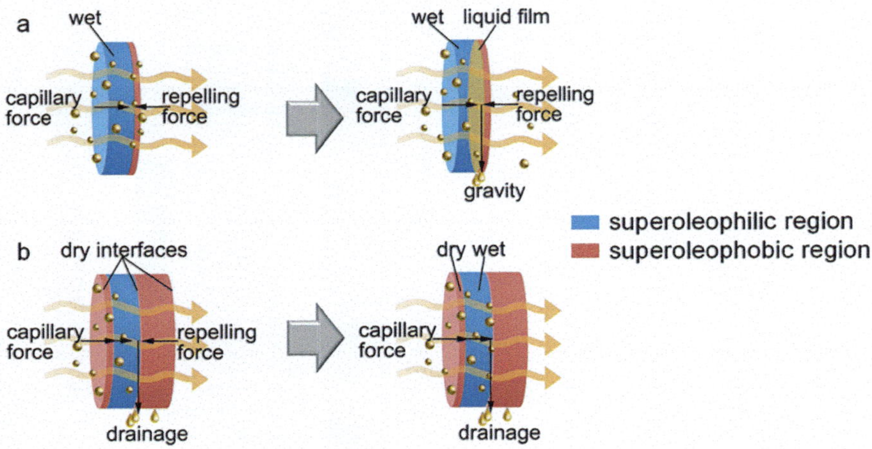

Fig. 4.27 **a** Filtration mechanism of super oil-affinity-super oil-repellent membrane structure; and **b** filtration mechanism of super oil-repellent-super oil-affinity-super oil-repellent structure [115]

the aforementioned process. Figure 4.27a illustrates that oil droplets blocked on the super oil-repellent filter membrane surface can be drawn into the oil-affinity layer. Accumulated oil droplets are discharged at the super oil-affinity-super oil-repellent interface due to gravity. Moreover, the super oil-repellent-super oil-affinity-super oil-repellent asymmetric wetting membrane, unlike the standalone super oil-affinity-super oil-repellent membrane, maintains high permeability while capturing oil mist. The super oil-affinity layer acts as temporary oil storage, redistributing collected oil droplets to reduce the resistance of oil mist entering membrane pores. The aggregated oil droplets are repelled upon reaching the subsequent super oil-affinity-super oil-repellent interface, where they are discharged due to gravity (Fig. 4.27b). The results indicate that the super oil-repellent-super oil-affinity-super oil-repellent filter membrane achieves a filtration efficiency of 99.45% for small oil mist (particles 0.01–0.5 μm) and close to 100% for large oil mist particles (0.5–20 μm), with a filtration resistance of 9.29 kPa.

4.4 Carbon Nanotube Membranes

4.4.1 Overview

Carbon stands as one of the most pivotal elements on Earth, serving not only as the backbone of human life but also exerting significant influence in various aspects of human existence. Its presence in daily life dates back to the inception of human civilization, manifesting in forms such as charcoal, coal, and ink. As the world entered the industrial revolution during the seventeenth century, materials like coke, graphite,

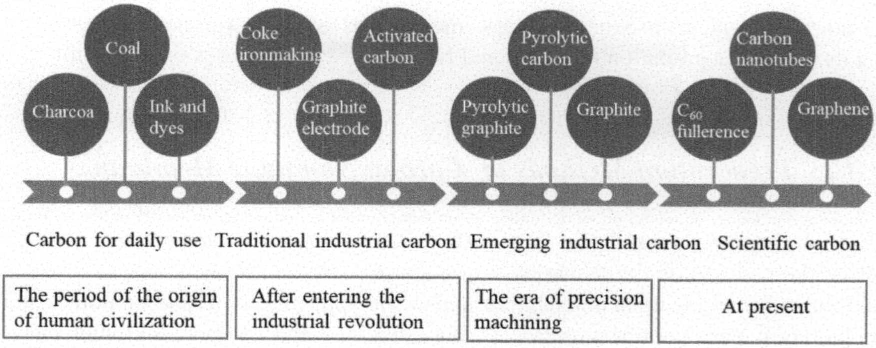

Fig. 4.28 Development history of carbon materials

and activated carbon emerged as the cornerstone of industrial applications. Further industrial advancements witnessed the pivotal role played by fine carbon materials like pyrolytic graphite and pyrolytic carbon in the early nineteenth century. As technology burgeoned, nano-carbon materials emerged as a focal point of contemporary research (as depicted in Fig. 4.28). Among these, Carbon Nanotubes (CNTs) have garnered immense attention. These cylindrical nanostructures feature a hollow center and enclosed ends, formed by rolling layers of graphene. Based on the number of graphene layers, they bifurcate into single-walled carbon nanotubes (comprising a single graphene layer) and multi-walled carbon nanotubes (with two or more graphene layers). Carbon nanotubes boast a considerable specific surface area, robust thermal stability, commendable alkali resistance, high Young's modulus, and exceptional mechanical strength. These properties render them extensively applicable in electrochemistry, energy storage (including hydrogen storage materials), catalysis, separation technologies, and various other domains.

Membranes composed of carbon nanotubes are collectively known as carbon nanotube membranes. These membranes can be categorized based on their intended applications into two main types: carbon nanotube gas–solid separation membranes and carbon nanotube catalytic membranes. Carbon nanotube gas–solid separation membranes are specifically designed for particulate filtration in gases, thereby facilitating gas cleaning. The high porosity and small diameter of these membranes enable efficient screening and interception of ultrafine solid particles or aerosols in gases. Despite this efficiency, they maintain low filtration resistance, making them commonly employed in the removal of solid particles from gases, such as $PM_{2.5}$, bacteria, viruses, and kitchen fumes. On the other hand, carbon nanotube catalytic membranes involve carbon nanotube membranes loaded with catalysts capable of catalytic degradation of nitrogen oxides, or volatile organic compounds (VOCs) present in gases. The significant surface area of carbon nanotubes, reaching a theoretical value of 2630 m^2/g, coupled with their substantial aspect ratio (micron-level tube length and nanometer-level diameter), allows for a higher catalyst load compared to other carriers. Consequently, these membranes exhibit promising potential in catalyzing the degradation of VOCs and NO_x.

4.4.2 *Preparation Methods of Carbon Nanotube Membranes*

Carbon nanotube membranes lack sufficient self-supporting properties, necessitating their integration with various supports to yield carbon nanotube composite membranes, making them practical for real-world applications. These supports often include polymer nanofibers, ceramics, glass fibers, metal meshes, and other materials. The preparation methods for different carbon nanotube composite membranes vary due to the unique characteristics of both the support materials and the carbon nanotubes themselves.

4.4.2.1 Preparation of Carbon Nanotube/Polymer Composite Membranes

Polymer nanofibers are frequently employed in fabricating carbon nanotube composite membranes. Due to the limited tolerance of polymer nanofibers to high temperatures, methods like chemical vapor deposition for carbon nanotube membrane preparation have become challenging. Therefore, the usual approach involves mixing prepared carbon nanotubes with polymer nanofibers to create composite membranes via hot pressing or electrospinning.

Wang et al. [123] introduced multi-walled carbon nanotubes into the polystyrene (PS) spinning solution and controlled electrospinning conditions to create CNTs/PS composite fiber membranes. These membranes exhibited 'wrinkles' and 'peak' protrusions on their surfaces. At a filtration speed of 85 L/min, the membrane displayed a filtration efficiency of 99.95% for NaCl aerosols with an average particle size of 260 nm but had a filtration pressure drop of 374.5 Pa. Despite its high filtration efficiency, this composite membrane had a drawback of elevated pressure drop. To address this, Yildiz et al. [124] proposed an edge hot pressing method for composite membrane preparation. They stacked a single-layer CNT board with uniform orientation onto a polypropylene (PP) fabric, layering another identical PP fabric atop the CNT layer, sandwiching the CNT layer between them. By stacking multiple CNT boards and PP fabric layers, then hot-pressing the entire structure, a robust CNT/PP composite membrane was formed. At a filtration speed of 10 cm/s, this filter displayed a dust filtration efficiency of 99.98% for 0.3 μm particles. However, the low bonding strength of this composite membrane resulted from limited connections between the carbon nanotubes and polymer fibers. Wang et al. [125] utilized polyetherimide nanofibers as an adhesive to stack four layers of continuously arranged CNT sheet material, followed by thermally bonding it with a polyimide (PI) nanofiber membrane. This method produced a high-temperature CNTs/PI composite membrane (Fig. 4.29) with outstanding properties. At a filtration speed of

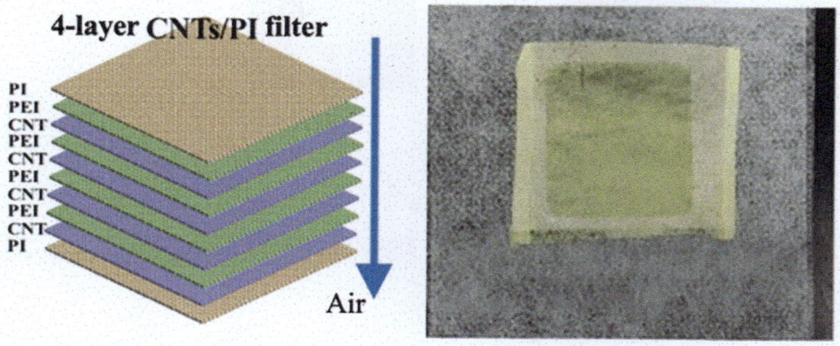

Fig. 4.29 Structure diagram and physical picture of the 4-layer CNTs/PI composite membrane [125]

5.3 cm/s, this membrane achieved a filtration efficiency of 99.99% for 0.3 μm particles, with a filtration pressure drop of 120 Pa. Additionally, the inclusion of CNT sheet material enhanced the tensile strength of the mixed filter by 9 MPa, enabling stable operation within the range of 200–250 °C. This hot-press spinning method resulted in a composite membrane boasting higher filtration efficiency, lower pressure drop, and improved mechanical strength. In another approach, Yildiz et al. [126] employed a novel electrospinning technique to produce carbon nanotube composite membranes. They coiled the CNTs sheet fabric around a rotating shaft and electrospun polyethylene oxide (PEO) onto the CNTs while the shaft rotated. By adjusting the CNT proportion from 0 to 100%, various CNTs/PEO ultra-high-efficiency filters were fabricated via hot pressing (Fig. 4.30). At a filtration speed of 15 cm/s, the composite filter exhibited a maximum filtration efficiency of 99.9981% (with 15% CNTs content) for aerosol particles ranging from 10 to 300 nm. This method not only reduces raw material costs but also offers controlled preparation of composite membranes, proving highly valuable for diverse applications.

4.4.2.2 Preparation of Carbon Nanotube/Inorganic Fiber Composite Membranes

Carbon nanotube/inorganic fiber composite membranes, such as carbon nanotube/glass fiber, carbon nanotube/metal mesh, and carbon nanotube/carbon fiber composites, are commonly used. Inorganic fibers are preferred due to their superior thermal stability and larger pore structures, often prepared using chemical vapor deposition methods. Park et al. [127] utilized a stainless steel mesh substrate with a fiber diameter of approximately 12 μm and a pore distribution ranging from 10 μm to 50 μm. Employing multiple sintering processes and a mixture of hydrogen/acetylene as reactants, they developed a CNTs/metal composite membrane. Incorporating carbon nanotubes effectively reduced the stainless steel mesh's pore size, enhancing particle capture rates and improving the membrane's capacity to handle particle pollutants,

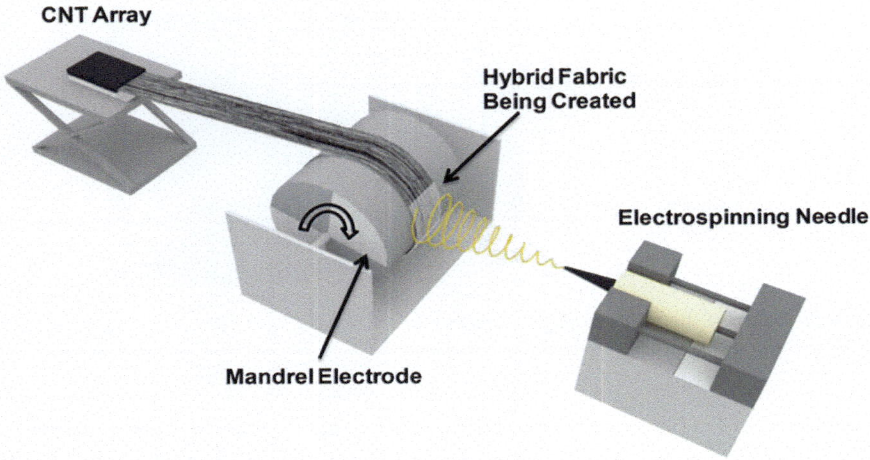

Fig. 4.30 Schematic diagram of CNTs/PEO composite membrane preparation [126]

resulting in a significant boost in filtration performance. Achieving a filtration efficiency of up to 98% for particles smaller than 200 nm, the carbon nanotube/metal mesh composite membrane exhibited notable mechanical strength. However, the bonding between the metal substrate and carbon nanotubes was relatively weak, leading to some shedding of carbon nanotubes during use. On the other hand, Li et al. [128] successfully engineered a CNTs/QF (quartz fiber) composite membrane with a gradient nanostructure on a glass fiber substrate at 760 °C, employing ethylene and xylene as reactants. Using aerosol technology combined with floating catalyst chemical vapor deposition, they created a three-layer structure in the composite membrane. With varying concentrations: 2.97% in the top layer, 0.68% in the middle layer, and 0.56% in the bottom layer, the multi-level structure notably enhanced the composite membrane's performance. It significantly improved filtration efficiency for the most penetrating particle size (MPPS) pollutants by an order of magnitude and substantially extended the membrane's service life.

4.4.2.3 Preparation of Carbon Nanotube/Ceramic Composite Membranes

Inorganic ceramic supports exhibit commendable chemical stability, robust mechanical strength, and resistance to high temperatures. However, their limitation lies in a relatively small specific surface area. As a consequence, carbon nanotube/ceramic composite membranes, showcasing superior performance attributes, are frequently synthesized by incorporating carbon nanotubes onto these supports. The primary method employed for the fabrication of these composite membranes remains chemical vapor deposition.

Xu et al. [129] employed foam nickel as the substrate and xylene, along with ferrocene, as precursor materials to fabricate a carbon nanotubes (CNTs) membrane through chemical vapor deposition. This membrane was further modified by dip-coating a layer of titanium dioxide catalyst to yield a TiO_2/CNTs foam nickel composite membrane. This method preserved the high gas flux characteristic of the foam nickel substrate while imparting favorable hydrophobicity and catalytic properties to the resulting composite membrane. Investigations substantiated a filtration efficiency of up to 99.6% for kitchen fumes encompassing particle sizes between 40 nm and 8 μm. Moreover, the composite membrane maintained a pressure drop below 2.0 kPa and exhibited a photocatalytic degradation rate of 76.7% for oleic acid. Halonen et al. [130] utilized porous Si/SiO_2 ceramics as the scaffolding material to fashion a uniformly oriented carbon nanotube membrane within the pore channels employing chemical vapor deposition conducted at temperatures ranging between 770 and 785 °C. Subsequently, they immersed the resulting composite membrane in a palladium acetate solution to create a Pd@CNTs/(Si/SiO_2) composite membrane (depicted in Fig. 4.31). This composite membrane demonstrated a filtration efficiency of up to 99% for particles measuring 0.3 μm in diameter and exhibited notable catalytic activity in propylene hydrogenation to propane. Zhao et al. [131] utilized xylene and ferrocene as reactants and conducted catalytic floating chemical vapor deposition at 850 °C to successfully synthesize a CNTs/Al_2O_3 composite membrane. Characterized by a multi-level pore structure, the composite membrane combined the gas permeability derived from the larger alumina pores with the solid particle filtration performance attributable to the dense surface carbon nanotube membrane layer. Research findings underscored the composite membrane's exceptional filtration efficiency of up to 99.9999% for dust particles measuring 300 nm in diameter, alongside a filtration pressure drop of 2.25 kPa. Compared to gas cleaning membranes prepared with polymer or inorganic fiber supports, ceramic-supported membranes exhibit relatively higher filtration pressure drops. However, their commendable heat resistance renders them more suitable for high-temperature, complex industrial flue gas treatments rather than household air cleaning materials.

Fig. 4.31 Schematic diagram (**a**) and electron microscope image and (**b**) of Pd@CNTs/(Si/SiO_2) composite membrane [130]

4.4.3 Application of Carbon Nanotube Membranes in Gas Cleaning

Carbon nanotube membranes boast unique structural attributes that confer several advantages, including a substantial specific surface area, enhanced gas permeability, and exceptional thermal resistance. Consequently, these membranes find widespread applications in air cleaning by eliminating solid particles, catalyzing oil fume separation, facilitating ozone degradation, among other uses. Furthermore, these membranes exhibit noteworthy antibacterial properties, effectively neutralizing airborne microbes.

4.4.3.1 Removal of Ultrafine Particles

As per the analysis on the Nusselt number and fiber diameter relationship [132], airflow patterns through fibers exhibit distinct characteristics: when fiber diameters fall below 13.2 nm, the airflow follows a laminar pattern; between 13.2 and 528 nm, the airflow transitions into a transitional flow state while passing through the fiber. Typically, carbon nanotubes possess diameters ranging from 2 to 500 nm, thereby enabling airflow through carbon nanotube fibers to predominantly exhibit molecular free flow and transitional flow. Consequently, the impedance offered by carbon nanotubes to airflow remains minimal, facilitating the maintenance of low filtration pressure drops in carbon nanotube membranes during the filtration of fine particles. Furthermore, the efficient filtration of gas particles by carbon nanotubes, achieved through inertial collision, direct interception, and other mechanisms (as illustrated in Fig. 4.32) [133], underscores their effectiveness in gas cleaning, thus broadening their applicability and conferring competitive advantages in this domain.

Li et al. [134] conducted a study on the fabrication of a highly stratified CNTs/QF composite membrane, tailored for the efficient filtration of submicron aerosols present in ambient air. The initial glass fiber support exhibited a pore size of 3.49 μm and a specific surface area of 3.12m^2/g. Its filtration pressure drop concerning 100 nm and 300 nm aerosols measured at 0.56 kPa, accompanied by filtration efficiencies of 99.99651% and 99.9862%, respectively. The integration of CNTs onto the glass fiber surface, manifesting as a brush-like structure, markedly augmented the support's performance. Notably, the composite membrane demonstrated a diminished pore size, ranging from 1.31 to 2.34 μm, with an amplified specific surface area escalating to 42.83 m^2/g, a 12-fold increase. Consequently, the aerosol filtration pressure increased to 0.84 kPa for both 100 and 300 nm particles. However, the filtration efficiencies markedly escalated to 99.9999785% and 99.999745%, respectively. This enhancement can be attributed to the presence of CNTs not only on the fiber surface but also within the pores. The slender and densely distributed carbon nanotubes mitigated gas flow resistance, averting a steep surge in pressure drop. Additionally, the uniformly distributed carbon nanotubes effectively captured NaCl aerosols on the composite membrane surface, exhibiting a high adherence rate to the

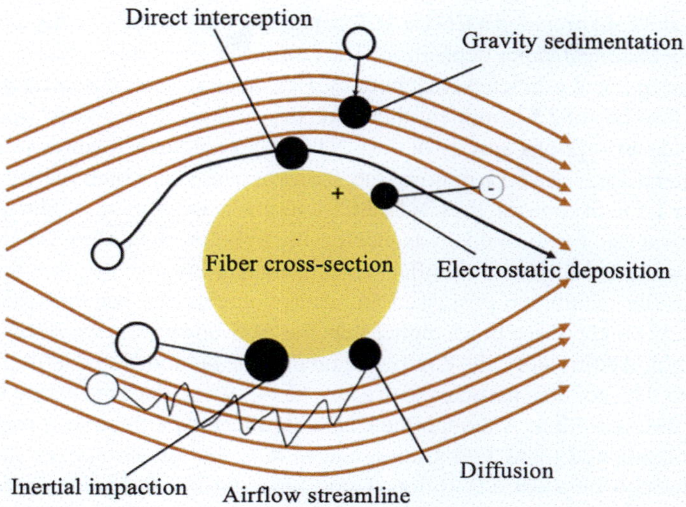

Fig. 4.32 Five mechanisms of nanotube filtration of particles [133]

carbon nanotubes. This underscored the advantageous role of carbon nanotubes with a substantial specific surface area in facilitating efficient particle removal.

Yang et al. [135] conducted a study on the preparation of a three-dimensional CNTs/stainless steel composite membrane for gas filtration and assessed its performance in filtering incense smoke particles. Initially, the stainless steel mesh exhibited negligible filtration effectiveness for both $PM_{2.5}$ and $PM_{2.5\sim10}$ particles. In contrast, the composite membrane displayed a filtration efficiency of 95.7% for $PM_{2.5}$ and an even higher 96.45% for $PM_{2.5\sim10}$ particles. Following three experimental cycles, the composite membrane maintained a consistent filtration efficiency of over 93% for particles. However, during the subsequent fourth filtration experiment, owing to the loss of carbon nanotubes due to multiple resistance heating cleanings (3.1 A and 12.7 W), the filtration efficiency for $PM_{2.5}$ particles decreased notably to 75.6%. This suggests that the composite membrane could reliably operate for at least 3 cycles, corresponding to approximately 30 h of stable performance.

4.4.3.2 Control of Gaseous Pollutants

The substantial specific surface area inherent in carbon nanotubes endows them with a considerable adsorption capacity for gaseous pollutants. Loading modifications with catalysts enable efficient removal of volatile organic compounds (VOCs), ozone (O_3), sulfur dioxide (SO_2), and various other gaseous pollutants using carbon nanotube membranes.

Yang et al. [136] prepared a CNTs/ACF composite membrane on activated carbon (ACF) using chemical vapor deposition. The composite membrane exhibited exceptional ozone removal efficiency, reaching 99.14% at an ozone concentration of 300 μg/L and a flow rate of 2.5 L/min. Furthermore, the same research group evaluated the formaldehyde adsorption capacity of the CNTs/ACF composite membrane. Initially, the ACF filter displayed a 50% removal rate for formaldehyde, which swiftly declined to zero after 1.5 h. In contrast, the CNTs/ACF composite membrane exhibited an 85% initial removal rate for formaldehyde, decreasing to below 30% after 1 h and losing its adsorption capability entirely after 6.5 h. Calculations revealed that the CNTs/ACF composite membrane possessed an adsorption capacity for formaldehyde of (62.49 ± 4.52) mg/g, three times higher than the ACF adsorption capacity of (19.11 ± 1.71) mg/g. Additionally, the composite membrane demonstrated a filtration efficiency of 95.97% for dust particles with a size of 145.9 nm, marking a 25% improvement over the ACF filter. Although the CNTs significantly enhanced formaldehyde adsorption compared to ACF, it's important to note that the composite membrane operates through physical adsorption. Upon reaching its adsorption capacity, the efficiency of formaldehyde removal is substantially reduced or completely nullified. Hence, further research should focus on developing carbon nanotube membranes capable of chemically degrading formaldehyde. Feng et al. [137] combined filtration coating technology with hydrothermal method to prepare UiO-66-NH_2@CNTs/PTFE composite membrane (Fig. 4.33).

At a flow rate of 2.0 m/min, the membrane exhibited an exceptional filtration efficiency of 99.997% for 300 nm dust particles, maintaining a pressure drop of 160 Pa. The original PTFE filter material, characterized by larger pores, lacked the capability to intercept smaller aerosol particles. However, the incorporation of UiO-66-NH_2 and CNTs effectively modified the pore structure of PTFE. This modification reduced the pore size, enabling the composite membrane to sieve a significant quantity of dust particles. Moreover, the high specific surface area of carbon nanotubes

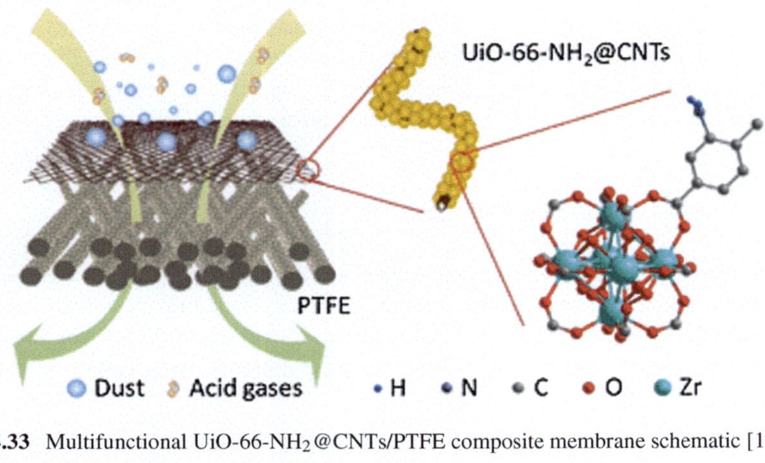

Fig. 4.33 Multifunctional UiO-66-NH_2@CNTs/PTFE composite membrane schematic [137]

within the composite membrane facilitated increased dopamine loading, enhancing the presence of amine free radicals for SO_2 adsorption. Consequently, the composite membrane displayed remarkable adsorption performance for SO_2, with the smaller pore size and larger specific surface area of UiO-66-NH_2 contributing to an enhanced SO_2 adsorption capacity of up to 0.6 mmol/g. This composite membrane showcases functional attributes for dust removal and synergistic treatment of gaseous pollutants, offering a novel research direction for indoor air cleaning design.

4.4.3.3 Removal of Microorganisms

The process through which carbon nanotubes eliminate microorganisms primarily operates via two mechanisms. Firstly, the dense structure of the carbon nanotube membrane facilitates the direct filtration and interception of microorganisms, thereby accomplishing their removal. Secondly, carbon nanotubes, being extended nanofibers, exhibit a certain toxicity towards bacteria and microorganisms [138]. This toxicity arises from the ability of long-fiber carbon nanotubes to penetrate the cell membrane of bacteria. Notably, the cell membrane struggles to wholly engulf these fibers, resulting in a delayed clearance of extended fibers (> 20 μm). Consequently, this delayed clearance disrupts the integrity of the cell membrane, triggering inflammation (as illustrated in Fig. 4.34), ultimately leading to the deactivation of bacteria and microorganisms, consequently achieving their removal from the environment.

Park et al. [139] developed a CNT/QF filter composite membrane via CNT coating onto a glass fiber filter employing an electrothermal deposition system. At a coating density of $1.5 \times 10^9/cm^2$, the composite membrane demonstrated an optimal average diameter of 200 nm for MS_2 bacteriophage filtration efficiency at 78.4% under filtration speeds ranging from 0 to 0.2 m/s. Simultaneously, the filtration quality factor peaked at 0.13. The principal strategy for eliminating bacteria and microbes from the environment involves treating the bacteriophage akin to particulate pollution, employing screening and interception mechanisms, while harnessing the sterilization properties inherent in carbon nanotubes for disinfection. Nonetheless, this sterilization method operates slowly and extended usage may potentially foster bacterial growth on the filter membrane.

Jung et al. [140] employed polyurethane (PU) fibers as a foundational substrate and utilized continuous atomization along with thermal evaporation/condensation to amalgamate CNTs particles and Ag powder, depositing them onto the fiber base within a heat pipe furnace. The original PU filter displayed limited antibacterial efficacy against E. coli and Gram-positive bacteria, with only 40% inactivation of Gram-positive bacteria and 60% of E. coli over 100 min. In contrast, the composite membrane exhibited a sterilization rate surpassing 90%. Upon investigating bacterial survival on the membrane for 1200 min, it was observed that the Ag@CNTs/PU composite membrane demonstrated a sterilization rate of 99.1% against Gram-positive bacteria and an inactivation rate of 99.8% against E. coli.

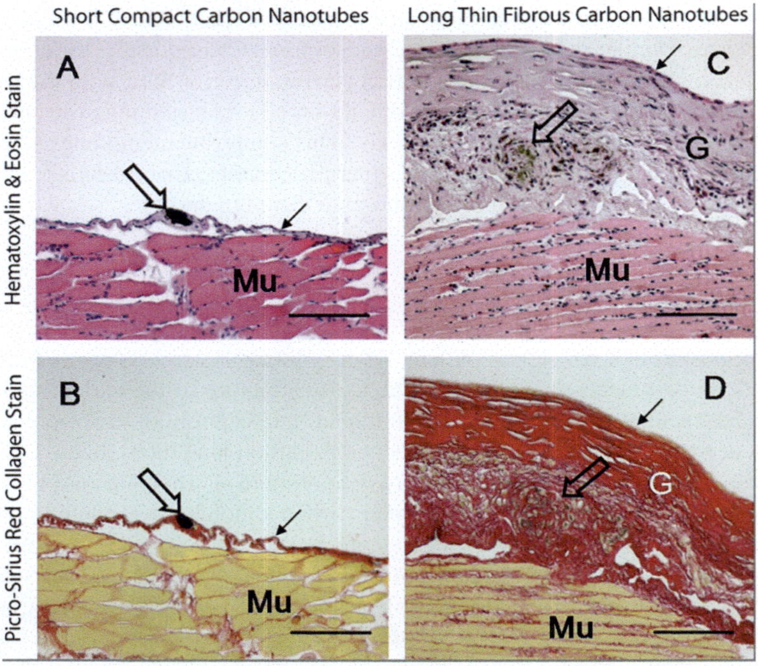

Fig. 4.34 Ratio of longitudinal and transverse damage to the diaphragmatic peritoneal surface after intraperitoneal injection of 2 types of carbon nanotubes [138]. (Mu is the diaphragm, G is granuloma, small arrows are mesothelium, large arrows are carbon nanotubes)

4.4.3.4 Synergistic Removal of Multiple Pollutants

Carbon nanotubes exhibit not only remarkably high filtration efficiency and effective removal capabilities for gaseous pollutants but also possess notable antibacterial properties. Consequently, they represent an optimal multifaceted gas cleaning membrane material capable of synergistically eliminating various airborne pollutants, including $PM_{2.5}$, VOCs, bacteria, among others.

Zhao et al. [141] utilized a macroporous alumina ceramic membrane as a foundational support for fabricating an Ag@CNTs/Al_2O_3 composite membrane. This composite membrane found applications in air cleaning, demonstrating proficiency in dust removal, formaldehyde degradation, and sterilization. Notably, it exhibited high removal rates for various microorganisms, including E. coli (99.97% removal rate), Bacillus subtilis (98.92% removal rate), and Aspergillus niger (98.75% removal rate), establishing its multifunctional efficacy in diverse applications (refer to Fig. 4.35).

A dense carbon nanotube separation membrane layer was synthetized through chemical vapor deposition, grown both on the surface and within the pores of an

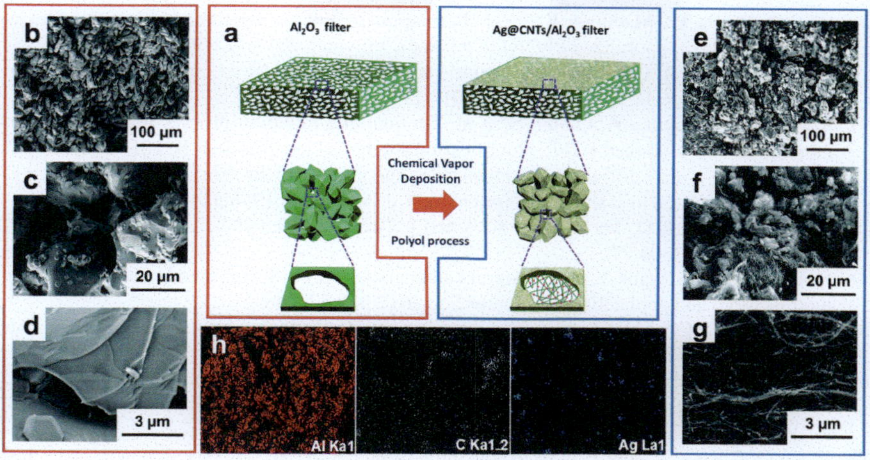

Fig. 4.35 Schematic diagram of the preparation process of the multifunctional $Ag@MWCNTs/Al_2O_3$ composite membrane with a deep hierarchical structure (**a**); Surface morphology of the Al_2O_3 composite membrane (**b–d**); Surface morphology of the $Ag@MWCNTs/Al_2O_3$ composite membrane (**e–g**); and element distribution on the surface of the $AgNPs@MWCNTs/Al_2O_3$ composite membrane (**h**) [141]

alumina ceramic substrate. This membrane's efficacy in dust filtration was evaluated using solid particles measuring 300 nm and 2.5 μm, revealing an exceptional dust filtration efficiency of 99.9999%. This efficiency far surpassed the established benchmarks for high-efficiency air filters (HEPA). Furthermore, the carbon nanotube membrane was augmented with silver nanoparticles via the polyol reduction method of silver nitrate. The incorporation of nano-silver significantly enhanced the composite membrane's performance in removing formaldehyde and bacteria. Notably, silver possesses catalytic properties for formaldehyde degradation, while carbon nanotubes exhibit substantial formaldehyde adsorption capacity. This synergistic action facilitated a composite membrane capable of achieving a 99% removal rate for formaldehyde at 55 °C. Simultaneously, the nano-silver component contributed to sterilization by disrupting bacterial cell membranes, thereby enhancing the overall sterilization efficiency of the membrane.

References

1. Liu Y, Liu YJ, Cui LN, et al. Effect of biaxial stretching process on microstructure and macro properties of nylon 6 membranes. China Plast Ind. 2020;48(6):35–42.
2. Wan C, Cao T, Chen X, et al. Fabrication of polyethylene nanofibrous membranes by biaxial stretching. Mater Today Commun. 2018;17:24–30.
3. Murphy MA, Horstemeyer MF, Gwaltney SR, et al. Nanomechanics of phospholipid bilayer failure under strip biaxial stretching using molecular dynamics. Modell Simul Mater Sci Eng. 2016;24(5):1–19.

4. Feng J, Zhang G, Macinnis K, et al. Formation of microporous membranes by biaxial orientation of compatibilized PP/Nylon 6 blends. Polymer. 2017;123:301–10.
5. Quarti C, Milani A, Castiglioni C. Ab initio calculation of the IR spectrum of PTFE: helical symmetry and defects. J Phys Chem B. 2013;117(2):706–18.
6. Kobayashi A, Oshima A, Okubo S, et al. Thermal and radiation process for nano-/micro-fabrication of crosslinked PTFE. Nucl Instrum Methods Phys Res Sect B. 2013;295:76–80.
7. Stein RS. The X-Ray diffraction, birefringence, and infrared dichroism of stretched polyethylene. 2. Generalized uniaxial crystal orientation. J Polym Sci. 1958;31(123):327–34.
8. Gore RW, Newark D. Very highly stretched polytetrafluoroethylene and process therefor: US3962153. 1976-04.27.
9. Feng S, Zhong Z, Wang Y, et al. Progress and perspectives in PTFE membrane: preparation, modification, and applications. J Membr Sci. 2017;549:332–49.
10. Dai Y, Wang XG, et al. Structure and properties of PTFE fiber membrane prepared by hot-rolling process. J Donghua Univ (Nat Sci). 2019;45(3):358–63.
11. Wang Y, Xu Y, Wang D, et al. Polytetrafluoroethylene/polyphenylene sulfide needle-punched triboelectric air filter for efficient particulate matter removal. ACS Appl Mater Interfaces. 2019;11(51):48437–49.
12. Wang G, Chen L, Zhang F, et al. Application of membrane technology in power plant dust removal. J Salt Chem Ind. 2015;2:50–3.
13. Liu L, Chen C, Wang Z, et al. Thermal-moisture comfort of multilayered fabric systems used as firefighting clothing. J Textile Res. 2019;40(5):119–23.
14. Ding L, Zhang D, Wu T, et al. Three-dimensional crystal structure evolution and micropore formation of β-iPP during biaxial stretching. Polymer. 2020;196: 122471.
15. Anton F. Process and apparatus for preparing artificial threads: US1975504. 1934.10-02.
16. Doshi J, Reneker DH. Electrospinning process and applications of electrospun fibers. In: Conference record of the 1993 IEEE industry applications conference twenty-eighth IAS annual meeting, Toronto, ON Canada; 1993. pp. 1698–1703
17. Fong H, Reneker DH. Elastomeric nanofibers of styrene-butadiene-styrene triblock copolymer. J Polym Sci Part B: Polym Phys. 1999;37(24):3488–93.
18. Reneker DH, Yarin AL. Electrospinning jets and polymer nanofibers. Polymer. 2008;49(10):2387–425.
19. Taylor G. Disintegration of water drops in an electric field. Edinburgh: The Royal Society; 1964. p. 383–97.
20. Wang N, Wang X, Ding B, et al. Tunable fabrication of three-dimensional polyamide-66 nano-fiber/nets for high efficiency fine particulate filtration. J Mater Chem. 2012;22(4):1445–52.
21. Wang N, Si YS, Wang N, et al. Multilevel structured polyacrylonitrile/silica nanofibrous membranes for high-performance air filtration. Sep Purif Technol. 2014; 126:44–51.
22. Darrell HR, Iksoo C. Nanometre diameter fibres of polymer, produced by electrospinning. Nanotechnology. 1996;7(3):216–23.
23. Liu H, Cao CY, Wei FF, et al. Flexible macroporous carbon nanofiber membrane with high oil adsorption capacity. J Mater Chem A. 2014;2(10):3557–62.
24. Zhang S, Liu H, Yu J, et al. Microwave structured polyamide-6 nanofiber/net membrane with embedded poly(m-phenylene isophthalamide) staple fibers for effective ultrafine particle filtration. J Mater Chem A. 2016;4(16):6149–57.
25. Wang Z, Zhao CC, Pan ZJ. Porous bead-on-string poly(lactic acid) fibrous membranes for air filtration. J Colloid Interface Sci. 2015;441:121–9.
26. Ahn YC, Park SK, Kim GT, et al. Development of high efficiency nanofilters made of nanofibers. Curr Appl Phys. 2006;6(6):1030–5.
27. Feng X, Wang B, Wang J, et al. Preparation and properties of polyacrylonitrile nanofiber membranes used for air filtering by electrospinning. J Textile Res. 2017;38(4):6–11.
28. Yun KM, Suryamas AB, Iskandar F, et al. Morphology optimization of polymer nanofiber for applications in aerosol particle filtration. Sep Purif Technol. 2010;75(3):340–5.
29. Hung CH, Leung WWF. Filtration of nano-aerosol using nanofiber filter under low peclet number and transitional flow regime. Sep Purif Technol. 2011;79(1):34–42.

30. Hu M, Zhong Z, Xing W. Development of nanofiber membrane for air purification. Chem Ind Eng Prog. 2018;37(4):1305–13.
31. Ogi T, Ono H, Bao L, et al. Morphology-controlled synthesis of electrospun nanofibers and their application for aerosol filtration. Kagaku Kogaku Ronbunshu. 2014;40(2):84–9.
32. Liu B, Zhang S, Wang X, et al. Efficient and reusable polyamide-56 nanofiber/nets membrane with bimodal structures for air filtration. J Colloid Interface Sci. 2015;457:203–11.
33. Li Z, Kang W, Zhao H, et al. Fabrication of a polyvinylidene fluoride tree-like nanofiber web for ultra high performance air filtration. RSC Adv. 2016;6(94):91243–9.
34. Li XQ, Wang N, Fan G, et al. Electreted polyetherimide-silica fibrous membranes for enhanced filtration of fine particles. J Colloid Interface Sci. 2015;439:12–20.
35. Wang Z, Pan ZJ, Wang JG, et al. A novel hierarchical structured poly(lactic acid)/titania fibrous membrane with excellent antibacterial activity and air filtration performance. J Nanomater. 2016;2016:1–17.
36. Wang Z, Pan ZJ. Preparation of hierarchical structured nano-sized/porous poly(lactic acid) composite fibrous membranes for air filtration. Appl Surf Sci. 2015;356:1168–79.
37. Yang Y, Zhang S, Zhao X, et al. Sandwich structured polyamide-6/polyacrylonitrile nanonets/bead-on-string composite membrane for effective air filtration. Sep Purif Technol. 2015;152(Supplement C):14–22.
38. Jeong S, Cho H, Han S, et al. High efficiency, transparent, reusable, and active $PM_{2.5}$ filters by hierarchical ag nanowire percolation network. Nano Lett. 2017;17(7):4339–46.
39. Ma LC, Wang JN, LiL, et al. Preparation of PET/CTS antibacterial composites nanofiber membranes used for air filter by electrospinning. Acta Polymerica Sinica. 2015;2:221–7.
40. Neeta L, Ramakrishnan R, Li B, et al. Fabrication of nanofibers with antimicrobial functionality used as filters: protection against bacterial contaminants. Biotechnol Bioeng. 2007;97(6):1357–65.
41. Montazer M, Malekzadeh SB. Electrospun antibacterial nylon nanofibers through in situ synthesis of nanosilver: preparation and characteristics. J Polym Res. 2012;19(10):1–6.
42. Choi J, Yang BJ, Bae GN, et al. Herbal extract incorporated nanofiber fabricated by an electrospinning technique and its application to antimicrobial air filtration. ACS Appl Mater Interfaces. 2015;7(45):25313–20.
43. Souzandeh H, Molki B, Zheng M, et al. Cross-linked protein nanofilter with antibacterial properties for multifunctional air filtration. ACS Appl Mater Interfaces. 2017;9(27):22846–55.
44. Modesti M, Roso M, Boaretti C, et al. Preparation of smart nano-engineered electrospun membranes for methanol gas-phase photoxidation. Appl Catal B-Environ. 2014;144:216–22.
45. Tian MJ, Liao F, Ke QF, et al. Synergetic effect of titanium dioxide ultralong nanofibers and activated carbon fibers on adsorption and photodegradation of toluene. Chem Eng J. 2017;328:962–76.
46. Zhu XC, Chen JH, Yu XN, et al. Controllable synthesis of novel hierarchical V_2O_5/TiO_2 nanofibers with improved acetone oxidation performance. RSC Adv. 2015;5(39):30416–24.
47. Zhu XC, Zhang S, Yu XN, et al. Controllable synthesis of hierarchical MnO_x/TiO_2 composite nanofibers for complete oxidation of low-concentration acetone. J Hazard Mater. 2017;337:105–14.
48. Chen CQ, Yu Y, Li W, et al. Mesoporous $Ce_{1-x}Zr_xO_2$ solid solution nanofibers as high efficiency catalysts for the catalytic combustion of VOCs. J Mater Chem. 2011;21(34):12836–41.
49. Oh GY, Ju YW, Jung HR, et al. Preparation of the novel manganese-embedded PAN-based activated carbon nanofibers by electrospinning and their toluene adsorption. J Anal Appl Pyrol. 2008;81(2):211–7.
50. Guo ZY, Huang JT, Xue ZH, et al. Electrospun graphene oxide/carbon composite nanofibers with well-developed mesoporous structure and their adsorption performance for benzene and butanone. Chem Eng J. 2016;306:99–106.
51. Hu M, Yin L, Zhou H, et al. Manganese dioxide-filled hierarchical porous nanofiber membrane for indoor air cleaning at room temperature. J Membr Sci. 2020;605.
52. Chattopadhyay S, Hatton TA, Rutledge GC. Aerosol filtration using electrospun cellulose acetate fibers. J Mater Sci. 2015;51(1):204–17.

53. Liu H, Cao C, Huang J, et al. Progress on particulate matter filtration technology: basic concepts, advanced materials, and performances. Nanoscale. 2020;12(2):437–53.
54. Feng S, Zhou M, Han F, et al. A bifunctional MnO_x@PTFE catalytic membrane for efficient low temperature NO_x-SCR and dust removal. Chin J Chem Eng. 2020;28(5):1260–7.
55. Jung S, An J, Na H, et al. Surface energy of filtration media influencing the filtration performance against solid particles, oily aerosol, and bacterial aerosol. Polymers (Basel). 2019;11(6):935.
56. Chen F, Ji Z, Qi Q. Effect of liquid surface tension on the filtration performance of coalescing filters. Sep Purif Technol. 2019;209:881–91.
57. Park SS, Kang MS, Hwang J. Oil mist collection and oil mist-to-gas conversion via dielectric barrier discharge at atmospheric pressure. Sep Purif Technol. 2015;151:324–31.
58. Jiang T, Guo Z, Liu W. Biomimetic superoleophobic surfaces: focusing on their fabrication and applications. J Mater Chem A. 2015;3(5):1811–27.
59. Liu M, Wang S, Jiang L. Nature-inspired superwettability systems. Nat Rev Mater. 2017;2(7):17036.
60. Ghaffari S, Aliofkhazraei M, Darband GB, et al. Review of superoleophobic surfaces: evaluation, fabrication methods, and industrial applications. Surf Interfaces. 2019;17: 100340.
61. Wu J, Ding Y, Wang J, et al. Facile fabrication of nanofiber- and micro/nanosphere-coordinated PVDF membrane with ultrahigh permeability of viscous water-in-oil emulsions. J Mater Chem A. 2018;6(16):7014–20.
62. Zhang D, Cheng Z, Kang H, et al. A smart superwetting surface with responsivity in both surface chemistry and microstructure. Angew Chem Int Ed Engl. 2018;57(14):3701–5.
63. Yong J, Chen F, Yang Q, et al. Superoleophobic surfaces. Chem Soc Rev. 2017;46(14):4168–217.
64. Wang T, Cui J, Ouyang S, et al. A new approach to understand the Cassie state of liquids on superamphiphobic materials. Nanoscale. 2016;8(5):3031–9.
65. Lin J, Tian F, Ding B, et al. Facile synthesis of robust amphiphobic nanofibrous membranes. Appl Surf Sci. 2013;276:750–5.
66. Zhu X, Feng S, Zhao S, et al. Perfluorinated superhydrophobic and oleophobic SiO_2@PTFE nanofiber membrane with hierarchical nanostructures for oily fume purification. J Membr Sci. 2020;594: 117473.
67. Yasuda H, Gazicki M. Biomedical applications of plasma polymerization and plasma treatment of polymer surfaces. Biomaterials. 1982;3(2):68–77.
68. Poncin-Epaillard F, Chevet B, Brosse JC. Functionalization of polypropylene by a microwave (433 MHz) cold plasma of carbon dioxide. Surface modification or surface degradation. Eur Polym J. 1990;26(3):333–9.
69. Riekerink MBO, Terlingen JGA, Engbers GHM, et al. Selective etching of semicrystalline polymers: CF_4 gas plasma treatment of poly(ethylene). Langmuir. 1999;15(14):4847–56.
70. Yu HY, He XC, Liu LQ, et al. Surface modification of poly(propylene) microporous membrane to improve its antifouling characteristics in an SMBR: O_2 plasma treatment. Plasma Process Polym. 2010;5(1):84–91.
71. Feng S, Zhong Z, Wang Y, et al. Progress and perspectives in PTFE membrane: preparation, modification, and applications. J Membr Sci. 2018;549:332–49.
72. Feng SS, Zhong ZX, Zhang F, et al. Amphiphobic polytetrafluoroethylene membranes for efficient organic aerosol removal. ACS Appl Mater Interfaces. 2016;8(13):8773–81.
73. Fan H, Gao A, Zhang G, et al. A facile strategy towards developing amphiphobic polysulfone membrane with double Re-entrant structure for membrane distillation. J Membr Sci. 2020;602: 117933.
74. Lin S, Nejati S, Boo C, et al. Omniphobic membrane for robust membrane distillation. Environ Sci Technol Lett. 2014;1(11):443–7.
75. Khan AA, Siyal MI, Lee CK, et al. Hybrid organic-inorganic functionalized polyethersulfone membrane for hyper-saline feed with humic acid in direct contact membrane distillation. Sep Purif Technol. 2019;210:20–8.

76. Xu C, Fang J, Low ZX, et al. Amphiphobic PFTMS@nano-SiO_2/ePTFE membrane for oil aerosol removal. Ind Eng Chem Res. 2018;57(31):10431–8.
77. Mao X, Chen Y, Si Y, et al. Novel fluorinated polyurethane decorated electrospun silica nanofibrous membranes exhibiting robust waterproof and breathable performances. RSC Adv. 2013;3(20):7562.
78. Chen LH, Huang A, Chen YR, et al. Omniphobic membranes for direct contact membrane distillation: effective deposition of zinc oxide nanoparticles. Desalination. 2018;428:255–63.
79. Li J, Yan L, Ouyang Q, et al. Facile fabrication of translucent superamphiphobic coating on paper to prevent liquid pollution. Chem Eng J. 2014;246:238–43.
80. Wu X, Wyman I, Zhang G, et al. Preparation of superamphiphobic polymer-based coatings via spray- and dip-coating strategies. Prog Org Coat. 2016;90:463–71.
81. Zhou H, Wang H, Niu H, et al. A waterborne coating system for preparing robust, self-healing, superamphiphobic surfaces. Adv Func Mater. 2017;27(14):1604261.
82. Li X, Shan H, Cao M, et al. Facile fabrication of omniphobic PVDF composite membrane via a waterborne coating for anti-wetting and anti-fouling membrane distillation. J Membr Sci. 2019;589: 117262.
83. Dizge N, Shaulsky E, Karanikola V. Electrospun cellulose nanofibers for superhydrophobic and oleophobic membranes. J Membr Sci. 2019;590: 117271.
84. Wu XQ, Wu X, Wang TY, et al. Omniphobic surface modification of electrospun nanofiber membrane via vapor deposition for enhanced anti-wetting property in membrane distillation. J Membr Sci. 2020;606: 118075.
85. Huang YX, Wang ZX, Hou DY, et al. Coaxially electrospun super-amphiphobic silica-based membrane for anti-surfactant-wetting membrane distillation. J Membr Sci. 2017;531:122–8.
86. An X, Liu Z, Hu Y. Amphiphobic surface modification of electrospun nanofibrous membranes for anti-wetting performance in membrane distillation. Desalination. 2018;432:23–31.
87. Woo YC, Chen Y, Tijing LD, et al. CF_4 plasma-modified omniphobic electrospun nanofiber membrane for produced water brine treatment by membrane distillation. J Membr Sci. 2017;529:234–42.
88. Zhang P, Lu W, Wang Y, et al. Fabrication of flexible and amphiphobic alumina mats by electrospinning. J Sol-Gel Sci Technol. 2016;80(3):690–6.
89. Zhang W, Chen S, Hu W, et al. Facile fabrication of flexible magnetic nanohybrid membrane with amphiphobic surface based on bacterial cellulose. Carbohyd Polym. 2011;86(4):1760–7.
90. Feng X. Modification of porous PTFE membrane with multifunction and its applications in air purification. Nanjing: Nanjing University of Technology; 2018.
91. Wei X, Chen F, Wang H, et al. Efficient removal of aerosol oil-mists using superoleophobic filters. J Mater Chem A. 2018;6(3):871–7.
92. Kampa D, Wurster S, Meyer J, et al. Validation of a new phenomenological "jump-and-channel" model for the wet pressure drop of oil mist filters. Chem Eng Sci. 2015;122:150–60.
93. Charvet A, Gonthier Y, Gonze E, et al. Experimental and modelled efficiencies during the filtration of a liquid aerosol with a fibrous medium. Chem Eng Sci. 2010;65(5):1875–86.
94. Mead-Hunter R, King AJC, Mullins BJ. Aerosol-mist coalescing filters: a review. Sep Purif Technol. 2014;133:484–506.
95. Zhang R, Liu B, Yang A, et al. In Situ investigation on the nanoscale capture and evolution of aerosols on nanofibers. Nano Lett. 2018;18(2):1130–8.
96. Dawar S, Chase GG. Correlations for transverse motion of liquid drops on fibers. Sep Purif Technol. 2010;72(3):282–7.
97. Kampa D, Wurster S, Buzengeiger J, et al. Pressure drop and liquid transport through coalescence filter media used for oil mist filtration. Int J Multiph Flow. 2014;58:313–24.
98. Liu J, Zhou H, Wu X, et al. Superoleophobic filters: improvement of filtration performance by front attachment of oil-guiding fabric. Adv Mater Interfaces. 2019;7(2):1901808.
99. Yang HC, Hou J, Chen V, et al. Janus membranes: exploring duality for advanced separation. Angew Chem Int Ed Engl. 2016;55(43):13398–407.
100. Wu J, Wang N, Wang L, et al. Unidirectional water-penetration composite fibrous membranes via electrospinning. Soft Matter. 2012;8(22):5996–9.

101. Zhao Y, Wang H, Zhou H, et al. Directional fluid transport in thin porous materials and its functional applications. Small. 2017;13(4):201601070.
102. Babar AA, Wang X, Iqbal N, et al. Tailoring differential moisture transfer performance of nonwoven/polyacrylonitrile-SiO_2 nanofiber composite membranes. Adv Mater Interfaces. 2017;4(15):1700062.
103. Babar AA, Miao D, Ali N, et al. Breathable and colorful cellulose acetate-based nanofibrous membranes for directional moisture transport. ACS Appl Mater Interfaces. 2018;10(26):22866–75.
104. Yan W, Miao D, Babar AA, et al. Multi-scaled interconnected inter-and intra-fiber porous Janus membranes for enhanced directional moisture transport. J Colloid Interface Sci. 2020;565:426–35.
105. Dong Y, Thomas NL, Lu X. Electrospun dual-layer mats with covalently bonded ZnO nanoparticles for moisture wicking and antibacterial textiles. Mater Des. 2017;134:54–63.
106. Tang S, Pi H, Zhang Y, et al. Novel Janus fibrous membranes with enhanced directional water vapor transmission. Appl Sci. 2019;9(16):3302.
107. Ju J, Shi Z, Deng N, et al. Designing waterproof breathable material with moisture unidirectional transport characteristics based on a TPU/TBAC tree-like and TPU nanofiber double-layer membrane fabricated by electrospinning. RSC Adv. 2017;7(51):32155–63.
108. Hu L, Gao S, Zhu Y, et al. An ultrathin bilayer membrane with asymmetric wettability for pressure responsive oil/water emulsion separation. J Mater Chem A. 2015;3(46):23477–82.
109. Zhang J, Yang Y, Zhang Z, et al. Biomimetic multifunctional nanochannels based on the asymmetric wettability of heterogeneous nanowire membranes. Adv Mater. 2014;26(7):1071–5.
110. Contal P, Simao J, Thomas D, et al. Clogging of fibre filters by submicron droplets: phenomena and influence of operating conditions. J Aerosol Sci. 2004;35(2):263–78.
111. Zhang Y, Barboiu M. Dynameric asymmetric membranes for directional water transport. Chem Commun (Camb). 2015;51(88):15925–32.
112. Essalhi M, Khayet M. Surface segregation of fluorinated modifying macromolecule for hydrophobic/hydrophilic membrane preparation and application in air gap and direct contact membrane distillation. J Membr Sci. 2012;417–418:163–73.
113. Mates JE, Schutzius TM, Qin J, et al. The fluid diode: tunable unidirectional flow through porous substrates. ACS Appl Mater Interfaces. 2014;6(15):12837–43.
114. Zeng C, Wang H, Zhou H, et al. Directional water transport fabrics with durable ultra-high one-way transport capacity. Adv Mater Interfaces. 2016;3(14):1600036.
115. Wei X, Zhou H, Chen F, et al. High-efficiency low-resistance oil-mist coalescence filtration using fibrous filters with thickness-direction asymmetric wettability. Adv Func Mater. 2019;29(1):1806302.
116. Wang H, Ding J, Dai L, et al. Directional water-transfer through fabrics induced by asymmetric wettability. J Mater Chem. 2010;20(37):7938–40.
117. Tian X, Jin H, Sainio J, et al. Droplet and fluid gating by biomimetic Janus membranes. Adv Func Mater. 2014;24(38):6023–8.
118. Zhao X, Li Y, Hua T, et al. Cleanable air filter transferring moisture and effectively capturing $PM_{2.5}$. Small. 2017;13(11):1603306.
119. Miao D, Huang Z, Wang X, et al. Continuous, spontaneous, and directional water transport in the trilayered fibrous membranes for functional moisture wicking textiles. Small. 2018;14(32):1801527.
120. Lao L, Shou D, Wu YS, et al. “Skin-like” fabric for personal moisture management. Sci Adv. 2020;6(14):eaaz0013.
121. Cao M, Xiao J, Yu C, et al. Hydrophobic/hydrophilic cooperative Janus system for enhancement of fog collection. Small. 2015;11(34):4379–84.
122. Liu J, Zhou H, Wu X, et al. Superoleophobic filters: improvement of filtration performance by front attachment of oil-guiding fabric. Adv Mater Interfaces. 2019;7(2):1901808.
123. Wang C, Li X, Cheng C, et al. Preparation of ultrafine fibrous membranes of polystyrene/MWCNTs composites as air filter by electrospinning. J Mater Sci Eng. 2016;34(6):960–6.

124. Yildiz O, Bradford PD. Aligned carbon nanotube sheet high efficiency particulate air filters. Carbon. 2013;64:295–304.
125. Wang Q, Yildiz O, Li A, et al. High temperature carbon nanotube: nanofiber hybrid filters. Sep Purif Technol. 2020;236: 116255.
126. Yildiz O, Stano K, Faraji S, et al. High performance carbon nanotube: polymer nanofiber hybrid fabrics. Nanoscale. 2015;7(40):16744–54.
127. Park SJ, Lee DG. Performance improvement of micron-sized fibrous metal filters by direct growth of carbon nanotubes. Carbon. 2006;44(10):1930–5.
128. Li P, Wang C, Li Z, et al. Hierarchical carbon-nanotube/quartz-fiber membranes with gradient nanostructures for high efficiency and long service life air filters. RSC Adv. 2014;4(96):54115–21.
129. Xu C, Xie W, Yu Y, et al. Photocatalytic and filtration performance study of TiO_2/CNTs: filter for oil particle. Process Saf Environ Prot. 2019;123:72–8.
130. Halonen N, Rautio A, Leino AR, et al. Three-dimensional carbon nanotube scaffolds as particulate filters and catalyst support membranes. ACS Nano. 2010;4(4):2003–8.
131. Zhao Y, Zhong Z, Low ZX, et al. A multifunctional multi-walled carbon nanotubes/ceramic membrane composite filter for air purification. RSC Adv. 2015;5(112):91951–9.
132. Zhang RF, Wei F. High-efficiency particulate air filters based on carbon nanotubes. Amsterdam: Elsevier; 2019. p. 643–66.
133. Wang CS, Otani Y. Removal of nanoparticles from gas streams by fibrous filters: a review. Ind Eng Chem Res. 2013;52(1):5–17.
134. Li P, Zong Y, Zhang Y, et al. In situ fabrication of depth-type hierarchical CNT/quartz fiber filters for high efficiency filtration of sub-micron aerosols and high water repellency. Nanoscale. 2013;5(8):3367–72.
135. Yang K, Yu Z, Yu C, et al. An electrically renewable air filter with integrated 3D nanowire networks. Adv Mater Technol. 2019;4(7):1900101.
136. Yang S, Zhu Z, Wei F, et al. Carbon nanotubes/activated carbon fiber based air filter media for simultaneous removal of particulate matter and ozone. Build Environ. 2017;125:60–6.
137. Feng S, Li X, Zhao S, et al. Multifunctional metal organic framework and carbon nanotube-modified filter for combined ultrafine dust capture and SO_2 dynamic adsorption. Environ Sci Nano. 2018;5(12):3023–31.
138. Donaldson K, Murphy FA, Duffin R, et al. Asbestos, carbon nanotubes and the pleural mesothelium: a review of the hypothesis regarding the role of long fibre retention in the parietal pleura, inflammation and mesothelioma. Part Fibre Toxicol. 2010;7:5.
139. Park KT, Hwang J. Filtration and inactivation of aerosolized bacteriophage MS_2 by a CNT air filter fabricated using electro-aerodynamic deposition. Carbon. 2014;75:401–10.
140. Jung JH, Hwang GB, Lee JE, et al. Preparation of airborne Ag/CNT hybrid nanoparticles using an aerosol process and their application to antimicrobial air filtration. Langmuir. 2011;27(16):10256–64.
141. Zhao Y, Low ZX, Feng S, et al. Multifunctional hybrid porous filters with hierarchical structures for simultaneous removal of indoor VOCs, dusts and microorganisms. Nanoscale. 2017;9(17):5433–44.

Chapter 5
High Temperature Gas Cleaning Membranes

5.1 Ceramic Membranes

Porous ceramic membranes, due to their outstanding advantages such as high temperature resistance, corrosion resistance, erosion resistance, high mechanical strength, wide pore size range, and wide application range, are one of the best choices for high temperature gas filtration materials [1–3]. Ceramic membranes have been used in gas–solid separation for many years, with a wide range of applications and significant effects. The porous ceramic membranes currently used in the field of high temperature dust removal are mainly divided into two types, namely oxide ceramic materials and non-oxide ceramic materials. The former includes alumina, spinel, cordierite and mullite, and the latter includes silicon carbide, silicon nitride, etc. At present, ceramic membrane filters have been applied in many fields such as coal gasification, waste incineration and pyrolysis, petrochemicals, regeneration of black metal melting, precious metal recovery, thermal soil reforming, fluidized bed metal purification, boiler devices, chemical manufacturing, and glass melting [4, 5]. The basic properties of several common porous ceramic materials are shown in Table 5.1.

Compared to other ceramics, porous silicon carbide ceramics have advantages such as high mechanical strength and good thermal shock resistance, and have become the mainstream material used for high-temperature gas filtration membranes. This section will introduce the preparation methods, properties, and application systems of silicon carbide ceramic membranes.

5.1.1 Overview of Silicon Carbide Ceramic Membranes

Silicon carbide (SiC) is a typical covalent compound, with the covalency of the Si–C bond in SiC accounting for 88%. The SiC crystal structure is composed of identical tetrahedra, with Si atoms at the center, and all structures are formed by

Z. Zhong and W. Xing, *Membrane Technology for Air Cleaning*,
https://doi.org/10.1007/978-981-96-1144-7_5

Table 5.1 Basic properties of porous ceramic materials [6]

Material	Chemical composition	Linear expansion coefficient/ ($\times 10^{-6}$/ °C)	Thermal shock resistance	Suitable operating temperature/ °C	Oxidation resistance	Mechanical strength
Corundum	Al_2O_3	8.8	Low	≤ 500	Low	Relatively high
Mullite	$3Al_2O_3{\cdot}2SiO_2$	3.5	Good	≤ 1100	Good	Relatively high
Cordierite	$2Al_2O_3{\cdot}5SiO_2{\cdot}2MgO$	1.8	Good	≤ 1100	Good	Average
Alumino silicate fiber	$3Al_2O_3{\cdot}2SiO_2$	—	Good	≤ 1100	Good	Poor
Silicon carbide	SiC	4.7	Good	≤ 950	Poor	High

parallel or antiparallel stacking of SiC tetrahedra [7]. SiC is divided into green, gray, and dark green colors according to purity, and its decomposition temperature under normal pressure is 2380 °C. Common crystal types include α-SiC, β-SiC, 6H-SiC, 4H-SiC, and 15R-SiC, among which α-type and β-type are the main crystal types. Silicon carbide ceramic membranes are mainly formed by high-temperature sintering of stacked silicon carbide particles. The sintered material has a certain mechanical strength and is a porous material [8]. Structurally, silicon carbide ceramic membranes consist of a silicon carbide support with large pore size and high strength, and a separation layer with small pore size and high separation efficiency, as shown in Fig. 5.1. Due to the advantages of silicon carbide materials, such as good thermal conductivity, high strength, small thermal expansion coefficient, and good thermal shock resistance [9], coupled with the excellent thermal shock resistance, corrosion resistance, creep resistance, and high mechanical strength of silicon carbide ceramic membranes at high temperatures, it is one of the most widely used materials in the field of high-temperature gas–solid separation [10].

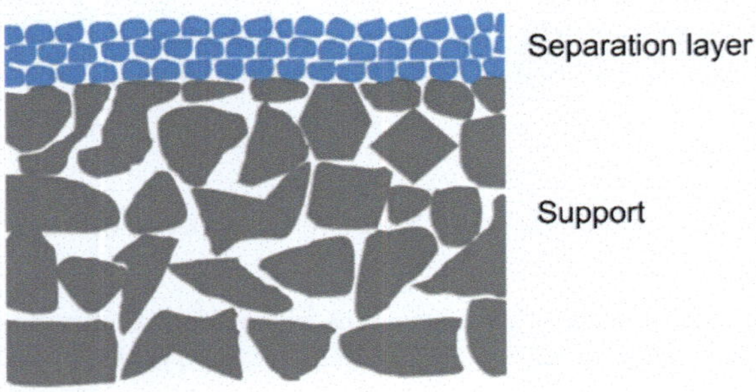

Fig. 5.1 Schematic diagram of the structure of silicon carbide ceramic membranes

5.1.2 *Preparation Methods of Silicon Carbide Ceramic Membrane Supports*

The silicon carbide ceramic membrane support used for high-temperature gas cleaning needs to have characteristics such as high porosity, large pore size, and high mechanical strength to solve the problem of high resistance of ceramic membranes in practical applications. To solve these problems, innovative improvements need to be made in the preparation methods. The preparation process of silicon carbide ceramic membrane support is complex and technically difficult, and the sintering mechanism is different from that of general oxide ceramics. It is usually made by using silicon carbide particles as aggregate, adding feldspar, clay and other sintering aids, as well as activated carbon, wood fiber and other pore-forming agents, using extrusion process, tape casting process, isostatic pressing process for molding, and then high-temperature sintering. The prepared support's surface needs to be smooth and free of large defects to avoid affecting the subsequent coating process. In addition, since silicon carbide is a covalent compound, the sintering temperature is high. In order to obtain pure silicon carbide membrane products, sintering needs to be carried out in a reducing atmosphere, and the sintering temperature usually needs to be above 2100 °C, which results in high production costs [11, 12]. In response to these preparation difficulties, researchers have made improvements in the preparation methods in order to reduce the sintering temperature and improve its gas permeability performance within a reasonable strength range. Common preparation techniques for silicon carbide ceramic supports mainly include the following, including in-situ reaction sintering, partial sintering, addition of pore-forming agents, template method, direct foaming method, etc. [8–13].

5.1.2.1 In-Situ Reaction Sintering Method

In order to reduce the product cost of silicon carbide ceramic membranes, reducing the sintering temperature has become one of the necessary means. In recent years, researchers have proposed the use of in-situ reaction sintering method to prepare silicon carbide ceramic membranes, which has quickly developed and achieved industrial production [14].

In-situ reaction sintering technology is a process in which the ceramic raw material body undergoes chemical reactions between solid, liquid, and gas phases at a certain temperature, simultaneously densifying and synthesizing specified components to obtain the predetermined sintered body. The product has almost no size shrinkage before and after firing. The performance of the SiC porous ceramic membranes prepared by this method is closely related to the choice of sintering aids. The types of sintering aids include metal oxides, alkaline earth metal oxides, cordierite ($2MgO{\cdot}2Al_2O_3{\cdot}5SiO_2$), mullite ($3Al_2O_3{\cdot}2SiO_2$), SiO_2, Si, Si_3N_4, AlN and natural clay, etc. [15–17]. The addition of these aids can reduce the temperature of SiC ceramic membrane preparation, save energy, and reduce production costs, and

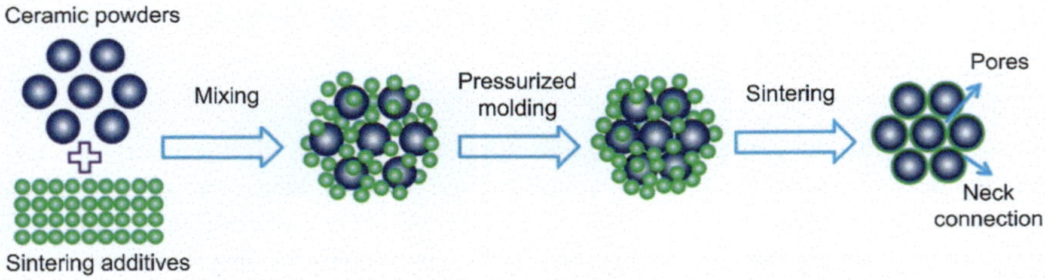

Fig. 5.2 Schematic diagram of in-situ reaction sintering preparation process

the material has excellent thermal shock resistance, chemical stability, and higher strength. The schematic diagram of the sintering process of this method is shown in Fig. 5.2.

In terms of low-temperature sintering to prepare high-strength silicon carbide membrane supports, Han et al. [18–20] used zirconia as a sintering aid and silicon carbide whiskers as a reinforcement, and successfully prepared high porosity, high strength, and thermally stable silicon carbide ceramic membrane supports by in-situ reaction sintering at 1300–1550 °C. The performance changes after doping with silicon carbide whiskers are shown in Figs. 5.3 and 5.4. It can be seen that as the amount of SiC whiskers increases, the flexural strength of the support firstly increases and then decreases, the porosity remains basically unchanged, and the addition of SiC whiskers improves the pore size and gas permeability of the support. When the addition amount of SiC whiskers is 3.3%, the flexural strength of the support is 28.7 MPa, which is nearly 3 times higher than that without SiC whiskers, the porosity is still as high as 43%, and the gas permance reaches 270 $m^3/(m^2\ h\ kPa)$, which is 2 times higher than that without whiskers. This is because the addition of SiC whiskers promotes the oxidation of silicon carbide, forming a new zircon phase, and enhancing the neck connection of SiC particles (as shown in Fig. 5.5).

Based on this work, mullite fibers were used to replace silicon carbide whiskers (as shown in Fig. 5.6), which further increased the porosity of the support while ensuring mechanical strength. The addition of mullite effectively ensures the flexural strength of the support and maintains a high porosity. The reinforcement of mullite fibers mainly increases the strength of ceramics through three mechanisms: fiber bridging, crack deflection, and fiber pull-out (Fig. 5.7). When 4% of mullite fibers are doped, the porosity and flexural strength reach 46.8% and 15.7 MPa respectively, the average pore size is 20 μm, and the highest gas permance reaches 1600 $m^3/(m^2\ h\ kPa)$, which is close to the performance of foam ceramic materials [21].

Xu et al. [21] and Yang et al. [22] investigated the effect of the addition of sintering aids such as sodium dodecylbenzenesulfonate (SDBS), NaOH, and sodium-A zeolite (NaA) on the performance of silicon carbide ceramic membranes, further reducing the sintering temperature to below 1300 °C. It was found that the porosity of the prepared support decreases with the increase of the sintering aid content, and the average pore size, gas permeability coefficient, and strength all show a trend of first increasing and then decreasing with the increase of the sintering aid content.

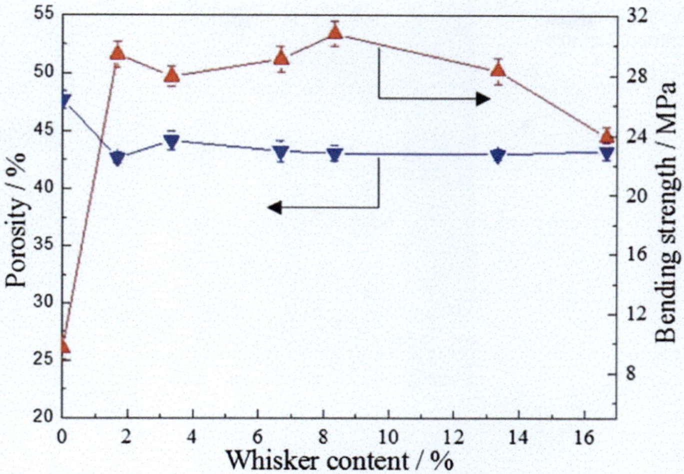

Fig. 5.3 Changes in porosity and flexural strength with SiC whisker content [18]

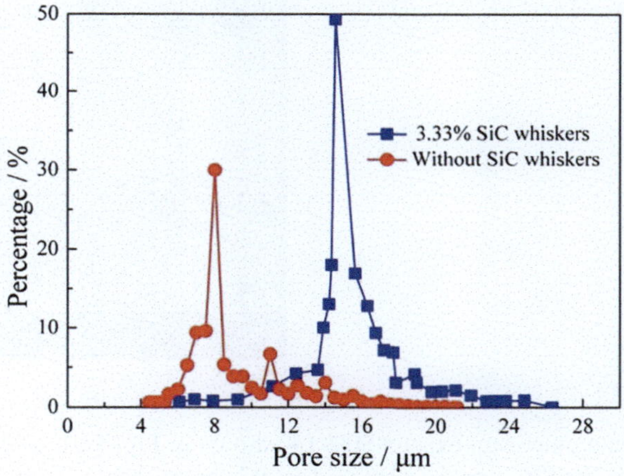

Fig. 5.4 The effect of the addition of SiC whiskers on the pore size distribution of the silicon carbide support [18]

After adding the sintering aid, the gas permeability and strength of the support are greatly improved, and the connection between silicon carbide particles is significantly strengthened (Fig. 5.8). The optimal amounts of SDBS, NaOH, and NaA as additives are 8 wt%, 1.5 wt%, and 10 wt%, respectively. The pore sizes of the supports were 20 μm, 20 μm, and 24 μm, the gas permeances were 900 $m^3/(m^2\ h\ kPa)$, 850 $m^3/(m^2\ h\ kPa)$, and 1300 $m^3/(m^2\ h\ kPa)$, and the mechanical strengths were 16 MPa, 16 MPa, and 27 MPa, respectively. Using Na source as a sintering aid reduces the sintering temperature, improves the gas permeability, and the support also shows excellent acid corrosion resistance. However, when used in an alkaline environment for a long time, the mechanical strength of the support will decay, and the reduction

Fig. 5.5 Silicon carbide ceramic membrane support microstructure [18]

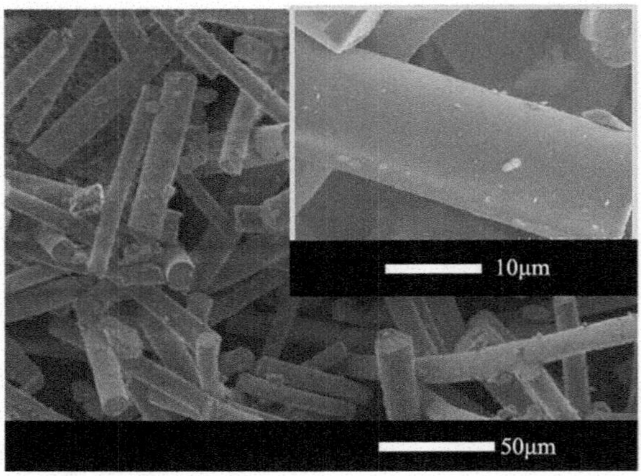

Fig. 5.6 Microscopic morphology of mullite ceramic short fibers [19]

rate may even exceed 50% (as shown in Fig. 5.9), so the alkali corrosion resistance of the SiC support prepared by this method needs to be improved [23].

The in-situ reaction sintering technology for preparing SiC porous ceramic membranes has the characteristics of low sintering temperature, low cost, simple preparation process, and easy to industrialize production. Under the control of different SiC aggregate particle sizes, sintering aids types, pore-forming agent content, and sintering processes, the sintering temperature range for the general reaction sintering method is between 950 °C and 1550 °C. But for the self-bonded SiC porous ceramics, the sintering temperature is higher than 1600 °C.

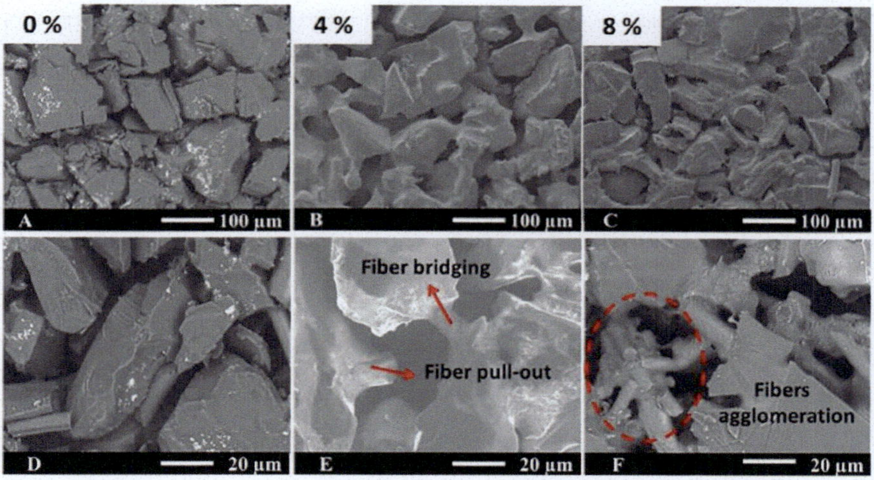

Fig. 5.7 Changes in the microstructure of porous silicon carbide ceramics with the content of mullite fibers [19]

Fig. 5.8 The effect of Na sintering aid types on the microstructure of the support [22]

5.1.2.2 Partial Sintering Method

The partial sintering method is one of the simplest and most typical preparation methods. At high temperatures, processes such as diffusion, dissolution, precipitation, or recrystallization occur on the surface of SiC particles, resulting in the formation of neck connections and pores between adjacent particles. The ceramic membrane prepared by this method has a uniform pore structure, with an average pore size ranging from approximately 1/5 to 1/2 of the diameter of the SiC particles [24], and its strength is mainly determined by the quality of the neck connections.

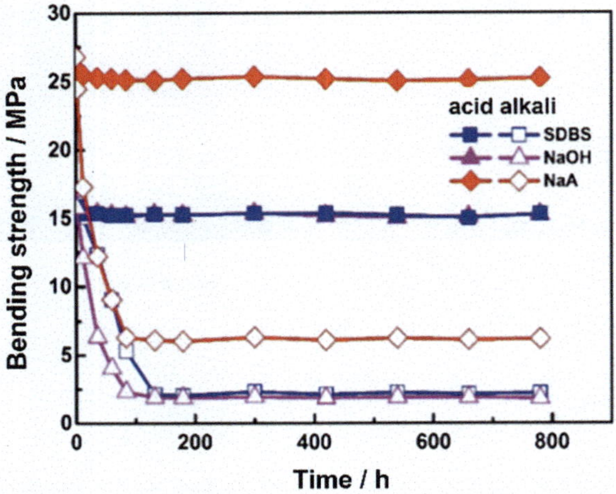

Fig. 5.9 The effect of acid and alkali corrosion on the mechanical strength of the support prepared by adding SDBS, NaOH, NaA [23]

The sintering temperature of the partial sintering method is generally higher, above 2000 °C. Wang et al. [25] used the solid-phase sintering method, using PMMA microspheres as pore-forming agents, and sintered at 2100 °C to obtain a porous silicon carbide support with a three-level pore size distribution (see Fig. 5.10), and a porosity between 51.5% and 81.3%. They proposed that when the temperature is above 2150 °C, the growth of silicon carbide grains accelerates. Although this method has a high porosity, the connectivity between the pores is relatively low, and the strength is not high. Yao et al. [26] used this method to sinter at 2150 °C to prepare a silicon carbide ceramic membrane with a three-point bending strength of up to 400 MPa, and proposed that the microstructure of the silicon carbide ceramic membrane can be controlled by adjusting the carbon source. Fukushima et al. from Japan [27] successfully prepared high-strength SiC porous ceramics at 1500–1800 °C by doping SiC nanoparticles and using cold isostatic pressing method. When the sintering temperature increased from 1500 °C to 1800 °C, the compressive strength of the material increased from 122 to 513 MPa. It is believed that the addition of nano-SiC particles promotes the mass transfer rate at high temperatures, increases the neck connections, and thus improves the overall strength of the ceramic.

Usually, although the ceramics prepared by the partial sintering method have higher strength, the open porosity is lower. The densification of the material can be prevented by using low-temperature liquid phase sintering, mixing of coarse and fine particles, pressure-less sintering, and recrystallization, and the pore size distribution of the resulting SiC porous ceramics is also relatively narrow. The biggest advantage of the partial sintering method is that it is very easy to control the size of the product, and the properties of the product prepared are closest to the characteristics of SiC material itself with a pore size of 0.1–10 μm; the disadvantage of this method is that the permeability is poor, so improving the porosity of the material to enhance the permeability is a research trend in future.

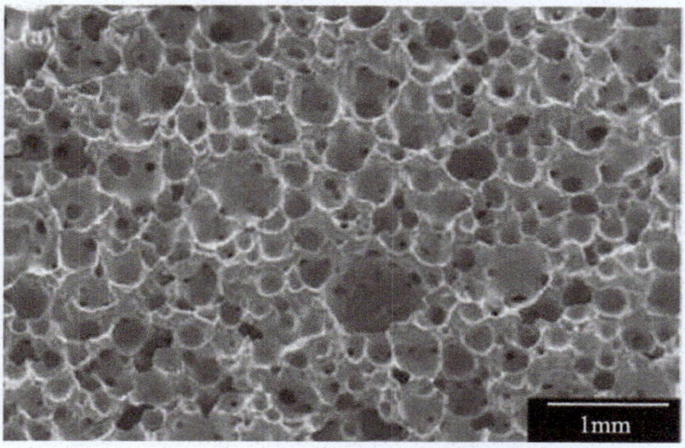

Fig. 5.10 Microstructure of silicon carbide porous ceramics prepared by partial sintering method [25]

5.1.2.3 Template Method

The template method was studied early on, with scientists using sponges as templates to prepare honeycomb ceramics as early as the 1960s. The main process of the template method is shown in Fig. 5.11. The preparation of silicon carbide porous materials by impregnating polyurethane sponge involves the following steps: (1) Impregnating the polyurethane sponge with SiC suspension or precursor solution, (2) Centrifuging to remove excess suspension, (3) Drying the impregnated polyurethane sponge, (4) Heat treatment at a certain temperature to decompose this polymer template, (5) Increasing the temperature, sinter in air atmosphere to obtain silicon carbide porous ceramics [28].

The porosity of ceramics prepared by the template method can reach more than 80%–90%, and a relatively regular network structure can be obtained. Zhu et al. [29] used polyurethane sponge as a template, SiC, Al_2O_3 and bentonite as main materials, silicon sol as a binder, and sintered at 1400 °C for 5 h. The obtained silicon carbide ceramic has a three-dimensional network structure with a flexural strength of about 1 MPa. Singh [30] used maple and peach wood as templates and obtained silicon carbide porous ceramics by vapor phase infiltration method in a high temperature

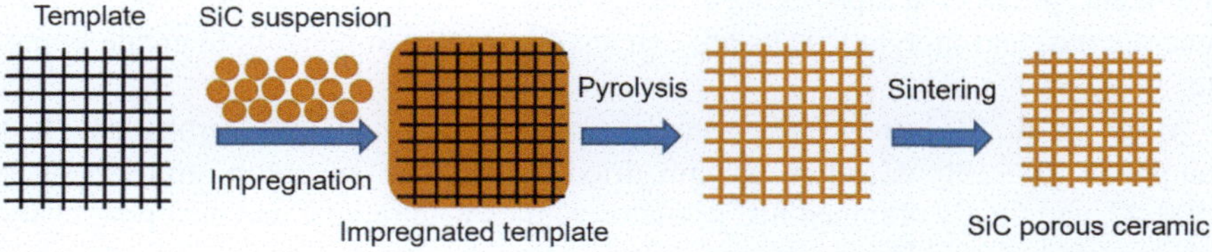

Fig. 5.11 Schematic diagram of the process of preparing silicon carbide porous ceramics using the template method [28]

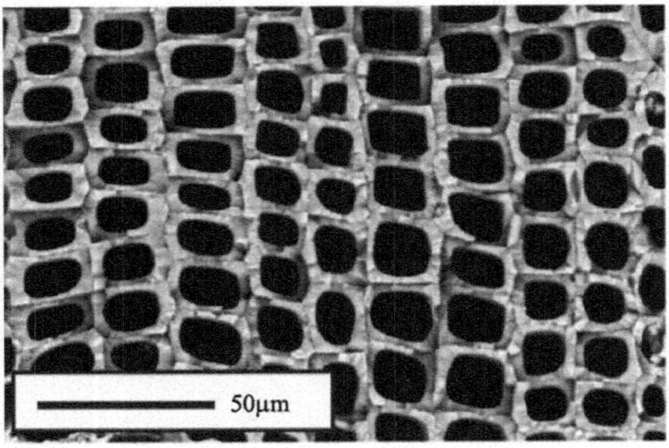

Fig. 5.12 Microstructure of silicon carbide porous ceramics prepared using pine wood [31]

and vacuum environment. The flexural strength of the ceramic was up to 200 MPa, but the porosity was relatively low. Vogli et al. [31] infiltrated silicon into pine wood and heated it at 1600 °C. The reaction resulted in biomorphic SiC porous ceramics, showing regular geometric pores (Fig. 5.12), with an average pore size of 20 μm, a porosity of up to 71%, and a flexural strength of only 12 MPa.

5.1.2.4 Pore-Forming Additive Method

The pore-forming additive method is to mix a substance that can be decomposed easily during sintering with silicon carbide. When the temperature rises, the substance is decomposed and discharged, leaving pores to obtain porous ceramics. Pore-forming agents are generally divided into organic and inorganic materials. Organic pore-forming agents include natural fibers and synthetic polymers, such as sawdust, starch, polystyrene (PS), polymethyl methacrylate (PMMA) and so on. Inorganic pore-forming agents include salts that can be decomposed at high temperatures such as ammonium carbonate, ammonium bicarbonate, ammonium chloride and so on, and inorganic carbon such as graphite, coal powder, and so on [32]. Usually, these template agents decompose at temperatures between 200 °C and 1000 °C. The form of pore formation is also diverse, liquid phase water or organic material templates form pores through freeze-drying or sublimation, inorganic carbon or organic polymer templates form pores through high-temperature oxidation to generate gas, and inorganic salts are extracted by water immersion to create pores, etc.

Yoon et al. [33] froze SiC samples impregnated with polycarbosilane and camphene to −196 °C, then vacuum dried, sintered in an argon atmosphere at 1400 °C, and finally obtained SiC porous ceramics with regular and neat pore channels (see Fig. 5.13). Fukushima et al. [34] using ice as a pore-forming agent, the water-containing silicon carbide green membrane was first cooled to −10 °C to −70 °C, then vacuum dried, and finally sintered at 1800 °C to obtain silicon carbide

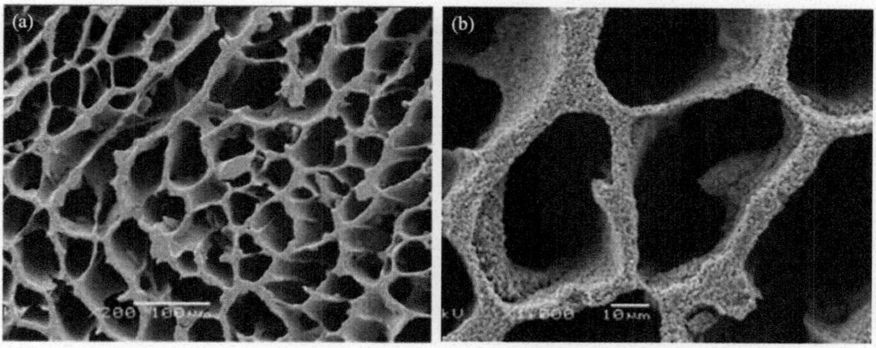

Fig. 5.13 Microstructure of silicon carbide porous ceramics prepared by freeze-drying process [33]

porous ceramic with unidirectional cylindrical pores. The total porosity exceeded 86.5%, the pore size was 34–147 μm, the compressive strength was 5.2–16.6 MPa, the flexural strength was less than 5.2 MPa, and the gas permeability was $2.27 \times 10^{-11} - 1.04 \times 10^{-10}$ m^2. Du et al. [35] used camphene as a freezing agent, added it to ZrB_2 and SiC powders in a certain ratio (20%, 25%, and 30%), and then ball milled at 50 °C for 12 h. After molding, curing, freezing, sublimation and finally sintered in argon atmosphere at 1950 °C for 2 h, a porous ceramic with 3D network structure was prepared. The strength was between 173 and 364 MPa, the porosity range was 42–66%, and the pore size decreased with the decrease of the freezing temperature.

The advantage of the pore-forming additive method is that it can precisely control the shape, size, and porosity of the pores by optimizing the shape, particle size, distribution, and preparation conditions of the pore-forming agent. The products prepared by this method have a uniform pore size distribution, high mechanical strength, small shrinkage, and are not easy to deform. The pores of the products prepared by this method are basically open pores, which is beneficial for the passage of gas or liquid. The disadvantage is that the pore distribution is not uniform, the porosity of the obtained product is not high (generally less than 50%), and the ceramic strength will be greatly affected by the change in the amount of pore-forming agent.

5.1.2.5 Direct Foaming Method

The direct foaming method forms a large number of pores by adding substances that can generate a large number of bubbles when heated to the ceramic slurry, generating bubbles and discharging them during heat treatment, and finally drying and sintering to obtain porous ceramics (see Fig. 5.14). Foaming agents include organic foaming agents, inorganic foaming agents, physical foaming agents, and surfactants [36, 37]. Foaming methods include gas injection, stirring to generate bubbles, liquid evaporation, and gas generation by chemical reaction. The characteristics of ceramics prepared by the foaming method are high porosity, up to 95%. Bao et al. [38] controlled the porosity by adding a volatile agent to foam the SiC slurry, which is relatively simple to operate. Fukushima et al. [39] used azodicarbonamide as a foaming agent and polycarbosilane (PCS) as a ceramic precursor. The mixed powder was foamed at 250–260 °C, cross-linked and cured at 200 °C for 2 h, and finally pyrolyzed at 1000 °C to obtain SiC porous ceramics (see Fig. 5.15). The porosity can reach 59–85%, and the pore size is 416–1455 μm.

Compared to other preparation processes, methods such as the addition of pore-forming agents and organic foam impregnation can only produce net-like open-pore SiC ceramics, while the foaming process is the only effective method for preparing closed-pore porous SiC ceramics. At present, this type of closed-porous material is increasingly used in environmental protection, chemical, and energy industries. It plays a significant role in insulation and weight reduction structures, so further research, development, application, and promotion will bring great economic and social benefits.

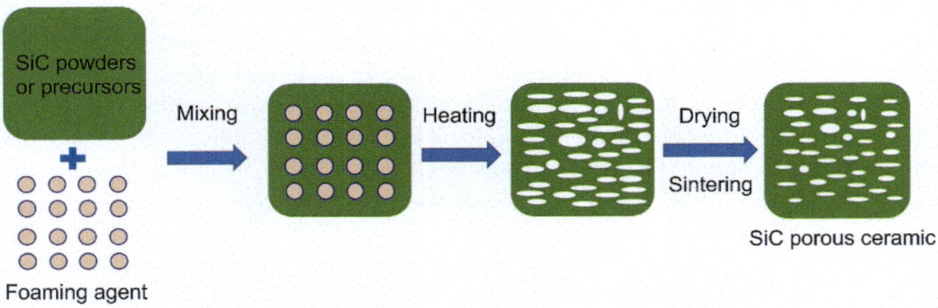

Fig. 5.14 Schematic of the process of preparing silicon carbide porous ceramics by the foaming method [38]

Fig. 5.15 Microstructure of silicon carbide porous ceramics prepared with azodicarbonamide [39]

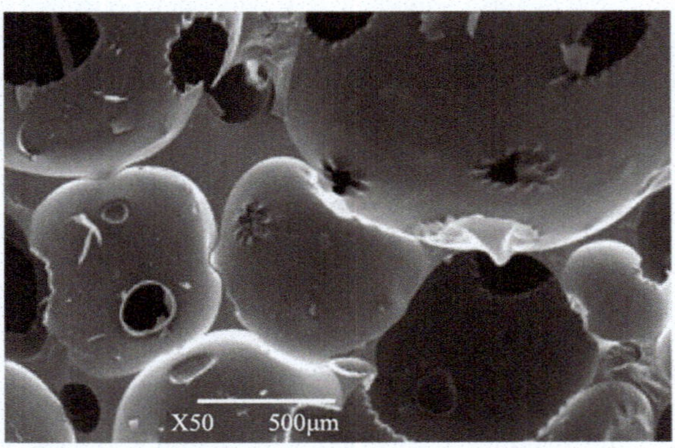

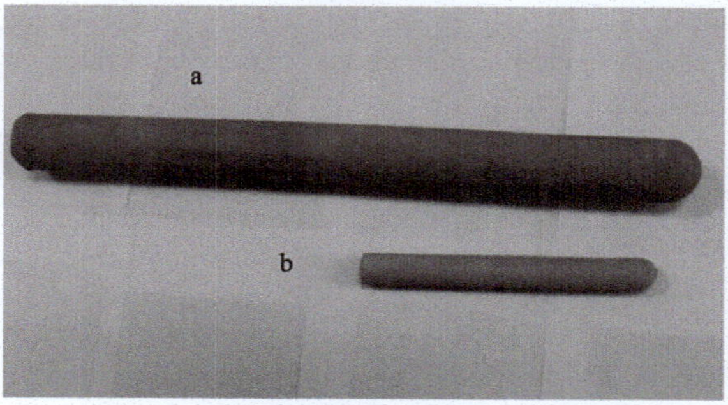

Fig. 5.16 Silicon carbide membrane support components **a** L = 500 mm, φ 48 mm × 32 mm and **b** L = 200 mm, φ 24 mm × 16 mm prepared by gel casting [43]

5.1.3 Molding Technology of Silicon Carbide Ceramic Supports Molding

Common molding methods for silicon carbide supports mainly include isostatic pressing, extrusion molding, slip casting, gel casting, etc. The common shapes of the supports are tubular, rod-shaped, flat, and honeycomb.

5.1.3.1 Gel Casting

The gel casting process was first proposed by the Oak Ridge National Laboratory in the United States in the 1990s. The principle is to form the slurry directly into a blank through an in-situ polymerization reaction, and then make ceramics through high-temperature sintering. The advantage of this process is that the ceramic green body formed has high strength, being less prone to collapse in the sintering process, and the operation is simple, and ceramics with relatively complex shapes can be prepared. Zhang et al. [40] used gel casting to prepare ceramic blanks with alumina as the raw material, with a strength of up to 14 MPa and a green density of up to 68%. Wang et al. [41] used gel casting to prepare silicon carbide ceramics, which were efficient but had poor mechanical strength, with a strength of only 3 MPa. Mouazer et al. [42] combined gel casting and foaming to prepare silicon carbide ceramics with a porosity of up to 78–88%. Most of the existing gel systems are toxic acrylamide systems, and a large amount of organic additives need to be added to adjust the viscosity of the slurry to improve the strength of the green body. There are many problems in the sintering process, such as large shrinkage deformation of the fired body and poor mechanical strength. Figure 5.16 shows two different diameters and lengths of silicon carbide membrane support prepared by the gel casting [43].

5.1.3.2 Slip Casting

Slip casting [44] is a process in which ceramic slurry is spread on a plane under the action of a scraper. The principle is to add a large amount of organic plasticizer to the powder to make the powder into a slurry with good viscosity. The slurry flows out on the slip casting machine and is scraped and coated on the base tape by the scraper. After drying and curing, it is formed. The cured blank can be processed, and finally, the ceramic product is obtained by sintering at high temperature. The advantage of slip casting is that the equipment is simple and the production efficiency is high, but the disadvantage is that this method can only prepare flat ceramic membranes. Guo et al. [45] used this method to prepare aluminum nitride ceramics with a volume density of 3.32 g/cm^3 and a thermal conductivity of up to 178 W/(m K). Zhou et al. [46] used the tape casting method to prepare electronic ceramics and optimized the preparation process. This method also has problems such as a large amount of organic plasticizer added, high requirements for drying and dewaxing. Tape casting is currently mature in the preparation of piezoelectric ceramic thin membranes, and further development is needed in the preparation of large-scale silicon carbide membranes. This is mainly because this molding method is sensitive to the choice of organics, and the type and ratio of organics used must be reasonably selected according to the powder used and the thickness of the required thin sheet. In addition, silicon carbide particles are hard and wear-resistant, which will wear the scraper. It is believed that through continuous improvement of methods by researchers, tape casting will become a silicon carbide membrane molding method with a very broad application prospect.

5.1.3.3 Isostatic Pressing

Isostatic pressing is a method of using the characteristics of incompressibility and uniform pressure transmission of liquid medium to press the ceramic powder in the mold. The advantage of isostatic pressing is that it can prepare products with complex shapes, with small sintering shrinkage, being suitable for producing tubular and flat supports [47]. Liu et al. [48] used isostatic pressing to prepare a silicon carbide support with an outer diameter of 60 mm, which was sintered at 1300 °C. Its average pore size can reach 150 μm, the porosity is small, only 36%, and the flexural strength is only about 15 MPa. Fukushima et al. [49] used 300 nm silicon carbide as aggregate to prepare high-strength supports by isostatic pressing. The strength reached up to 513 MPa, and the flexural strength increased with the increase of molding pressure. Because no pore-forming agent was added, the porosity of the prepared support was less than 40%. The microstructure of this silicon carbide support is shown in Fig. 5.17. There are many fine particles and small pore sizes. Isostatic pressing can basically prepare silicon carbide membrane supports in one near-net size molding, reducing the number of sintering times, but there are also some problems with isostatic pressing, such as difficult demolding, challenges in continuous production, and low level of automation.

Fig. 5.17 Microstructure of the silicon carbide support prepared by isostatic pressing [49]

5.1.3.4 Extrusion Molding

Extrusion molding is a type of plastic molding, one of the molding methods for preparing long wires and tubes, and is the most suitable low-cost process for producing equal-section products. It refers to the process of obtaining green bodies by spiral extrusion or piston extrusion of the prepared wet material after vacuum refining to obtain a better plastic mud. However, it has high requirements for the mud. Firstly, the powder must have sufficient fineness and smooth shape to ensure the fluidity of the mud, and secondly, a suitable amount of solvent and plasticizer must be added to ensure the plasticity of the mud.

Extrusion molding is very suitable for continuous production, with high efficiency, easy-to-automate operations, and different shapes of blanks can be obtained by simply changing different molds. The disadvantage is that in addition to the strict requirements for the mud, it also requires high precision of equipment processing and large sintering shrinkage.

Extrusion molding can extrude tubular or rod-shaped supports with diameters from 1 to 800 mm, 100 to 200 holes/cm^2 of honeycomb or plate-like supports. Isobe et al. [50] used alumina material with a particle size of 0.7 μm, carbon fiber with a diameter of 14 μm as a pore-forming agent, and prepared porous alumina ceramics through extrusion molding. Its three-point bending strength reached up to 171 MPa, the pore size was 14 μm, and the pore size had the characteristic of unidirectional arrangement. Jiang et al. [51] used high specific surface area silicon nitride as raw material to extrude and prepare silicon nitride ceramics. The pore size range was between 15 and 25 μm, the porosity was as high as 67%, and the gas permeability coefficient was as high as 1.2×10^{-12} m^2. It showed outstanding performance in microstructure, with no obvious defects and high bending strength. Xu et al. [52] used silicon carbide with a particle size of 100 μm as aggregate, sodium dodecyl benzene sulfonate and zirconia as sintering aids, and carbon powder as a

Fig. 5.18 (**a**) The green bodies of the multi-channel tubular silicon carbide support and (**b**) final product prepared by extrusion molding method[52]

pore-forming agent. They prepared a multi-channel tubular porous silicon carbide ceramic support by extrusion molding. The bending strength of the support was above 20 MPa. Figure 5.18 shows the green body and finished product of the multi-channel silicon carbide support tube prepared by extrusion molding.

5.1.4 Preparation Methods of Silicon Carbide Ceramic Separation Membrane Layer

Asymmetric ceramic membranes generally use impregnation and lifting, spin coating, spray coating, and chemical vapor deposition to coat a separation membrane layer on the support. After a certain heat treatment process, the final product is prepared.

The impregnation and lifting method [53] is common in the sol coating process. The sol refers to a transparent colloidal dispersion system formed by uniformly dispersing nano-sized (1–100 nm) dispersed particles in another continuous phase dispersion medium. The impregnation and lifting method mainly includes impregnation, lifting, and heat treatment. First, the support is immersed in the prepared sol, and then it is uniformly pulled out of the liquid surface at a constant speed. The speed of lifting is particularly critical. The faster the lifting speed, the thinner the membrane layer, and the longer the preparation time required for the membrane layer; the slower the lifting speed, the thicker the membrane layer, making it easier for the sol to agglomerate on the surface of the support to form gel particles. Then, the support covered with a uniform liquid membrane is dried at a constant temperature, the solvent quickly evaporates, and a gel membrane is formed on the surface of the support. Repeat the above steps until an appropriate thickness is obtained for a gel layer. Finally, different pore size range separation membranes are obtained through different sintering programs. Figure 5.19. shows the effect of the solid content of the membrane-forming liquid on the thickness of the membrane. When the solid content

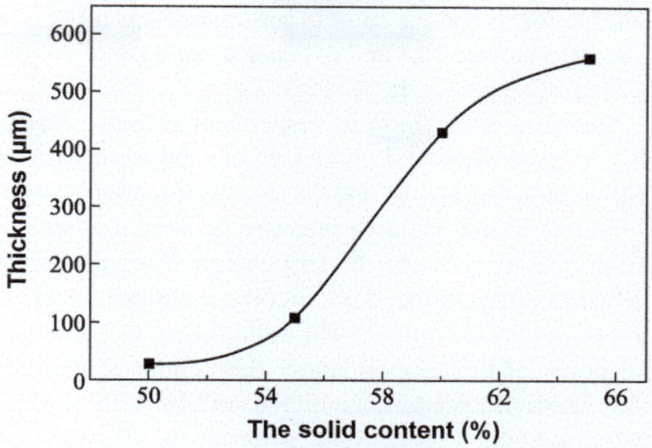

Fig. 5.19 The effect of the solid content of the membrane-forming liquid on the thickness of the membrane [54]

increases from 55 to 60%, the membrane thickness increases rapidly and then tends to be flat. Therefore, when preparing silicon carbide membranes by the impregnation and lifting method, the relationship between membrane thickness and the solid content of the membrane-forming liquid should be considered, and an appropriate solid content should be selected [54].

The thickness of the membrane prepared by spin-coating is precisely controllable between 30 and 2000 nm [55], and the preparation process mainly consists of three steps: drop-coating, spraying, and drying. First, the coating solution is sprayed onto the surface of support, which rotates at a certain angular speed without stopping, and the sol is uniformly coated on the surface of the support. Then, it is dried at a constant temperature, where the solvent evaporates quickly and forms a layer of gel membrane on the surface of the support. Finally, a complete separation membrane is obtained by sintering under a set sintering program. Among them, rotation speed and drying are the key factors and steps affecting the structure and thickness of the membrane layer. Spin-coating method has the advantages of simple equipment, easy maintenance, simple preparation process, short time consumption, precise and controllable membrane thickness, cost-effective, energy-saving and environmentally friendly. Spin-coating technology is not only suitable for coating on flat support bodies, but also suitable for preparing separation membrane layers on complex-shaped supports (such as tubular ceramic supports).

Chemical vapor deposition [56] is the deposition of a source material on the surface of the support or in the pores by chemical methods at low temperatures, with the help of gas-phase reaction on the surface of the support and deposition to generate a separating membrane layer. This method must satisfy the following conditions: (i) the reactants must have a sufficiently high vapor pressure, and the deposits must have a sufficiently low vapor pressure to ensure that the generated deposits remain on the heated support throughout the course of the reaction; (ii) except for the required deposits, all other reaction products must be gaseous. When preparing a separation membrane layer using this method, the control of the deposition time is very important. The equipment for chemical vapor deposition is expensive and not suitable for the preparation of large-area separation membrane layers. The types of raw materials are few and expensive, so its application to industrial production still needs further research.

Spraying is common in suspension systems. Suspension refers to micron-sized (0.1–10 μm) dispersed phase particles uniformly dispersed in a dispersion medium with a certain viscosity, and the suspension exists stably for a certain period. Spraying is the use of the negative pressure generated when compressed air passes through the nozzle to suck out the suspension. The suspension is atomized with the rapid diffusion of compressed air, and the atomized droplets are deposited on the surface of the support to form a uniform liquid layer. Therein, the dispersion medium enters the pores of the support under the action of capillary force, while the dispersed phase particles accumulate on the surface of the support to form a membrane. After sintering under a set sintering program, a separation layer is obtained. The spraying process requires low equipment and operating costs, which is simple to operate, is suitable for continuous production, is capable of laminating membranes on the surface of the support with complex shapes, and the prepared membrane layer has good densification and high bonding strength. The pore structure of the membrane layer is mainly affected by the particle size and its distribution, composition of the dispersed particles in the suspension, thickness of the membrane layer, structure of the support sintering procedure and other factors. Among them, the dispersed phase particles with suitable particle size distribution are the basis of the spraying process, and obtaining a suspension with good dispersion stability is one of the key conditions and prerequisites for the preparation of separation membranes with small average pore sizes, narrow pore size distribution, and no any defects.

Figure 5.20a shows the effect of spraying time on the thickness of silicon carbide membrane. The membrane thickness increases rapidly when the spraying time is less than 10 s, and then it tends to level off. Given that the membrane thickness greatly affects gas permeability, a short spraying time may result in membrane defects. Therefore, it is generally accepted that a spraying time between 8 and 10 s is more suitable. Figure 5.20b is a microscopic structure diagram of the membrane layer when sprayed for 8 s, with a membrane thickness of about 90 μm [57]. Wei et al. [58] successfully prepared a tubular silicon carbide separation membrane layer by rotating spraying, with a membrane thickness of 250 μm and a gas permance of 45.2 $m^3/(m^2$ h kPa). Figure 5.21 is a photo of the appearance of the tubular silicon carbide ceramic membrane and an electron microscope photo reflecting its microscopic structure.

5.1.5 Application of Silicon Carbide Ceramic Membranes

The shape of the hanging candle silicon carbide ceramic membrane currently used in industrial processes is shown in Fig. 5.22 [10], where the hanging candle filter element is closed at one end and open at the other with a flange configuration.

Developed countries such as the United States, Japan, and Germany have initially implemented silicon carbide porous ceramic dust removal technology in the field

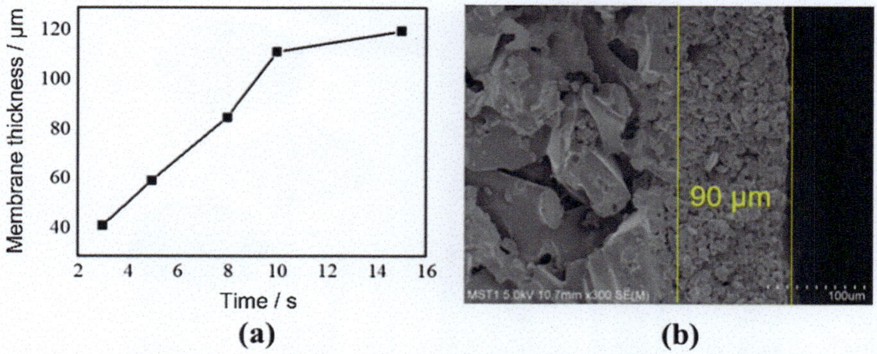

Fig. 5.20 **a** The effect of spraying time on membrane thickness and **b** the microscopic structure of the membrane layer after 8 s of spraying [57]

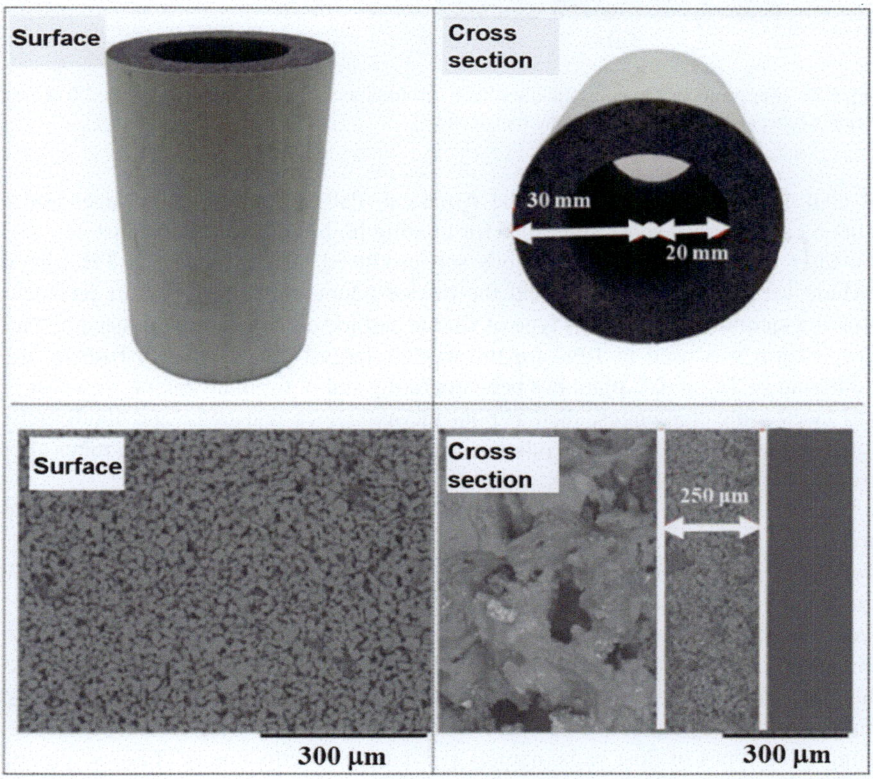

Fig. 5.21 Macroscopic photographs and microelectron microscopy of tubular silicon carbide ceramic membrane [58]

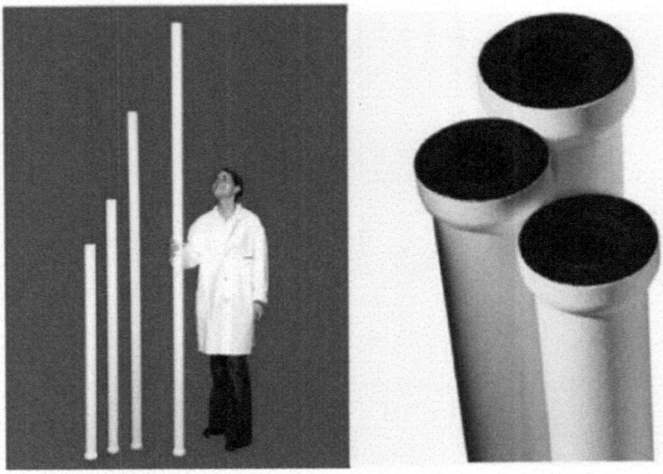

(a) (b)

Fig. 5.22 Appearance of hanging candle silicon carbide ceramic membrane [10]: **a** 1.5–2.0 m long ceramic tube, and **b** tubular membrane flange end

of coal chemical industry [59]. The American Pall Corporation has developed a silicon carbide porous ceramic filter for treating high-temperature exhaust gas, and the filter tube used is the company's self-produced DIA-SCHUMALITH silicon carbide ceramic tubular filter, and the device adopts advanced Venturi jet back-blowing technology [60]. This type of silicon carbide ceramic tube produced by Pall Corporation was used in 1994 for the world's largest IGCC plant in Bijnem, the Netherlands. The IGCC plant has been operating stably for many years. However, it still faces some challenges, such as poor thermal shock resistance and alkali metal corrosion resistance of ceramic tubes, as well as neck breakage of ceramic tubes after prolonged operation [61].

Pastila et al. [62] used commercial SiC ceramic high-temperature filter tubes to investigate the effect of prolonged filtration and dedusting on the microstructure as well as the strength of the ceramics under the harsh conditions of high-temperature water vapor and gaseous sodium at 870 °C. The results indicate that the microscopic structure of the SiC filter tubes was altered, resulting in decreased strength and compromised filtration performance. Further investigation is required to assess their filtration efficiency and oxidation-resistant properties at high temperatures. Lupión et al. [63] compared the performance of baghouse with ceramic filtration, using three types of dust bags, namely polytetrafluoroethylene (PTFE), polyimide (P84) and glass fiber (FB700). The ceramic tubular membrane used was the DIA-SCHUMALITH type filter tube produced by Pall Corporation. After considering the filtration effect and economic benefits, it was believed that PTFE filtration material was more competitive below 235 °C, and silicon carbide porous ceramic membrane filter has a competitive advantage in high temperature (>300 °C) situations. However, the use of silicon carbide porous ceramic filtration results in higher pressure drop, lower gas handling capacity, and increased energy consumption.

In recent years, rapid progress in the preparation and application research of silicon carbide porous ceramics has been made in China and mainly in universities and research institutes, such as Nanjing Tech University, University of Science and Technology of China, Hainan University, Institute of Metals of the Chinese Academy of Sciences, Jiangsu Ceramic Research Institute and Shandong Industrial Ceramic Design Institute, etc. These units have also each carried out research on the design and preparation of new silicon carbide porous ceramics and composite materials, developing from early alumina, zirconia, titanium oxide materials to silicon carbide composite gradient porous materials, and carrying out applied research in high-temperature gas filtration, which lays a good foundation for promoting industrial applications [64]. Shandong Industrial Ceramic Design Institute has successfully developed a series of high-temperature ceramic membrane filtration materials and supporting filtration and dedusting technology of silicon carbide and mullite-silicon carbide. Cao et al. [65] used porous silicon carbide ceramic to filter the fly ash emitted from the Shell gasification device, and the results showed that in the coal gas cleaning process at temperatures above 300 °C, fine particles with a particle size of 0.3 μm can be effectively removed, and the fly ash content in the purified coal gas is reduced to 1–2 mg/m^3.

In recent years, Nanjing Tech University has used its own research and development advantages to develop a high-temperature dust-removing silicon carbide membrane with superior performance after many years of technical research. This technology uses in-situ reaction sintering to prepare a double-layer structure membrane, with the inner layer being a support with a larger average pore size to ensure the strength of the filter tube, and a silicon carbide membrane layer with a smaller average pore size is coated on the outer surface of the support to achieve surface filtration (Fig. 5.23). The pore size of the support body is relatively large, the porosity is generally greater than 40%, and the average pore size is 40–60 μm, and the average pore size of the separation membrane layer is adjustable in the range of 1–10 μm, so that the appropriate membrane pore size can be selected according to the actual dust particle size, and the detailed performance parameters are shown in Table 5.2 [66].

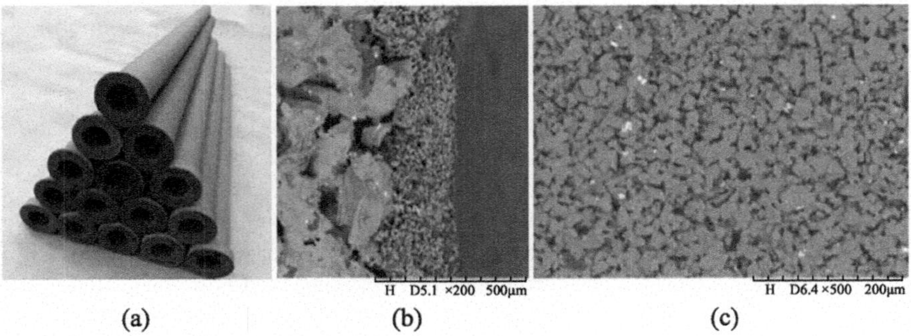

Fig. 5.23 **a** Digital image of silicon carbide ceramic tubular membrane, **b** cross-section scanning electron microscope image **c** membrane surface scanning electron microscope image

Table 5.2 Performance parameters of silicon carbide ceramic membrane elements [66]

Properties	Value	Properties	Value
Filter material	Silicon Carbide	Bending strength/MPa	15–20
Membrane pore size/μm	1–10	Filtration accuracy/μm	0.3
Porosity/%	> 40	Maximum tolerance pressure/MPa	10
Acid resistance (Mass retention)/%	99.9	Maximum working temperature/°C	1000
Alkali resistance (Mass retention)/%	99.8		

5.2 Other Membrane Materials

5.2.1 Ceramic Fiber Membranes

5.2.1.1 Overview

Ceramic fiber membrane refers to the membrane layer or support material that performs separation and filtration. It is a type of membrane material made by configuring casting solution, multiple coating, drying, and high-temperature sintering of inorganic ceramic short fibers. Compared with particles, ceramic fibers significantly improve the heat shock resistance and permeability of the membrane layer, and to some extent solve the problem of high filtration resistance faced by traditional ceramic membrane materials in gas–solid separation applications [67]. The structure of the ceramic fiber membrane is quite diverse, with some directly using ceramic fibers to make filter materials similar to cloth bags, mainly made of silicon carbide, alumina, mullite, and aluminum silicate fibers. For example, Buell Company, Westinghouse Company, and American Electric Power Research Institute in the United

States use ceramic fibers with a diameter of 10–12 μm (composed of 62% Al_2O_3, 24% SiO_2, and 14% B_2O_3) to weave ceramic cloth bags for filtration purposes. These bags were tested under filtration conditions of 816 °C and 0.98 MPa at a filtration wind speed of 2 m/min. The results showed an impressive dust removal efficiency of up to 99.7%, with the highest pressure drop reaching 1489 Pa after stabilization. Pulse air is used for back-blowing during the dust removal process [68]. The American company ACUREX uses ceramic fibers with a diameter of 3 μm to weave into blanket-style filter materials, which are then covered with a layer of stainless steel wire mesh on both sides. In tests conducted under conditions of 800 °C and 0.98 MPa, the filtration speed reached 6 m/min, and the dust removal efficiency could reach 99.9%. Dust removal is achieved by pulse gas back-blowing, and after 5000 back-blows at high temperatures, the overall strength of the filter material still meets the requirements [69]. In addition, the early ceramic fiber filter bags developed by the American company 3 M, which contain elements such as aluminum, boron, and silicon, have the advantages of better breathability and lighter weight. The maximum filtration speed can reach 5 m/min, but the membrane is prone to break during the usage process. The bag-shaped ceramic fiber filter unit KE-85 developed by the German company BWF can withstand temperatures up to 850 °C. The rigid ceramic fiber filter bags developed by the British company TENMAT are heat-resistant (up to 1600 °C), non-combustible, and can filter dust particles with a diameter of less than 1 μm. The dust emission concentration can be reduced to below 1 mg/m^3, suitable for industries such as metallurgy, waste incineration, and thermal power plants, with the advantages of low filtration resistance, good dust removal effect, and convenient operation and maintenance.

5.2.1.2 Preparation Method

Ceramic fiber membranes are usually prepared by vacuum suction molding process. First, ceramic fibers and silica sol are mixed in a high-speed stirrer, and the fibers are dispersed and sheared at high speed. During the stirring process, a sodium carboxymethyl cellulose solution with a mass fraction of 0.5% (based on the total mass of the slurry) is continuously added to the mixture. After stirring, the slurry is poured into a mold with a filter net at the bottom, and then the excess liquid in the slurry is removed by vacuum suction, and the fibers and a small amount of liquid aggregate in the mold to form a green body. The green body is dried for 24 h and finally sintered at high temperatures. Zhang et al. [70] used mullite nanofibers instead of mullite microfibers to prepare mullite nanofiber porous ceramics by gel casting. The density of the prepared mullite nano-porous ceramics (0.202 g/cm^3) is only 3/4 of the density of mullite micro-porous ceramics (0.266 g/cm^3), and the flexural strength of mullite nano-porous ceramics (0.837 MPa) is also higher than that of mullite micro-porous ceramics (0.515 MPa); the obtained material has a double pore size distribution, respectively 1–5 μm and 5–15 μm. The preparation process is shown in Fig. 5.24.

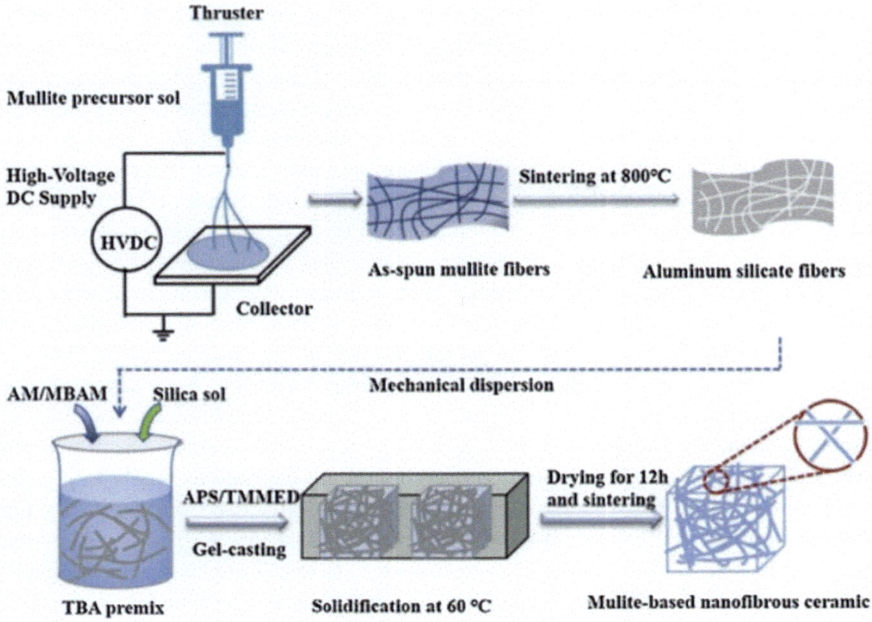

Fig. 5.24 The preparation flow chart of mullite fiber membrane [70]

The various shapes of filter materials made from the above-mentioned ceramic fibers have relatively low filtration resistance and higher filtration air speed than silicon carbide ceramic filter materials. However, the strength of the filter tubes or fiber bags made from ceramic fibers is not high. When encountering conditions with large temperature fluctuations, containing HF gas components and high water vapor content during use, the tubes are prone to breakage, which is a common problem at present. In recent years, some scholars have proposed using porous ceramics as carriers and coating the surface with a ceramic fiber membrane to filter high-temperature dust-containing gases. The filter material prepared in this way has high mechanical strength, and the surface membrane layer can ensure that the material has high filtration accuracy. The American 3M company uses silicon carbide as the support material and loads the surface with an alumina fiber membrane layer [71]. DuPont in the United States has developed a filter material with alumina as the fiber membrane layer and silicon carbide porous ceramics as the support body. B&W company has developed alumina fiber-alumina porous ceramic filter material [72]. The developed filter materials have a fiber membrane layer porosity of basically more than 75%, the mechanical strength mainly depends on the porous support body, the porosity of the porous support body is generally 30–40%, and the flexural strength is generally more than 15 MPa.

5.2.1.3 Application Context

Currently, the most successful application of high-temperature ceramic fiber membranes is the Cerafill 2H10 ceramic fiber composite membrane produced by the German Schumacher company using silicon carbide and alumina fibers as the main raw materials. The membrane layer has a porosity of more than 90% and can be used for a long time at 950 °C. The air permeability resistance is generally 1/3 to 1/2 of that of ceramic membranes [73]. In addition, as early as the 1970s, Asahi Glass Co., Ltd. of Japan developed a cordierite high-temperature ceramic fiber membrane filter element and promoted its application in the ironmaking furnace and other smelting furnace industries. The tubular porous ceramic membrane (LOTEC-V) produced by Asahi Glass Co., Ltd. of Japan has an average pore size of 40–65 μm, no gradient in the height direction, and the support body is directly sintered from β-cordierite powder with a small amount of sintering aid. Its high-temperature resistance is excellent and can handle high-temperature dust smoke at 1000 °C, but because the porosity is only 16–22%, the filtration resistance is large, and the flexural strength is only 15 to 18 MPa, the application range is small.

In addition to the traditional terminal surface filtration form, Westinghouse Company in the United States developed an alumina-mullite fiber composite ceramic membrane in 1988 and developed a cross-flow sheet ceramic. Hitachi Metals Co., Ltd. and Atsuhiro Noguchi and others developed a cylindrical ceramic honeycomb structure body composed of cordierite with multiple flow paths (as shown in Fig. 5.25). The peripheral walls of the ceramic honeycomb structure are made of a porous material. The adjacent peripheral walls are traversed by the gas-containing particles, which allows for filtration of the particulate matter as it passes through the honeycomb structure. At present, this structure's particle-trapping ceramic honeycomb filter is mainly used on diesel particulate filters (DPF) [74]. The structural advantage of the honeycomb filter element lies in the parallel arrangement of its gas cleaning channel and dust-laden gas channel, making it easy to assemble and compact in structure. However, compared to the tubular ceramic membrane, the cleaning process of the honeycomb filter element is more complex. When applied in high-dust systems, this type of filter structure is prone to dust clogging, leading to a continuous increase in pressure drop.

5.2.1.4 Current Status of Development in China

In recent years, Tsinghua University, Nanjing Tech University, and Shandong Industrial Ceramic Design Institute in China have all carried out research using ceramic fibers as transition layers or membrane layers, and have made a series of progress. For example, Li and coworkers [75, 76] in Tsinghua University used silicon carbide porous ceramics as the support body, coated a layer of alumina fiber transition layer on the surface, and then sprayed a layer of finer particle accumulation membrane

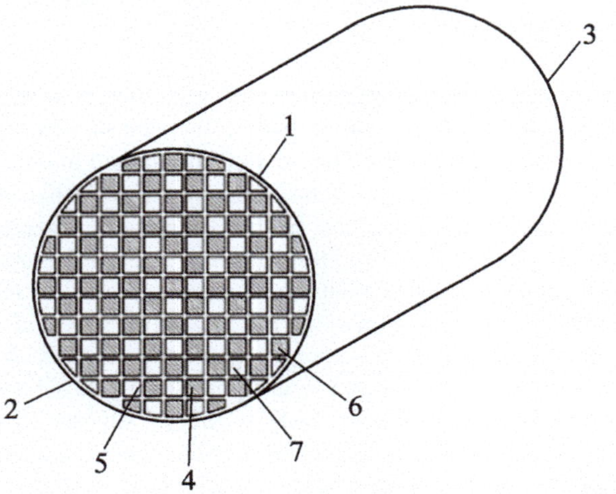

Fig. 5.25 Honeycomb structure ceramic filter element [74]: 1—Outer peripheral wall, 2—Inflow side end face, 3—Outflow side end face, 4—Mesh sealing part, 5—Inner peripheral wall, 6—Exhaust channel, 7—Intake channel

layer on the fiber transition layer. The fiber layer effectively prevents particle infiltration, and reduces membrane thickness and filtration pressure drop, as shown in Fig. 5.26.

Liu and co-workers from Nanjing Tech University [77] used the different lengths and diameters of alumina ceramic fibers and wollastonite fibers to make fiber porous ceramic samples with a composite structure by a two-step forming method and analyzed and discussed various factors affecting material performance. The microstructure of the ceramic fiber tube was analyzed and measured by scanning electron microscopy (SEM) and metallographic microscopy. The research found that the porosity of the prepared ceramic fiber tube reached up to 78%, and the resistance was only 228 Pa even at a flow rate of 4 m/min, but the flexural strength of

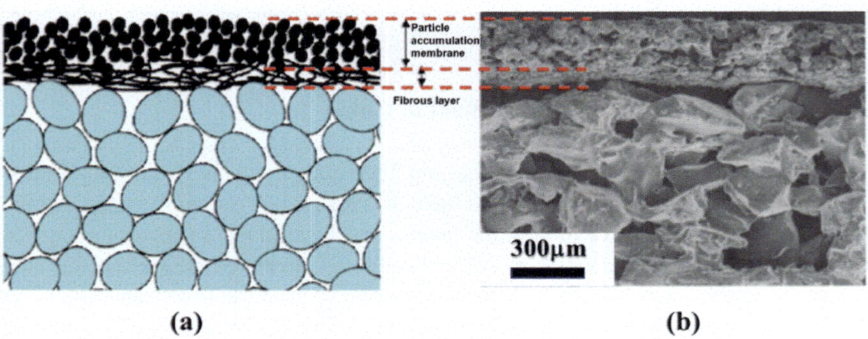

Fig. 5.26 Ceramic fiber composite membrane structure [75, 76]

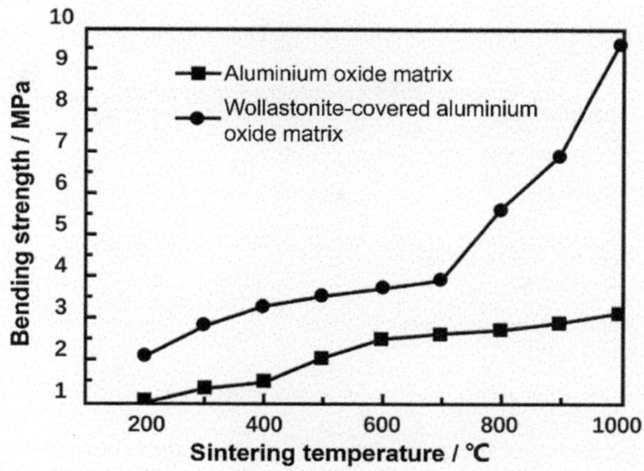

Fig. 5.27 Change in flexural strength with sintering temperature [77]

the material was low, and even when the sintering temperature reached 1000 °C, the flexural strength was only 9.7 MPa. The change in flexural strength with sintering temperature is shown in Fig. 5.27.

When considering membrane layer anti-infiltration, Xing and co-workers from Nanjing Tech University [78] utilized a surface modification method to alleviate the infiltration phenomenon by decorating the surface of a porous support material with silicon carbide whiskers. Then, a layer of 50–150 μm membrane coating was sprayed onto the surface. After high-temperature sintering, the intermediate transition layer is sacrificed through a solid-phase reaction, leaving only two layers. Furthermore, the membrane layer particles do not infiltrate, further reducing the filtration resistance of the membrane material (< 800 Pa).

Xue and co-workers [79, 80] in Shandong Industrial Ceramic Design Institute prepared continuous fiber-reinforced ceramic fiber filter materials using cordierite, continuous high-silica oxygen fiber, and mullite fiber as the main raw materials and a silicon-boron system high-temperature binder. The porosity of the fiber filter material produced at a firing temperature of 850 °C is not less than 70%, the mechanical strength is not less than 5 MPa, the filtration resistance at a wind speed of 1 m/min is not more than 200 Pa, and the fiber filter material has good fracture toughness. Table 5.3 shows the performance parameters of the cordierite ceramic fiber membrane element.

In recent years, Miao and others [81] in China University of Petroleum successfully prepared alumina silicate fiber-based ceramic filter elements using different drying methods and calcination temperatures. Studies show that there are differences in the microstructure of filter elements prepared by three drying methods, and the microstructure significantly affects the filtration strength and initial pressure drop of the filter elements. Compared with air drying and vacuum drying, microwave drying inhibits the migration of the binder. The filter elements prepared by microwave drying have evenly distributed binders, and the pore structure in the middle part is basically

Table 5.3 Cordierite ceramic fiber membrane element performance parameters

Properties	Value	Properties	Value
Filter material	Cordierite/Ceramic Fiber	Flexural strength/MPa	4–6
Membrane pore size/μm	20, 30	Filtration accuracy/μm	1.0
Porosity/%	38–42	Maximum working pressure/MPa	0.1
Acid resistance (Mass retention)/%	99.8	Maximum working temperature/°C	650
Alkali resistance (Mass retention)/%	99.5		

the same as the pore structure in the edge part (Fig. 5.28), thereby improving the uniformity of the structure and increasing the mechanical strength.

In practical applications, industrial high-temperature exhaust gases mostly contain corrosive gases, such as SO_2 and alkali metal vapors, etc. The membrane on the filter material surface and support body will be corroded by those corrosive gases, leading to a decrease in filtration efficiency and filtration accuracy, and a shortening of the material's service life. This greatly limits the application of oxide ceramics (fibers) in the field of high-temperature gas cleaning. Therefore, most of the current oxide ceramic membrane materials are used in the field of liquid–solid separation.

As the second generation of advanced high-temperature gas filtration materials, ceramic fiber membrane filter materials have shown great superiority and application prospects [82]. However, due to the limitations of the types of high-performance continuous fiber materials and high manufacturing costs, their application in the field of hot gas cleaning has not reached the level of popularity people expect. Furthermore,

Fig. 5.28 Morphology of filter elements prepared by different drying methods [81]: F1—Air drying, F2—Vacuum drying, F3—Microwave drying, a—Edge part, b—Middle part

the uniformity control of filtration material performance is still a technical challenge that needs to be overcome, mainly due to the influence of the dispersion of high-concentration short fiber slurries. In addition, the differences in the internal and external structure and performance of the material caused by vacuum filtration during the casting process of the fiber pulp are also technical issues that cannot be ignored. If it is to be widely promoted, further research needs to be done in areas such as the development of low-cost continuous fibers and the optimization of composite membrane preparation processes.

5.2.2 *Powder Sintered Metal Membranes*

Powder sintered metal membrane materials are rigid structured porous materials made from metal or alloy powders as raw materials, through molding and high-temperature sintering. They are characterized by a large number of interconnected or semi-connected pores inside. The pore structure is formed by regular and irregular powder particle stacking, and the size and distribution of the pores and the porosity depend on the powder particle size composition and preparation process. Common sintered metal powder porous materials are made of materials such as bronze, stainless steel, iron, nickel, titanium, tungsten, molybdenum, and refractory metal compounds. For example, metal membranes made from 310 stainless steel, Inconel600 (Fe–Cr–Ni), Monel (Ni–Cu–Fe–Mn), and Hastelloy (Ni–Cr) have advantages such as large filtration area, high filtration accuracy, low-pressure loss, good sealing performance, and fewer consumables.

The preparation of powder-sintered metal membrane carriers generally adopts traditional processes such as mold pressing and sintering, isostatic pressing, loose sintering, powder rolling, powder plastic extrusion, and slurry casting. Depending on the material, the sintering temperature range varies greatly (750–2000 °C). The preparation of porous metal membrane materials on the surface usually adopts a solid particle sintering method, phase separation method, thin membrane deposition method, etc. The use of the solid particle sintering method is more common, and its process includes suspension preparation, membrane coating, drying, and sintering. Factors affecting the pore size and pore size distribution during preparation include the metal powder system, powder particle shape and size, particle size distribution, binder viscosity, additives, and sintering temperature. The powders used in the suspension preparation process include titanium powder, stainless steel powder, and nano titanium dioxide powder. For nano titanium dioxide powder, water or alcohol can be used for uniform dispersion; for titanium powder and stainless steel powder, a high-viscosity dispersant should be selected, and polyvinyl alcohol colloid can be used as a dispersant. The membrane coating process generally includes brushing, impregnation, filtration, spraying, centrifugation, etc., and finally drying and sintering to obtain metal membrane products. The powder-sintered metal membrane materials used in the field of high-temperature flue gas cleaning are of pore size gradient composite structure, mainly composed of a substrate and a membrane layer. The substrate is a stainless steel porous metal material, and the membrane layer generally chooses intermetallic compounds that are easy to composite with the substrate, or porous metal membranes that composite ceramics on the porous metal substrate. Table 5.4 gives the performance parameters of typical high-temperature filter metal membranes. Compared with ceramic membranes, it can be found that the mechanical strength of metal materials is significantly higher than that of ceramic membrane materials.

In recent years, many experts have been dedicated to developing porous metal hollow fiber membranes to improve the filling density and separation efficiency of porous metal membranes. This not only takes advantage of the high strength and high-temperature resistance of porous metal membranes but also combines the large specific surface area and high utilization rate of hollow fiber membranes [84]. Therefore, porous metal hollow fiber membranes are currently a hot topic in metal membrane research. Rui and others [85] investigated the impact of the gas environment (air, CO_2, N_2, He, H_2) during the sintering process of stainless steel hollow fiber membranes on the microstructure and performance of the membrane. The results found that the air and CO_2 environments would reduce the mechanical performance of the membrane, N_2 and He environments would reduce the corrosion resistance of the membrane, and only the membrane obtained in the H_2 environment has excellent mechanical strength, toughness, and stability. Chi and others [86] A graphene-coated nickel hollow fiber (G–Ni–HF) membrane was prepared using phase transformation and chemical vapor deposition (CVD) processes. Studies have found that nickel hollow fiber membranes without graphene protection dissolve after several hours in nitric acid, while G–Ni–HF membranes with graphene protection remain intact. Figure 5.29 is a cross-sectional and fiber structure diagram of the hollow fiber metal Ni membrane (a protective layer can be seen on its surface from Fig. 5.29d, e).

In terms of high-temperature gas–solid separation membranes, Northwest Nonferrous Metal Research Institute and Antai Technology Co., Ltd. are the earliest in China to start researching sintered metal microporous membranes. These two institutions have formed commercialized porous metal membrane products with pore sizes of 0.05–0.10 μm. However, powder-sintered metal membranes are still in the early stages in China, and foreign companies GKN and Hyflux have completed the

Table 5.4 Basic performance parameters of typical powder sintered metal membrane materials [83]

Filtration material	Average pore size/μm	Permeability/ [L/ (min cm^2 Pa)]	Porosity/ %	Strength/ MPa	Elongation	Filtration efficiency/ %
AT&M-FeAl	10.1	2.45×10^{-4}	45.3	120	0.5–2	> 99.9
AT&M-310S	22.5	3.55×10^{-3}	35	160	> 30	> 99.5
FeAl	10	2.25×10^{-4}	30	122	– 1	> 99.9
Hastelloy	10	2.0×10^{-4}	23	240	> 8	> 99.9
Schumacher-T10(ceramic)	10	1.25×10^{-4}	38	13	< 1	> 99.9

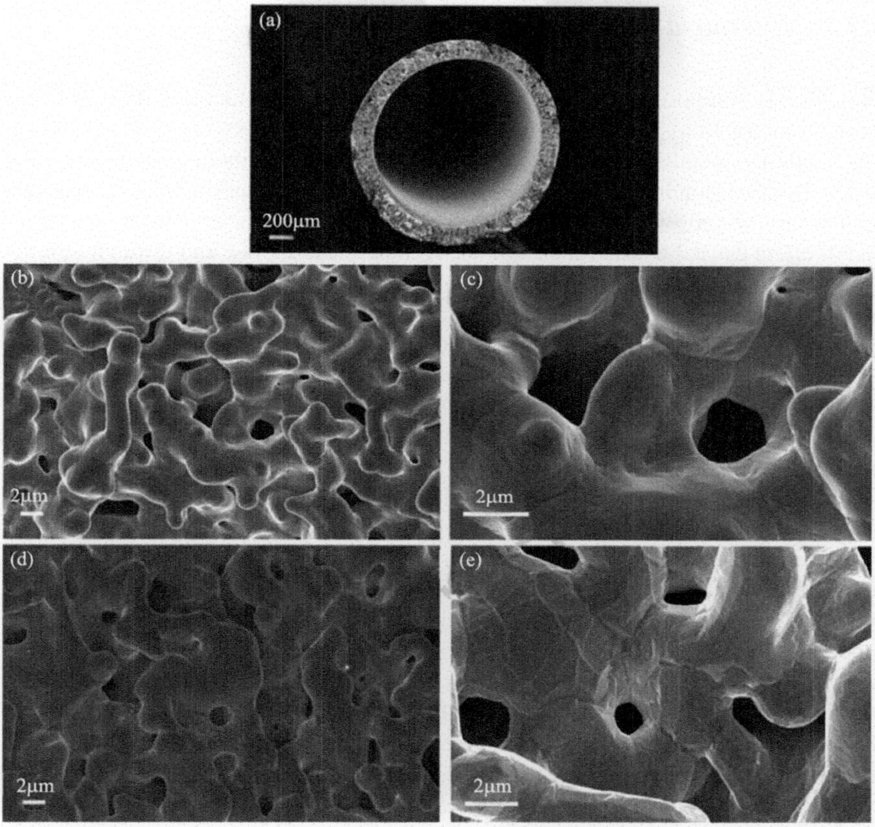

Fig. 5.29 **a** SEM photo of the cross-section of the hollow Ni metal fiber membrane, **b**, **c** surface of the original Ni-HF membrane, and **d**, **e** surface of the G-Ni-HF membrane [86]

layout of powder-sintered metal membranes in China. In the application process, there are still some problems with powder-sintered metal membrane materials, such as a decrease in strength with increasing temperature above 600 °C. This is mainly because the metal membrane is prone to oxidation, causing changes in the pore size of the membrane and a decrease in filtration accuracy, and the strength and corrosion resistance are lower than ceramic materials. In addition, research has found that the resistance of metal porous materials increases rapidly with increasing flow rate and temperature during filtration. When the temperature reaches 600 °C, the resistance reaches 3000 Pa. These problems need to be further studied by researchers [87].

5.2.3 Intermetallic Compound Membranes

Compounds composed of metal elements and metalloid elements in integer ratios (stoichiometric ratios) are called intermetallic compounds. They are bonded by covalent bonds and metallic bonds and retain the characteristics of metals, such as metallic luster, thermal conductivity, and electrical conductivity. Intermetallic compound membranes are porous membrane materials and components sintered by the Kirkendall effect caused by interdiffusion between metal elements, and through atomic-scale biased diffusion reaction synthesis technology, precise control of membrane pore size at the micron and submicron level can be achieved, realizing the entire process of environmentally friendly production and precise control of intermetallic compound membrane preparation technology. Intermetallic compound membranes have good high-temperature chemical stability, high-temperature thermal stability, high-temperature mechanical and physical properties, and high-temperature filtration performance.

Intermetallic compound membranes have an asymmetric structure, similar to ceramic membrane structures, divided into a support layer and a separation layer, with pore sizes ranging from 0.5 to 50 μm, a diameter of 60 mm, and a maximum length of 2500 mm. Because the metal membrane is thin, with a wall thickness of only 1.5–3 mm, the intermetallic compound membrane has a large flux and low resistance (as low as 1000 Pa), can work at 450–650 °C, achieve high-precision filtration, and the clean gas dust content is less than 5 mg/m^3. Currently, Chengdu Yitai Technology Co., Ltd. in China has successfully developed industrial products based on the research team of Central South University and has been successfully applied in the fields of coal chemical industry, mine thermal furnace, ferroalloy, phosphorus chemical industry, and other high-temperature flue gas treatment.

Liu et al. [88] prepared a Ti–Si intermetallic compound porous membrane for the first time to replace the chemical vapor deposition or solid-phase sintering process, using titanium powder and quartz tube (SiO_2) as raw materials and using an efficient and low-cost in-situ reaction method. The preparation process involves loading Ti powder into a quartz tube, then forming it by cold isostatic pressing, and finally demolding and sintering. The prepared membrane exhibits a pore size of less than 0.5 μm, a thickness of 2 μm, and a uniform pore size distribution. The preparation process is shown in Fig. 5.30, and the microstructure of the product is shown in Fig. 5.31. The Ti-Si membrane prepared by this method is low-cost, simple, and easy to scale up, and it has broad application prospects in the fields of filters, heat exchangers, and ion batteries in the future.

In terms of the application of intermetallic compounds, the Fe–Al metal membrane product produced by Chengdu Yitai Technology Company in China has been successfully applied in the treatment of yellow phosphorus tail gas. During the production of yellow phosphorus, coke is used to reduce phosphate ore in an electric furnace, and the gaseous yellow phosphorus produced is discharged from the electric furnace together with phosphorus furnace gas. The furnace gas directly enters the scrubbing tower, and the yellow phosphorus in the furnace gas is layered and purified in hot

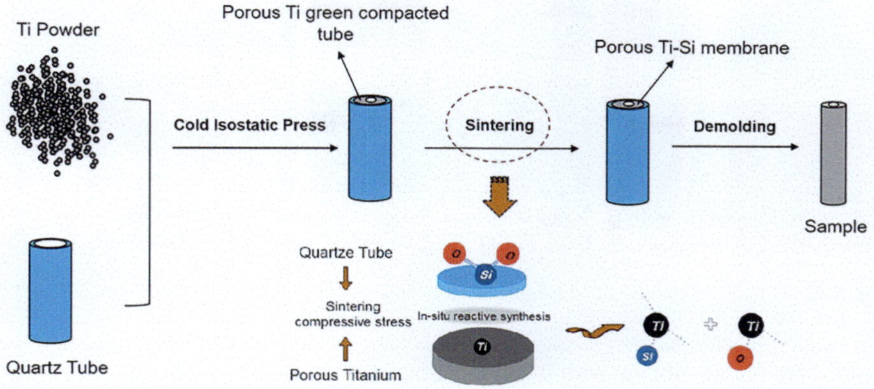

Fig. 5.30 Schematic of the in-situ sintering process for preparing Ti–Si intermetallic compound membrane [88]

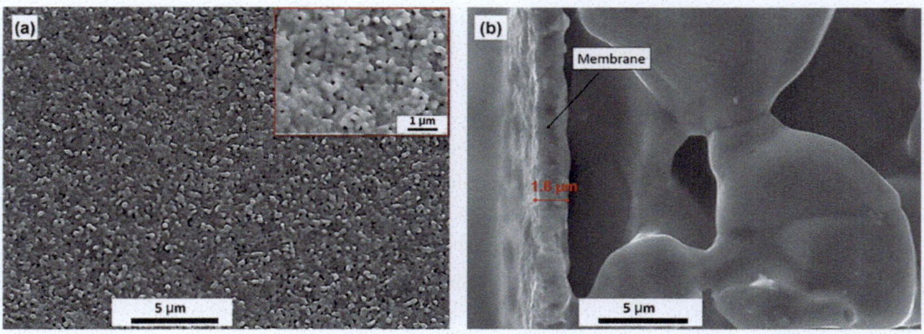

Fig. 5.31 SEM image of Ti-Si intermetallic compound membrane [88]. **a** Surface; and **b** cross-section

water, and dust particles settle. Because the furnace gas contains a large amount of dust, a large amount of phosphorus sludge is produced, consuming water resources. Yitai Technology uses Fe-Al metal membrane to purify yellow phosphorus furnace gas, which improves the purity and recovery rate of yellow phosphorus, avoids the production of phosphorus sludge, and solves the bottleneck problems of long process and high energy consumption in this industry [89].

5.2.4 Ceramic/Stainless Steel Composite Membranes

The ceramic/stainless steel composite membrane is formed by loading a ceramic membrane on the surface of a stainless steel substrate, combining the high precision

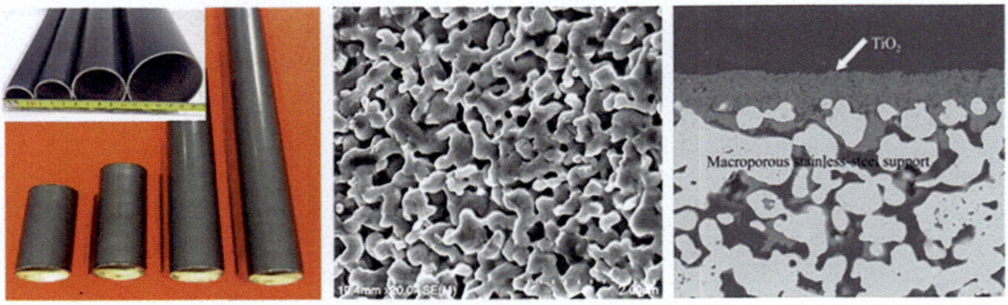

Fig. 5.32 Digital photos and SEM images of TiO_2/Stainless steel composite membrane [90, 91]

of the ceramic membrane with the high strength of the metal membrane. The filtration precision can reach 50 nm. The membrane material can be TiO_2, Al_2O_3, ZrO_2, YSZ, etc., with a thickness of 100–200 μm [90, 91]. The development of ceramic/stainless steel composite membranes is difficult, the fundamental reason being that the properties of metal and ceramic are very different, with weak affinity and a large difference in thermal expansion coefficients. Compared with the loaded stainless steel membrane, the ceramic/stainless steel composite membrane also needs to solve the problem of ceramic layer strength and sintering stress. Therefore, when preparing a ceramic/stainless steel composite membrane, a layer of stainless steel micro-powder is deposited on the stainless steel substrate as a transition layer before the ceramic coating process is implemented, and the ceramic coating is directly carried out without sintering the transition layer. During the sintering process, the stainless steel micro-powder can effectively buffer the sintering shrinkage force of the ceramic layer like a 'roller skate'. In addition, the stainless steel transition layer has a larger contact area with the ceramic layer, which also improves the adhesion of the ceramic, thus solving the problem of easy peeling and poor strength of the ceramic layer. Figure 5.32 is the digital photos of TiO_2/stainless steel composite membrane products and their microstructures.

Zhu et al. [92] prepared Ni-$BaZr_{0.1}Ce_{0.7}Y_{0.2}O_{3-\delta}$ asymmetric metal-ceramic membrane (Ni-BZCY ACM) through cold isostatic pressing and high-temperature sintering. The sintering process was at 900 °C, pre-sintered for 30 min, then sintered for 5 h under an H_2/Ar atmosphere at 1400 °C, finally obtaining the Ni-BZCY asymmetric metal/ceramic composite membrane. Figure 5.33 shows the SEM image of the cross-section structure of the Ni-BZCY asymmetric metal/ceramic composite membrane with a membrane thickness of 39.6 μm and the change in H_2 permeation flux with temperature. Under humid conditions, as the temperature rises from 700 °C to 900 °C, the H_2 permeation flux increases from 6.52×10^{-8} to 1.48×10^{-7} mol/(cm^2 s), while under dry conditions, it increases from 2.25×10^{-8} to 5.93×10^{-8} mol/(cm^2 s). The permeation flux of H_2 under humid conditions is 2.5–2.9 times that under dry conditions.

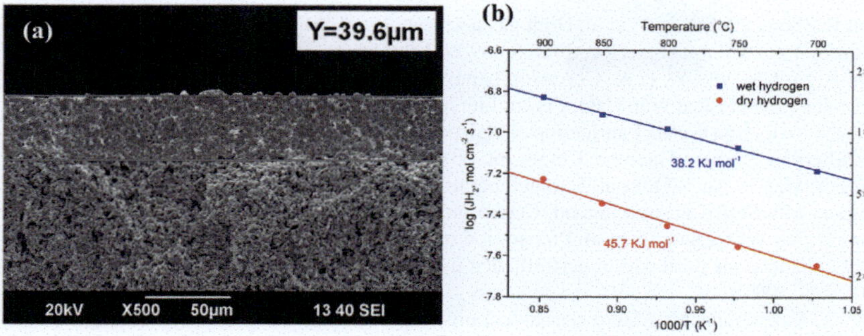

Fig. 5.33 **a** Cross-sectional SEM photo of Ni-BZCY asymmetric membrane and **b** hydrogen permeation flux diagram [92]

References

1. Dachamir H, Marco DL, Michaela W, et al. Silicon carbide filters and porous membranes: a review of processing, properties, performance and application. J Membr Sci. 2020;610:118193–207.
2. Lupion M, Rodriguez GM, Alonso FB, et al. Investigation into the influence on dust cake porosity in hot gas filtration. Powder Technol. 2014;264:592–8.
3. Lehtovaara A. Ceramic filter behavior in gasification. Bio-resource Technol. 1993;46:113–8.
4. Startin A, Elliott G. Treating industrial hot gases with ceramic filters. Filtr Sep. 2001;11:39–40.
5. Qiu JF. Design of high-temperature ceramic filters and transient flow model analysis for reverse pulse cleaning. Xi'an: Xi'an Jiaotong University; 2004.
6. Zhu JJ. Research on the high-temperature stability of ceramic membrane materials. Nanjing: Nanjing Tech University; 2013.
7. Li SP. Special ceramic technology. Wuhan: Wuhan University of Technology Press; 1990.
8. Fan YQ, Qi H, Xu NP. Advance in preparation techniques of porous ceramic membranes. J Chem Ind. 2013;64(1):107–15.
9. Li CR, Xie ZQ, Kang GX, et al. Research and application progress of SiC ceramics: a review. Bull Silicates. 2020;39(5):1353–70.
10. Steffen H. Hot gas filtration: a review. Fuel. 2013;104:83–94.
11. Kim YS, Min KS, Shim JI, et al. Formation of porous SiC ceramics via recrystallization. J Eur Ceram Soc. 2012;32:3611–5.
12. Li YS, Yan YJ, Yin J, et al. Processing, microstructures and mechanical properties of aqueous gelcasted and solid-state-sintered porous SiC ceramics. J Eur Ceram Soc. 2014;34:3469–78.
13. Jiang B, Wang YJ, Li ZM. Research progress on preparation methods of porous SiC Ceramics. China Ceram. 2012;11:1–3.
14. She JH, Deng ZY, Doni JD, et al. Oxidation bonding of porous silicon carbide ceramics. J Mater Sci. 2002;37:3615–22.
15. Ding SQ, Zeng YP, Jiang DL. Thermal shock resistance of in situ reaction bonded porous silicon carbide ceramics. Mater Sci Eng A. 2006;425:326–9.
16. Bai CY, Deng XY, Li JB, et al. Fabrication and properties of cordierite-mullite bonded porous SiC ceramics. Ceram Int. 2014;40:6225–31.
17. Liu Q, Ye F, Hou ZP, et al. A new approach for the net-shape fabrication of porous Si_3N_4 bonded SiC ceramics with high strength. J Eur Ceram Soc. 2013;33:2421–7.
18. Han F, Zhong ZX, Zhang F, et al. Preparation and characterization of SiC whisker-reinforced SiC porous ceramics for hot gas filtration. Ind Eng Chem Res. 2015;54:226–32.

19. Han F, Zhong ZX, Yang Y, et al. High gas permeability of SiC porous ceramics reinforced by mullite fibers. J Eur Ceram Soc. 2016;36:3909–17.
20. Han F, Xu CN, Wei W, et al. Corrosion behaviors of porous reaction-bonded silicon carbide ceramics incorporated with CaO. Ceram Int. 2018;44(11):12225–32.
21. Xu CN, Xu C, Han F, et al. Fabrication of high performance macroporous tubular silicon carbide gas filters by extrusion method. Ceram Int. 2018;44(15):17792–9.
22. Yang Y, Han F, Xu WQ, et al. Low-temperature sintering of porous silicon carbide ceramic support with SDBS as sintering aid. Ceram Int. 2017;43(3):3377–83.
23. Yang Y, Xu WQ, Zhang F, et al. Preparation of highly stable porous SiC membrane supports with enhanced air purification performance by recycling NaA zeolite residue. J Membr Sci. 2017;541:500–9.
24. Ohji T, Fukushima M. Macro-porous ceramics: processing and properties. Int Mater Rev. 2012;57(2):115–31.
25. Wang F, Gao ZF, Xu JQ, et al. Preparation of silicon carbide ceramics with multi-level pore structure by gelcasting combined with solid phase sintering. J Inorg Mater. 2016;3:305–10.
26. Yao XM, Liang HQ, Liu XJ, et al. Effect of carbon source and addition ratio on the microstructure and properties of solid phase sintered silicon carbide ceramics. J Inorg Mater. 2013;9:1009–13.
27. Fukushima M, Zhou Y, Yoshizawa Y. Fabrication and microstructural characterization of porous silicon carbide with nano-sized powders. Mater Sci Eng B. 2008;148:211–4.
28. Yao X, Tan S, Huang Z, et al. Effect of recoating slurry viscosity on the properties of reticulated porous silicon carbide ceramics. Ceram Int. 2006;32:137–42.
29. Zhu XW, Jiang DL, Tan SH. Preparation of silicon carbide reticulated porous ceramics. Mater Sci Eng A. 2002;323:232–8.
30. Singh M, Salem JA. Mechanical properties and microstructure of biomorphic silicon carbide ceramics fabricated from wood precursors. J Eur Ceram Soc. 2002;22(14):2709–17.
31. Vogli E, Sieber H, Greil P. Biomorphic SiC-ceramic prepared by Si-vapor phase infiltration of wood. J Eur Ceram Soc. 2002;22:2663–8.
32. Dong GX, Qi H, Xu NP. The effect of activated carbon doping on the structure and performance of porous alumina ceramic supports. J Silicate. 2012;40(6):844–50.
33. Yoon BH, Park CH, Kim HE, et al. In situ synthesis of porous silicon carbide (SiC) ceramics decorated with SiC nanowires. J Am Ceram Soc. 2007;90:3759–66.
34. Fukushima M, Nakata M, Zhou Y, et al. Fabrication and properties of ultrahighly porous silicon carbide by the gelation-freezing method. J Eur Ceram Soc. 2010;30:2889–96.
35. Du JC, Zhang XH, Hong CQ, et al. Microstructure and mechanical properties of ZrB2-SiC porous ceramic by camphene-based freeze casting. Ceram Int. 2013;39:953–7.
36. Pi XM, Xu LF. Research on the preparation of mullite porous ceramics by direct foaming method. Ceramics. 2017;7:50–4.
37. Li QX. Preparation of alumina porous ceramics by direct foaming method. Refractories Lime. 2009;4:52–4.
38. Bao X, Nangrejo MR, Edirisinghe MJ. Synthesis of silicon carbide foams from polymeric precursors and their blends. J Mater Sci. 1999;34(11):2495–505.
39. Fukushima M, Colombo P. Silicon carbide-based foams from direct blowing of polycarbosilane. J Eur Ceram Soc. 2012;32(2):503–10.
40. Zhang CY, Qi P, Li ZJ, et al. Research on the gel casting process of alumina-based ceramics-(I) preparation and rheological properties of the suspension. J Inorg Mater. 1999;4:623–8.
41. Wang X, Xie ZP, Huang Y, et al. Gelcasting of silicon carbide based on gelation of sodium alginate. Ceram Int. 2002;28(8):865–71.
42. Mouazer R, Thijs I, Mullens S, et al. SiC foams produced by gel casting: synthesis and characterization. Adv Eng Mater. 2004;6(5):340–3.
43. Bian Q. Research on the preparation of asymmetric porous silicon carbide ceramic membranes. Nanjing: Nanjing University of Technology; 2013.
44. Xie YZ, Peng CQ, Wang XF, et al. Research progress in tape casting technology. J Nonferrous Metals. 2015;7:1846–57.

45. Guo J, Sun YJ, Zhang HW, et al. Preparation and performance study of AlN benzene-free slurry for tape casting. Electronic Components Mater. 2015;8:69–72.
46. Zhou JM, Wang YD, Wang SX, et al. Preparation of electronic ceramic substrates by tape casting process. Bull Silicate. 2010;5:1114–8.
47. Eklund A. Hot and cold isostatic pressing of ceramic. Ceram Ind. 2016;166:21–4.
48. Liu YF, Liu KQ, Zeng ZY, et al. Microstructure analysis of SiC ceramic support by isostatic processing. Key Eng Mater. 2015;633:45–8.
49. Fukushima M, Zhou Y, Miyazaki H, et al. Microstructural characterization of porous silicon carbide membrane support with and without alumina additive. J Am Ceram Soc. 2006;89(5):1523–9.
50. Isobe T, Tomita T, Kameshima Y, et al. Preparation and properties of porous alumina ceramics with oriented cylindrical pores produced by an extrusion method. J Eur Ceram Soc. 2006;26(6):957–60.
51. Jiang G, Yang J. Extrusion of highly porous silicon nitride ceramics with bimodal pore structure and improved gas permeability. J Am Ceram Soc. 2018;101(2):520–4.
52. Xu CN. Fabrication of high performance macroporous tubular silicon carbide gas filters by extrusion method. Nanjing: Nanjing Tech University; 2018.
53. Zou D, Qiu MH, Chen X, et al. One-step preparation of high-performance bilayer α-alumina ultrafiltration membranes via co-sintering process. J Membr Sci. 2017;52(4):141–50.
54. Shen YJ, Bian Q, Fan YQ, et al. Fabrication and characterization of porous SiC support and Al_2O_3/SiC composite membrane. Membr Sci Technol. 2013;33(1):22–6.
55. Wang D, Liu HY, He JH, et al. Progress in preparation of functional membranes by spin-coating. Imaging Sci Photochem. 2012;30(2):27–9.
56. Sea BK, Kusakabe K, Morooka S. Separation of hydrogen from steam using a SiC-based membrane formed by chemical vapor deposition of triisopropylsilane. J Membr Sci. 1998;14(6):73–82.
57. Bian Q, Fan YQ. Preparation and properties of porous silicon carbide ceramic by gelcasting. Membr Sci Technol. 2014;34(1):29–33.
58. Wei W. Preparation and performance study of high-temperature dust removal silicon carbide separation membrane. Nanjing: Nanjing Tech University; 2017.
59. Heidenreich S, Simeone E, Haag W, et al. Hot gas filtration for syngas cleaning in biomass gasification. Gefahrstoffe Reinhalt Luft. 2011;71(6):281–5.
60. Salinger M, Heidenreich S. Ceramic hot gas filters for IGCC coal gasification-experiences and new developments. In: Advanced gas cleaning technology. Tokyo; 2005.
61. Heidenreich S, Wolter C. Hot gas filter contributes to IGCC power plant's reliable operation. Filtr Sep. 2004;41(5):22–4.
62. Pastila P, Helanti V, Nikkia AP, et al. Environmental effects on microstructure and strength of SiC-based hot gas filters. J Eur Ceram Soc. 2001;21(9):1261–8.
63. Lupión M, Francisco J, Navarrete B, et al. Assessment performance of high-temperature filtering elements. Fuel. 2010;89(4):848–54.
64. Ren XJ, Cheng ZY, Liu XQ. Research status and prospects of ceramic membranes on gas-solid separation. Membr Sci Technol. 2005;25(2):65–8.
65. Cao JC, Xue YX, Li XY, et al. The application of SiC material with ceramic membranes in shell coal gasfication plant. Adv Ceram. 2013;34(1):51–3.
66. Wei W, Zhang WQ, Jiang Q, et al. Preparation of non-oxide SiC membrane for gas cleaning by spray coating. J Membr Sci. 2017;540:381–90.
67. Wang XM. Design, fabrication and characteristics of porous gradient ceramics fiber membrane for purifying high temperature. Wuhan: Wuhan University of Technology; 2007.
68. Kanaoka J, Chikao R, Kishima LW, et al. Observation of the process of dust accumulation on a rigid ceramic filter surface and the mechanism of cleaning dust from the filter surface. Adv Powder Technol. 1999;10(4):417–26.
69. Ji HJ, Yang JK, Xiao B. Dynamic study of high-temperature dust removal by ceramic filters. Ind Safety Environ Protection. 2003;29(2):17–20.

70. Zhang Y, Wu YJ, Yang XK, et al. High-strength thermal insulating mullite nanofibrous porous ceramics. J Eur Ceram Soc. 2020;40:2090–6.
71. White LR, Sancki SM. High-temperature filtration using ceramic filter. 3M Company. 1981; 93(3): 265–75.
72. Xing Y, Kuang CJ. Research on high-temperature filtering materials. Filtr Sep. 2004;14(2):1–4.
73. Wither CJ. Improvement in the performance of ceramic media for filtration of hot gases. Filtr Sep. 1988;25(2):100–3.
74. Noguchi A, Soga K. Ceramic honeycomb filter and its manufacturing method: CN106714930B. 2019-04-12.
75. Li JF. Structure design and performance research of porous silicon carbide ceramics for high-temperature filtration. Beijing: Tsinghua University; 2011.
76. Deng XY, Bai CY, Li JB, et al. A coating method for a composite silicon carbide ceramic fiber transition layer: ZL201210510731.1. 2014-07-09.
77. Liu W, Cui YS, Jin J, et al. Preparation and properties of fibrous ceramic composite tube for purifying high temperature dust gas. Chin J Environ Eng. 2012;6(9):3248–52.
78. Xing WH, Zhong ZX, Wei W, et al. A method for preparing a silicon carbide separation membrane: CN 106083060A. 2018-08-17.
79. Xue YX, Wang X, Zhang JM, et al. Preparation of continuous fiber reinforced ceramic fiber filtration material. Adv Ceram. 2019;40(6):432–40.
80. Wang YM, Xu YX, Meng XQ, et al. Preparation and characterization of filtering tube of ceramic composite with pore-gradient structure. J Synth Crystals. 2007;36(5):1079–84.
81. Miao LF, Wu XL, Ji ZL, et al. Effects of heat-treatment conditions in the preparation of aluminum silicate fiber-based ceramic filter element for hot-gas filtration. Ceram Int. 2020;46:18193–9.
82. Judkins RR, Stinton DP, Smith RG, et al. Development of ceramic composite hot-gas filters. J Eng Gas Turbines Power. 1996;118(3):495–9.
83. Xiang XD. Theory, technology and application of dust fiber filtration. Beijing: Metallurgical Industry Press; 2007.
84. Hu WZ, Gao CL, Sun XH. Research progress on high-temperature resistant membrane materials. Water Treat Technol. 2019;45(8):10–5.
85. Rui WJ, Zhang C, Cai C, et al. Effects of sintering atmospheres on properties of stainless steel porous hollow fiber membranes. J Membr Sci. 2015;489:90–7.
86. Chi YS, Li T, Chong JY, et al. Graphene-protected nickel hollow fibre membrane and its application in the production of high-performance catalysts. J Membr Sci. 2019;597:117617–36.
87. Zhang WJ, Wei XL, Li T. Research on metal wire mesh dust removal technology for high temperature flue gas in industrial furnaces. In: Clean coal technology; 2020.
88. Liu ZJ, Liu ZM, Ji S, et al. Fabrication of Ti-Si intermetallic compound porous membrane using an in-situ reactive sintering process. Mater Lett. 2020;271:127786–122791.
89. Gao L, Wang T, Liu B. Yellow phosphorus filtration pretreatment device and its constituted yellow phosphorus purification system: ZL204824175U. 2015-12-02.
90. Wei L, Huang Y, et al. Preparation of porous TiO_2 stainless-steel membranes by carbon assisted solid-state particle sintering. J Inorg Mater. 2015;30(4):427–31.
91. Wei L, Yu J, Hu XJ, et al. Buffering of the sintering stress for fabrication of microporous Al_2O_3/stainless-steel membranes. J Inorg Mater. 2014;29(4):433–7.
92. Zhu ZW, Meng XG, Liu W, et al. Study on hydrogen permeation of Ni-$BaZr_{0.1}Ce_{0.7}Y_{0.2}O_3$ asymmetric cermet membrane. International Journal of Hydrogen Energy. 2019; 43: 4959–66.

Chapter 6
Gas Phase Pollutant Cleaning Membranes

6.1 VOCs Separation Membranes

VOCs, or volatile organic compounds, refer to organic substances with a melting point lower than room temperature and a saturated vapor pressure exceeding 13.3 Pa or a boiling point less than or equal to 250 °C under standard pressure of 101.3 kPa [1]. Common components include hydrocarbons, halogenated hydrocarbons, alcohols, esters, ketones, aromatic compounds, etc. [2]. The sources of VOCs are complex and can generally be divided into natural sources and anthropogenic sources [3]. The former includes forest fires, volcanic eruptions, vegetation and crop respiration, etc., while the latter includes emissions from motor vehicles, biomass combustion, industrial production, etc. [3]. In industrial processes, the evaporation and leakage of organic solvents are inevitable [4]. The emission of these VOCs not only results in significant economic losses and safety hazards during the industrial production process but also poses threats to the environment and human health. These emitted VOCs undergo reactions with nitrogen oxides in the atmosphere when exposed to sunlight, leading to the production of highly active free radicals. This process generates ozone, peroxyacetyl nitrate, aldehydes, and other photochemical smog, causing secondary pollution [5]. Additionally, it contributes to the formation of secondary organic aerosols (SOA), which not only impact atmospheric visibility but also present carcinogenic and teratogenic hazards to humans [6].

In order to comply with the laws and regulations regarding VOC emissions in different countries, researchers have developed various technologies to either eliminate or reduce these emissions. Based on their governance strategies, these technologies can be categorized into source control, process control, and end-of-pipe treatment [7].

Source control is considered the optimal technology for VOC management. It involves screening raw materials or added solution solvents used in the industry and employing green solvents, such as water, as substitutes for traditional organic

Z. Zhong and W. Xing, *Membrane Technology for Air Cleaning*,
https://doi.org/10.1007/978-981-96-1144-7_6

solvents in production, thereby reducing emissions [4]. This method still has significant obstacles at present, and the selection of raw materials for some processes is not sufficient to support the use of green solvents.

Process control refers to upgrading and transforming existing process routes or equipment to reduce VOC emissions during the production process [4]. This method has a long development cycle, high development costs, and is not universally applicable.

End-of-pipe treatment technology involves employing additional processes to treat already emitted VOCs, acknowledging the reality of VOC emissions and aiming to prevent further pollution caused by their release into the environment. End-of-pipe treatment technology does not need to change existing process conditions, it only needs to be integrated at the end of the process where VOCs are emitted. At the same time, the diversity of its technology can well meet the needs of VOC emission control, thus it has a high market share [4].

End-of-pipe treatment technology can be further divided into two categories, namely elimination technology and recovery technology, the former includes catalytic combustion, biodegradation, photocatalytic decomposition and plasma destruction [8, 9] etc., the latter includes adsorption, absorption, condensation and membrane separation technology [10–13]. Among them, membrane separation technology, due to its high separation selectivity, easy integration, green environmental protection, and low energy consumption, is considered one of the most promising separation technologies, and has been widely used in the chemical, pharmaceutical, energy, printing and other fields [14–16], playing an important role in replacing traditional separation technologies and upgrading traditional processes. Domestic and foreign research institutions actively cooperate with enterprises and have done a lot of work in the research and development of membrane materials, the scale-up of membrane fabrication, and the improvement of membrane separation coupling processes [17–24], playing an important role in the development and promotion of membrane VOCs separation technology.

This chapter provides an overview of materials used in VOC separation membranes, their preparation methods, performance parameters, mass transfer models, separation process flows, component design, and includes examples of industrial applications. It serves as a valuable resource, offering fundamental data and references for the control and treatment of VOCs originating from industrial sources.

6.1.1 Types of VOC Separation Membranes

The performance of membrane materials is key to the industrial application of separation membranes. VOC separation technology by membrane method can roughly divide membranes into two types according to different separation mechanisms: microporous polymer membranes that follow the molecular sieve separation mechanism and dense polymer membranes that follow the solution diffusion mechanism. The former separates based on the size of the molecules, with large VOC molecules being retained by the membrane and small nitrogen or air molecules preferentially passing through the separation membrane, thus achieving separation. Therefore, it can be called a VOC-sieving membrane. The latter separates based on different preferential adsorption affinities, with large molecular components preferentially passing through the membrane and small nitrogen or air molecules being retained, so it can be called a VOC-permeation-selective membrane [25].

6.1.1.1 VOC-Sieving Membranes

Molecular sieve is a common porous crystalline silicoaluminate material with a uniform pore structure [26]. It is generally considered to be a cage-like molecule composed of several identical types of metal ions in the structural unit center, and several silicon oxide, aluminum oxide molecules, and a small amount of bound water [26]. The skeleton of the molecular sieve is generally composed of SiO_4^{4-} and AlO_4^{5-} tetrahedra, which forms the net-like structure of the molecular sieve by sharing oxygen atoms. Due to its high mechanical strength and stable chemical properties, the molecular sieve is one of the popular materials for membrane fabrication. Zhou et al. [27] prepared a dense, continuous, uniform, and pore-free and crack-free 13X molecular sieve membrane layer on the surface of α-Al_2O_3 ceramic tube, which can remove three kinds of VOCs impurities such as dimethyl ether, methanol, and propionaldehyde in nitrogen to below 100 μL/L. The diameter of the cage-like pores of the molecular sieve is relatively fixed and is also limited by the size of the inorganic anions. For some molecules with kinetic diameters close to air, effective separation cannot be achieved [28]. Furthermore, the high cost of preparing molecular sieve membranes poses some difficulties in their large-scale promotion and use, which limits their application in the field of VOC separation.

Metal-Organic Frameworks (MOFs) are organic–inorganic hybrid materials with regular topological structures or regular pore channels, formed by transition metal ions or metal clusters as centers, combined with organic ligands in the form of coordination bonds [29]. Unlike molecular sieves, which have difficulty controlling pore size, MOFs can prepare VOC separation membranes with uniform pore size, diverse structures, and high specific surface area by changing the synthesis method and adjusting the central metal ions [28]. Wu et al. [30] prepared a ZIF-8 membrane on ZrO_2 using the solvothermal method, which was used for gasoline vapor recovery. The permeation rate of n-hexane is 7.16×10^{-8} mol/(m^2 s Pa), and under the conditions of a transmembrane pressure difference of 0.05MPa, a gas recovery rate of 29.3% and a volume fraction of n-hexane of 30%, the separation factor of nitrogen/n-hexane is 2.46. Although the MOFs membrane shows good VOC separation and recovery effects under laboratory conditions, its large-scale preparation is challenging, and the membrane formation is prone to grain boundary defects, limiting its industrial application [28].

Microporous polymers are a type of porous organic polymer formed by the disordered stacking of rigid twisted macromolecular chains in space, and this disordered chain stacking results in inherent microporous structures in the polymer [31]. By adjusting this part of the microporous structure, or modifying the polymer, separation membranes suitable for trapping VOCs can be prepared. However, when this form of microporous polymer is used for VOC separation, the high affinity and swelling action between the chain segments and VOCs eliminate the function of trapping VOCs through the pores, and instead show a property of preferentially permeating VOCs and trapping small molecules [25].

In order to achieve VOC trapping and reduce swelling, researchers have proposed optimization schemes from both structural and material aspects. Replacing the traditional chain structure with a three-dimensional network structure can effectively improve the rigidity of the polymer, weaken the swelling effect of VOC molecules on the membrane, and improve the sieving selectivity; at the same time, choosing hydrophilic materials or organics with hydrophilic groups can reduce the adsorption of VOCs molecules on the membrane surface. Based on this, Zhou et al. [18] used trisubstituted 2, 6, 14-triaminotriptycene as a monomer, and reacted with acyl chlorides of different carbon chain lengths to prepare a series of network triptycene-based polyamide imide membranes for nitrogen and VOCs separation, as shown in Fig. 6.1. By optimizing the size of the network pores and the polymer stacking pores, the prepared microporous polymers have high retention rates for cyclohexane, n-hexane, and n-heptane, and also have a nitrogen permeation coefficient of over 2000 Barrer. This indicates that microporous polymer membranes have great potential in oil and gas recovery. However, the issue of the stability of the microporous polymer membrane itself also limits its industrial use. Therefore, enhancing the stability of microporous polymers is also one of the current research hotspots.

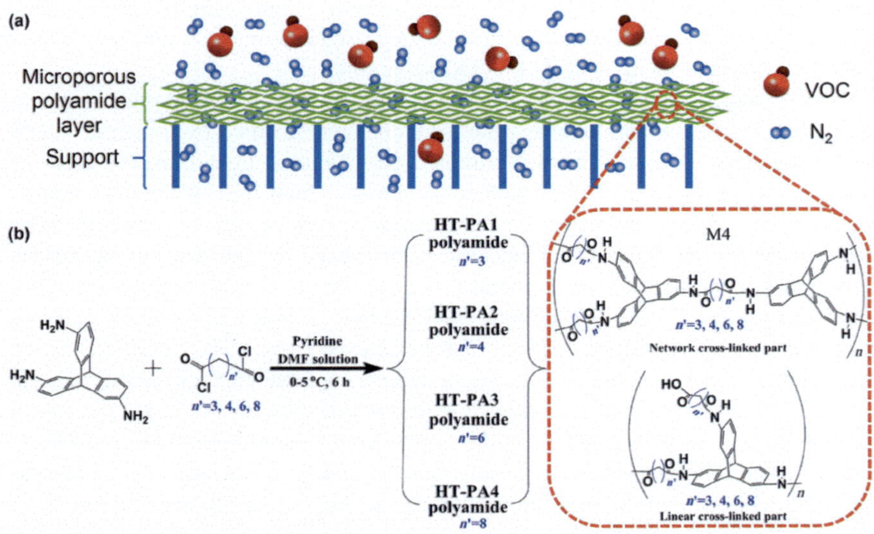

Fig. 6.1 Polyimide (PIM) membranes prepared from a trisphenylene monomer [18]

6.1.1.2 VOC-Permeation-Selective Membranes

VOC-permeation-selective membranes are often made of rubbery polymers, which have high gas permeability, but their selectivity and mechanical properties are relatively average. Polydimethylsiloxane (PDMS) is a polymer with a general formula structure of $–Si(CH_3)_2O–$, and it is one of the most widely used membrane materials in VOC separation membranes [32]. Because the bond length and bond angle of Si–O–Si are larger than those of C–C–C and C–O–C, the distance between non-bonded atoms increases, and the intermolecular interaction force decreases. This characteristic establishes that polydimethylsiloxane possesses exceptionally high gas permeability [33]. Polydimethylsiloxane has garnered significant attention due to its notable stability, ductility, thermal stability, chemical stability, high gas permeability, and excellent resistance to solvents [34]. Tubular membranes on surface of ceramic supports exhibit elevated gas permeability, robust mechanical strength, and outstanding chemical stability. The development of tubular membranes on wall of ceramic supports further enhances the mechanical properties of the membrane and reduces the damage rate during transportation. For different application systems and different production conditions, Chinese scholars have developed PDMS hollow fiber membranes and flat membranes suitable for different industrial processes, as shown in Fig. 6.2. These different configurations of PDMS composite membranes can handle VOC waste gas under different feed conditions, thereby reducing VOCs emissions from industrial sources and effectively recovering VOCs.

Polydimethylsiloxane also has certain problems, such as average solvent swelling resistance, average gas separation selectivity, and limited permeation rate. In order to effectively improve the film-forming performance of PDMS, researchers have made some modifications to PDMS. There are two types of modification methods: side chain modification and main chain modification [33]. Side chain modification uses

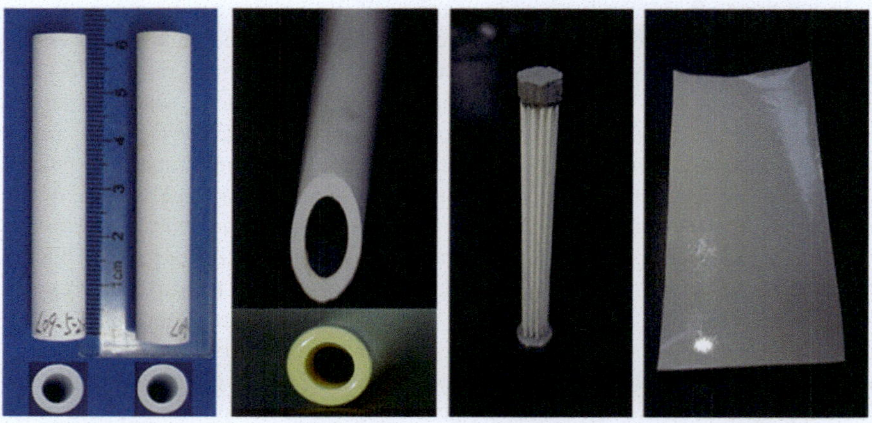

(1) Tubular outer membrane (2) Tubular intima (3) Hollow fiber (4) Flat membrane

Fig. 6.2 Different configurations of PDMS composite membranes

larger groups or polar groups to replace the $-CH_3$ on the side chain of polydimethylsiloxane molecules; main chain modification introduces larger groups on the main chain –Si–O–, or replaces –Si–O– with $-Si-CH_2-$. Side chain modification and main chain modification increase the glass transition temperature and molecular chain segment packing density of the polymer, thereby improving its chemical stability, thermal stability, permeability, and selectivity. Many researchers have produced new rubbery polymer materials through such means, such as polyoctylmethylsiloxane (POMS) [33], polyphenylmethylsiloxane (PPMS) [35], etc., and have achieved good results in separating various $VOCs/N_2$ systems.

Although the permeability and selectivity of VOC gas can be improved through modifications, traditional polymer membranes all exhibit a trade-off between gas permeability and selectivity (under the Robinson's upper bound). In contrast, inorganic materials (such as zeolites, MOFs, carbon materials, etc.), with their unique pore structure, have a higher level of selectivity than ordinary polymers, so a mixed matrix membrane can be prepared by introducing inorganic porous fillers into the polymer to improve the permeability of the membrane.

Yuan et al. [36] modified ZIF-8 particles with 5,6-dimethylbenzimidazole (DMBIM) to make them hydrophobic, and introduced them into the PDMS membrane, preparing a mixed matrix membrane with high separation performance. When the loading of ZIF-8-DMBIM was 10%, the separation performance of the PDMS mixed matrix membrane was the highest. In the separation of propane/nitrogen, the selectivity increased by 116% and the permeability increased by 91% compared to pure PDMS. Although the mixed matrix membrane showed good separation performance, in the process of industrial scale-up preparation, the added particles tend to agglomerate, causing the doped particles to be unevenly dispersed, which has a significant impact on the macroscopic performance of the mixed matrix membrane. Attaining an ideal particle-polymer interface and achieving uniform dispersion of nanoparticles in the surface layer of fibers remains a key focus of research.

In addition to PDMS, polyether block amide (PEBA) is also a polymer membrane material that performs well in the process of separating and recovering VOCs. PEBA refers to a polymer whose molecular formula contains -CO-PA-COO-PE-O- segments, where -PA- represents the rigid polyamide segment, which has strong polarity, and -PE- represents the flexible polyether segment [37]. The rigid polyamide segment enhances the mechanical properties of the polymer and imparts resistance to swelling, while the flexible polyether segment increases the free volume of the polymer, facilitating the passage of gas molecules. Under the joint action of the two types of segments, by adjusting the ratio of rigid and flexible segments, PEBA membranes can show good vapor permeability performance for different systems [38, 39]. Huang et al. [40] used PEBA membranes to separate heptane/nitrogen mixed gas, and investigated the separation performance of PEBA membranes for n-hexane/nitrogen and n-heptane/nitrogen systems under various operating conditions. The experimental results show that the permeability coefficient of n-heptane in the PEBA membrane reaches 5×10^{-8} mol/(m_2 s Pa), and the permeability coefficient of n-hexane reaches 1.6×10^{-8} mol/(m_2 s Pa), showing the good affinity of PEBA membranes for VOCs, and also indicating that PEBA has good selectivity in the process of oil and gas separation. The properties of PEBA membranes are affected by the polyamide and polyether segments. If the proportion of polyamide segments is too high, the reaction temperature for preparing the membrane will be too high, and it will not be possible to prepare a thin film with a large flux by traditional coating methods; if the proportion of polyether segments is too high, the polymer will be greatly affected by swelling, and the separation effect will decrease during the separation process. Hence, the development of various types of PEBA for distinct separation systems has become a widely addressed research direction.

6.1.2 Preparation Methods of VOC Separation Membranes

The VOC separation membrane prepared with PDMS has received widespread attention due to its good separation effect and chemical stability. The common preparation methods of PDMS membranes are shown in Fig. 6.3. The PDMS polymer monomer, catalyst, and crosslinking agent are dissolved in an organic solvent, and after reacting to a certain extent under heated conditions, they are coated on a porous support by spin coating or blade coating. After holding at room temperature for a period, the solvent is further volatilized by raising the temperature, and the PDMS coating further solidifies [41]. Han et al. [42] coated a layer of PDMS polymer on a PVDF support by such a coating method, and the prepared PDMS/PVDF composite flat membrane has a good separation effect for n-pentane, n-hexane, n-heptane, and nitrogen. At room temperature, after running for three months, the separation factor of aliphatic hydrocarbon/nitrogen is still greater than 100, and the flux of VOCs exceeds 22.5 mol/(m^2 h). Infrared spectroscopy characterization shows that the characteristic peaks of the composite membrane have not changed, indicating that the chemical structure of the membrane material has not been affected by the organic solvent. The PDMS membrane prepared by this coating method has good stability, strong binding force with the support, and the potential for industrial scale-up preparation.

On this basis, researchers have developed another representative method that can quickly solidify PDMS, control the reaction, and is solvent-free. Qin et al. [44] based on the previous work of photo-curing acrylate-thiol hydrogel, induce methyl acrylate-polydimethylsiloxane (MA-PDMS) polymerization and crosslinking under ultraviolet light. This process is easily achievable even in an atmospheric environment, and PDMS membranes can be continuously prepared on a large scale by this method, as shown in Fig. 6.4. However, this method is limited to certain polymers and can only be used for free radical reaction PDMS. Further development is still needed to achieve rapid crosslinking of other types of PDMS.

Using the excellent properties of ceramic supports, membranes with good stability, high mechanical strength, and good separation performance can be prepared. Therefore, the preparation method of ceramic tubular membranes has always attracted the attention of researchers. The common method of preparing tubular membranes is by dip-coating, in which the ceramic tubular membrane is immersed in the casting solution for a period, then taken out, and the casting solution is coated on the ceramic

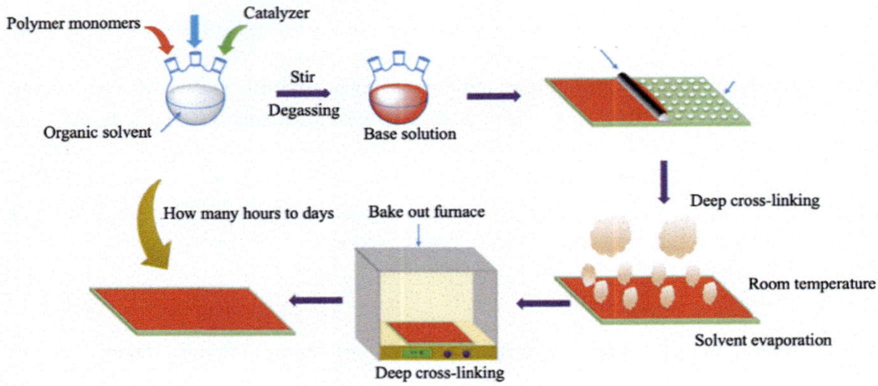

Fig. 6.3 Typical preparation process of PDMS composite membrane [43]

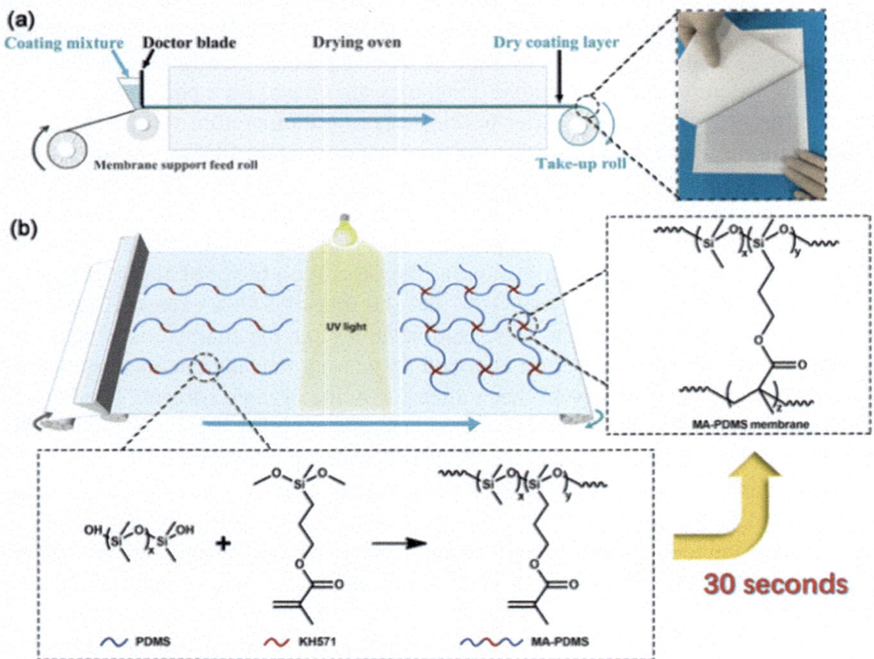

Fig. 6.4 Conventional preparation process of PDMS membranes: thermal crosslinking method (**a**) and ultraviolet polymerization method (**b**) [44]

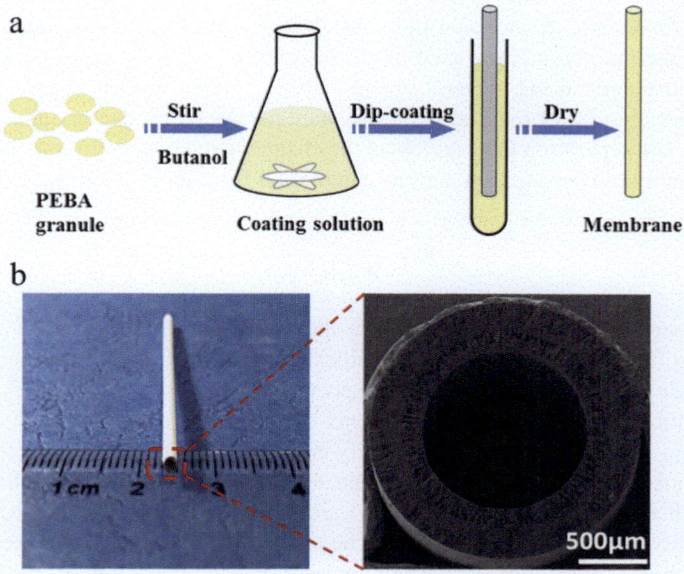

Fig. 6.5 Hollow fiber membrane prepared by the dip-coating method [45]

tubular membrane, followed by a drying process similar to that of flat membranes. Li et al. [45] coated a layer of PEBA membrane on a ceramic hollow fiber by the dip-coating method (Fig. 6.5). By adjusting the viscosity and concentration of the PEBA coating, a stable and excellent separation effect of PEBA ceramic hollow fiber composite membrane can be obtained. This membrane has stable separation performance in a 200-hour test and has good selectivity for butanol, laying the foundation for the industrial application of PEBA hollow fiber membranes.

6.1.3 Performance Parameters of VOCs Separation Membranes

In order to compare the separation effects between different separation membranes, the following parameters are commonly used to reflect the performance and separation ability of the membrane or membrane material. Flux (J) refers to the mass of the component to be separated per unit area of the membrane in unit time, and its calculation formula is as follows.

$$J = \frac{M}{At} \tag{6.1}$$

where, M represents the mass of the substance passing through the membrane; A represents the membrane area used in the separation process; t represents the time elapsed in the separation process.

The separation factor (β) is also an important parameter for evaluating separation efficiency. The separation factor refers to the change in the concentration ratio before and after membrane separation, and its calculation formula is as follows.

$$\beta = \frac{W_{f,A}/W_{f,B}}{W_{P,A}/W_{P,B}} \tag{6.2}$$

where, w represents the mass fraction of the component; the subscripts A and B represent different components in the substance to be separated; f and p represent the feed side and the permeation side, respectively.

The solubility coefficient (S) represents the size of the membrane's solubility for a certain component, and the solubility coefficient is related to the solubility of the component and the type of membrane. Usually, high-boiling-point liquefiable gases (with higher critical temperatures) are more soluble in the membrane, and the solubility coefficient is also larger. The diffusion coefficient (D) represents the diffusion ability of gas molecules in the membrane, and the diffusion coefficient is related to the dissolved gas and the structure of the membrane. Usually, the smaller the molecular weight, the faster the diffusion in the membrane, and the larger the diffusion coefficient.

The permeability coefficient (P) is used to represent the difficulty of gas passing through the membrane, and it also reflects the solubility and diffusion ability of the gas to the membrane. It generally refers to the permeation flux under unit time, unit membrane area, and unit driving force. Since the permeation process is generally considered to be a combination of the dissolution process and the diffusion process, the permeability coefficient is equal to the product of the diffusion coefficient and the solubility coefficient, as shown in the following formula. To simplify the unit, Barrer is defined as the commonly used unit of permeability coefficient.

$$P = SD \tag{6.3}$$

$$P_i = \frac{J_i}{\frac{\Delta p}{L}} = \frac{J_i L}{Pi_f - Pi_p} \tag{6.4}$$

$$Barrer = 10^{-10}\,\mathrm{cm^3(STP)\cdot cm \cdot cm^{-2}\cdot S^{-1}\cdot cmHg^{-1}} \tag{6.5}$$

where J_i represents the flux of component i; L represents the thickness of the membrane; P_i represents the partial pressure received by component i.

Gas selectivity (α) is used to represent the difference in affinity between the membrane material and different types of gases; typically, it is the ratio of permeability coefficients when two components pass through the membrane. The permeability coefficient can be obtained by multiplying the solubility coefficient and the diffusion coefficient. Similarly, selectivity can also be obtained by calculating the solubility selectivity (α_S) and diffusion selectivity (α_D), and the specific calculation formula is as follows.

$$a = \frac{P_A}{P_B} = a_s a_D \tag{6.6}$$

$$a_s = \frac{S_A}{S_B} \tag{6.7}$$

$$a_D = \frac{D_A}{D_B} \tag{6.8}$$

where permeability $\frac{P_i}{L}$ refers to the volume of gas passing through a unit membrane area under unit pressure difference per unit time. To simplify the unit, GPU is introduced as the unit of permeability.

$$GPU = 10^{-6}\ \mathrm{cm}^3(STP) \cdot \mathrm{cm}^{-2} \cdot \mathrm{s}^{-1} \cdot \mathrm{cmHg}^{-1} \tag{6.9}$$

6.1.4 Mass Transfer Models of VOC Separation Membranes

The membrane separation process is a typical rate separation process, which is realized by the difference in the adsorption diffusion rate of the separated substances in the membrane. The factors affecting the adsorption diffusion rate include the molecular size of the substances to be separated, hydrophilicity and hydrophobicity, relative volatility, molecular structure, and the structure and physicochemical properties of the membrane material [46–48]. Therefore, according to the different interactions between the separated substances and the membrane materials, the membrane VOCs separation mechanism is mainly divided into the separation mechanism for porous membranes (such as molecular sieve separation mechanism) and the solution diffusion separation mechanism for dense membranes [49].

6.1.4.1 Mass Transfer Mechanism of Porous Membranes

The separation mechanism of porous membranes can be divided into Knudsen diffusion, surface diffusion, capillary condensation, and molecular sieving, etc., according to the relationship between the free path or molecular size of the separated substances and the pore size of the membrane, as shown in Fig. 6.6 [50].

When the pore size of the porous membrane is smaller than the average free path of the gas molecules, the gas molecules collide with the pore wall during their passage through the membrane, and this motion behavior belongs to Knudsen diffusion [51].

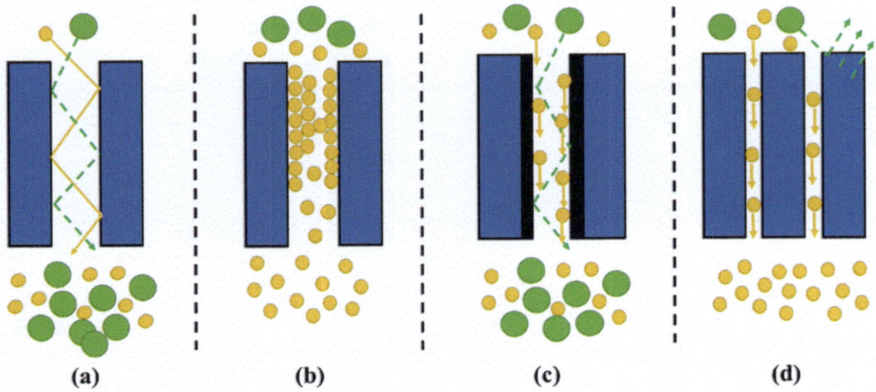

Fig. 6.6 Schematic diagram of the separation mechanism of porous membranes

However, the separation performance of the membrane using this separation mechanism is relatively low, and the theoretical selectivity of the membrane separation process comes from the molecular kinetic diameter and molecular weight, which are the properties of the molecules themselves, and the influence of the membrane itself in this process is relatively small.

When gas molecules are adsorbed on the surface of a porous medium, the concentration gradient on the membrane surface will cause the molecules to move, and this motion behavior belongs to surface diffusion [52]. Molecules with higher affinity for the membrane surface are preferentially adsorbed, occupying the pores, thereby reducing the adsorption of non-preferentially adsorbed selective molecules, achieving the purpose of separation. As more and more molecules are adsorbed on the membrane surface, multilayer diffusion and capillary condensation will occur, further hindering or blocking the membrane pores, reducing the diffusion of other gases, and further improving the separation performance. Controlling these processes is difficult, so there is less research based on the separation mechanism.

As the gap between the pore size of the membrane and the size of the separated substances further narrows, that is, when the pore size of the membrane is between the kinetic diameters of large and small molecules, large molecules are retained, while small molecules can pass through. This mechanism is called the molecular sieving mechanism [46]. Based on the screening mechanism, the adsorption selectivity of the membrane can be improved by controlling the surface properties of the membrane; the diffusion selectivity of the membrane can be controlled by adjusting the size of the pore diameter; a higher separation selectivity can be achieved by choosing suitable membrane preparation conditions. The actual separation operation process is simple and easy to control, which is the key research direction for future VOC separation membranes. The difficulty still lies in effectively controlling the size of the membrane pore diameter and mitigating the weakening of the screening effect caused by the swelling effect of VOCs on the membrane.

6.1.4.2 Dense Membrane Mass Transfer Mechanism

Unlike porous membranes, the separation mechanism of dense membranes has always been described by the dissolution-diffusion mechanism. As shown in Fig. 6.7, separation is achieved by relying on the difference in the preferential adsorption diffusion rate of the substance to be separated with the membrane. The substance to be separated undergoes adsorption on the membrane surface through diffusion from the solution's main body. Subsequently, the adsorbed substance diffuses within the membrane and is collected through desorption on the permeation side [53]. The dissolution-diffusion model simply introduces the process of membrane separation, without a detailed explanation of the influence of operating conditions (such as temperature, concentration, transmembrane driving force, etc.) during the separation process.

With the development of membrane materials, the characterization of the active layer has gradually attracted the attention of researchers. By analyzing the interaction between the separation layer and the feed flow, not only can guidance be provided for the synthesis process, but the entire process can also be better simulated and calculated, providing a theoretical calculation basis for industrialization. The permeation process is usually divided into three stages: dissolution, diffusion, and desorption. In the process of VOC separation, dissolution is often the rate-determining step, which has a large impact on the separation process and has been widely studied. The dissolution process of VOCs on the surface of glassy polymers and rubbery polymers is different. It is generally believed that the dissolution and diffusion process of VOCs on the surface of rubbery polymers can be described by Fick's law, while the dissolution process of VOCs on the surface of glassy polymers cannot be described by Fick's law. When VOCs dissolve on the surface of glassy polymers, they are affected by two factors [54–56]. The first influence is the slow change in the structure of the polymer

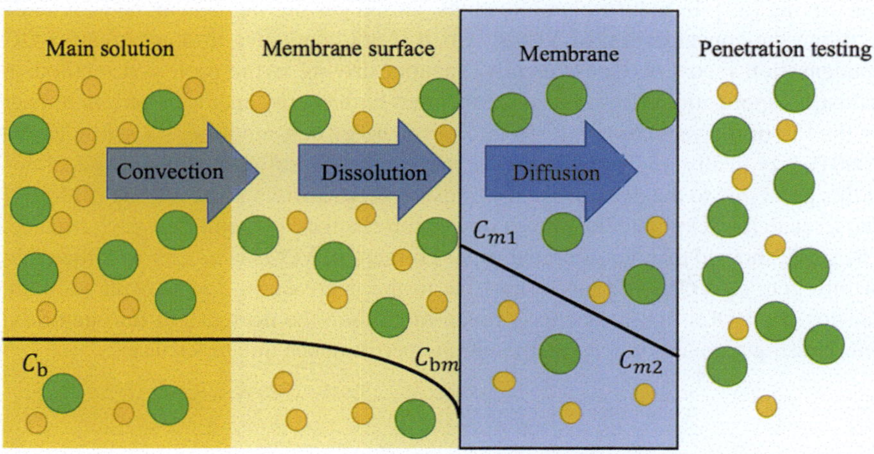

Fig. 6.7 Schematic diagram of the dissolution-diffusion principle

during the dissolution process. The slow expansion of the polymer chain segments causes tension between the chains, resulting in a pressure drop in the polymer, causing more molecules to be absorbed by the polymer. The second influence is that different parts of the polymer have different loads, causing different parts of the polymer to swell to different degrees, thereby producing different degrees of swelling potential or swelling stress, causing its adsorption behavior to differ from the Fick diffusion process [57].

Due to the relatively minor impact of the diffusion process on the VOC separation process and the difficulty in measuring related parameters, there has been limited research in this area. It is generally believed that the diffusion of molecules in the membrane refers to the diffusion of molecules in the free volume of the membrane. This theoretical model assumes that the amorphous region of the polymer is an ordered region with molecular chain segments, and the molecular chain segments exist in a parallel bundle form [58, 59]. This theory also assumes that molecular motion occurs in a specific chain-stacking manner or along parallel polymer chains. The molecular motion process can be divided into two parts. One part is the molecules jumping at a relatively fast speed from the space of the polymer chain segment, which requires low activation energy; the second part refers to the movement of the polymer chain segment allowing the molecules to transfer between the chain segments, which requires higher activation energy, this energy equals the minimum energy required for chain segment separation.

The desorption process is the least influential part of the separation process, and current researchers still have relatively little research on the specific process of desorption, and have not yet fully described the impact of related parameters. In the desorption process, the basis of membrane transfer is the transmembrane driving force, temperature, concentration, electrode potential, and pressure can all become transmembrane driving forces. VOC separation membranes are most commonly pressure driven. The driving methods mainly include vacuum driving and sweep gas driving [53]. In industrial applications, sweep gas driving is avoided as it tends to dilute the concentration of VOCs, which is not conducive to downstream VOC condensation recovery. Consequently, vacuum driving is the preferred method in industrial applications. Low vacuum consumes high energy and has a certain impact on the life of the vacuum pump. To minimize energy consumption, a vacuum above 2000 Pa is typically employed. However, the membrane performance sharply declines at this point due to the decrease in the transmembrane driving force. Zong et al. [60] and Li et al. [61] explored the impact of vacuum changes on membrane performance. The study showed that the saturated vapor pressure of VOCs is much lower than that of nitrogen, so VOCs are more sensitive to the decrease in downstream vacuum, causing the VOCs flux to decrease much more than the decrease in nitrogen flux, thus causing a greater impact on membrane performance due to vacuum.

6.1.5 Process and Module Design of VOC Separation Membranes

In the membrane separation process, the performance of the membrane affects the industrial application cost of VOC treatment by the membrane method. However, the time and cost required to develop new materials to prepare high-performance membranes are long and high, and it is difficult to achieve in a short time. Therefore, optimizing the component design for the membrane process to improve mass transfer performance is the first choice to improve separation effects and reduce processing costs. In addition, by designing a reasonable process flow for the treated gas, the effect of membrane separation can be optimized to the greatest extent, thereby achieving the purpose of reducing the membrane area and reducing costs.

In order to optimize the mass transfer process of gas or liquid flowing over the membrane surface, researchers conducted related tests on ceramic tubular membranes of different lengths and developed matching membrane components of different lengths, as shown in Fig. 6.8a. By optimizing the membrane preparation process and component design, the separation factor and flux of different length membrane components remain unchanged when separating the ethanol/water mixture system. It exhibits a higher level of competitiveness in performance compared to commercially available PDMS membranes.

Based on the development of large-scale ceramic tubular membrane fabrication technology, Jiusi High-tech Co., Ltd. has developed a membrane component with a length of 100 cm and a membrane filling area of $1m^2$, as shown in Fig. 6.8b. As the membrane component incorporates an increasing number of tubular membranes, researchers have begun to focus on the mass transfer behavior of the fluid within these tubular membranes. In order to improve the mass transfer behavior of porous channel components or tubular membranes in separation, it is necessary to reduce the dead zone of the fluid in the component and improve the mass transfer separation effect. Liu et al. [62] optimized the arrangement of tubes and feed conditions through computational fluid dynamics (CFD) simulations and comparative experiments on porous channel tubular membranes. Through the CFD simulation study of the ceramic-loaded PDMS composite membrane hollow fiber membrane, the internal

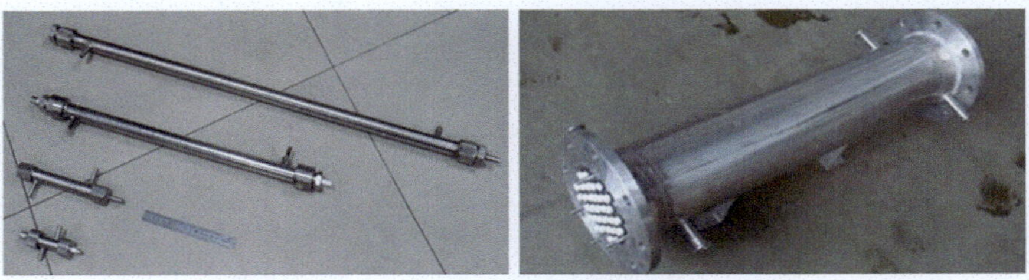

(a) (b)

Fig. 6.8 Tubular membrane matching components of different lengths and areas

flow field distribution of the module was analyzed, and it was concluded that the module design could be effectively optimized by controlling the packing density and cross-sectional arrangement, as shown in Fig. 6.9. By filling 7 bundles of hollow fiber membranes with high packing density, a high-performance hollow fiber module was obtained, which has good application prospects in the field of biofuel separation.

In addition to optimizing the membrane components, in the industrial production process, the VOC recovery process also depends on the properties, types, concentration, flow rate, material value, etc. of the VOCs. Process design for VOCs under different operating conditions is also an important part of membrane process development. It is commonly understood that the membrane process can yield

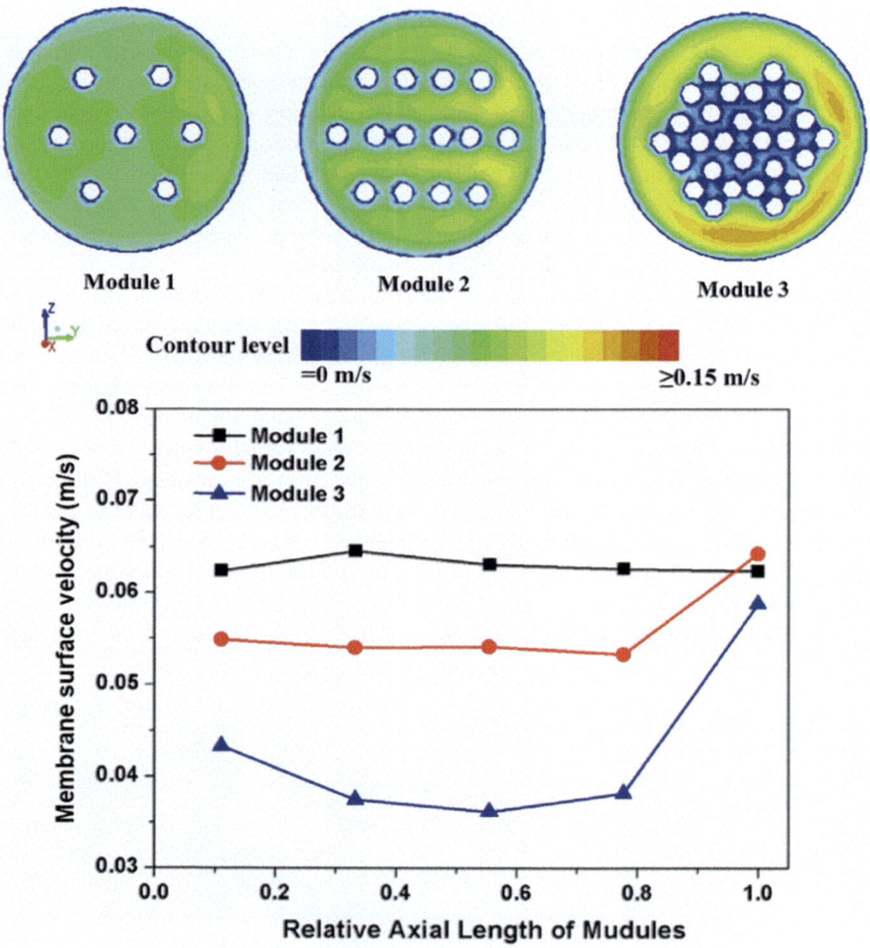

Fig. 6.9 Solution distribution of different configuration components

favorable results when the volume fraction of VOCs surpasses 0.1% and attains a certain recovery value. The membrane process coupled with adsorption, compression condensation, catalytic combustion and other processes can cope with different operating conditions and treatment requirements in the industrial production process.

Based on this, Nanjing Tech University in China has developed a pilot device and conducted related tests on the cyclohexane recovery process, as shown in Fig. 6.10. After testing various membrane areas to handle different volume fractions of cyclohexane feed, it was observed that with an increase in membrane area, the volume fraction of cyclohexane on the retention side decreases, leading to higher recovery amounts. However, this also results in an increase in operating costs. In addition, the recovery of cyclohexane under different condensation temperatures was investigated at the condensation port, and the results are shown in Fig. 6.11. It can be seen that as the temperature decreases, the saturated volume fraction of cyclohexane decreases. When the temperature is – 50 °C, the saturated volume fraction of cyclohexane drops to about 0.1%. This means that the energy consumption required for condensation is very high. However, after membrane concentration, the cyclohexane

Fig. 6.10 Digital image of the pilot device in the laboratory

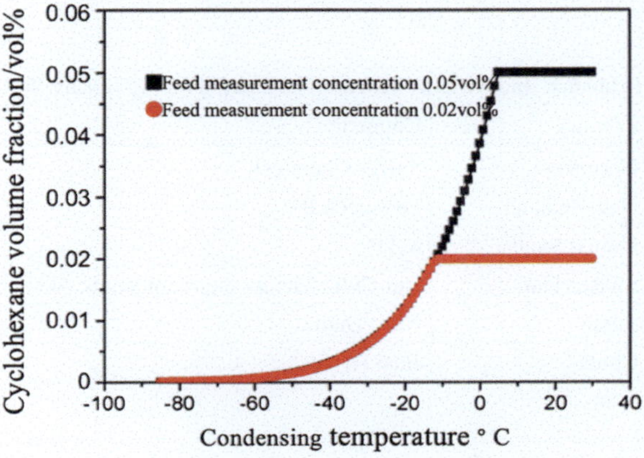

Fig. 6.11 The effect of condenser temperature on the saturated volume fraction of cyclohexane vapor

on the membrane permeation side is condensed to a volume fraction of 2–5% and then mixed with the raw gas for cyclic recovery. At this time, the temperature required by the condenser is about – 20–0 °C. Compared with cooling only by condensation, energy consumption is significantly reduced.

6.1.6 Industrial Application of VOC Separation Membranes

With its simplicity, high integrability, high recovery efficiency, and low energy consumption, membrane technology has steadily evolved membranes and supporting process technologies suitable for industries such as petrochemicals, coating printing, and biopharmaceuticals, as research in this field continues to deepen and expand. It has shown good separation effects when separating and recovering common organic solvents such as benzene, toluene, acetone, alcohols, hydrocarbons, and other VOCs [63, 64]. Table 6.1 lists the examples of the industrial application of membrane technology in various countries in recent years.

The Changchun Institute of Applied Chemistry, Lanzhou Institute of Chemical Physics, Dalian Institute of Chemical Physics of the Chinese Academy of Sciences, and the Membrane Science and Technology Research Institute of Nanjing Tech University actively cooperate with chemical enterprises, using developed gas separation membranes to couple different processes according to the on-site conditions to treat waste gas at different working conditions.

The Dalian Institute of Chemical Physics of the Chinese Academy of Sciences used a silicone rubber/polyamide polyether copolymer flat membrane to investigate the separation effect of VOCs/N_2. Using a polyamide polyether copolymer with better selectivity as the base membrane, a composite membrane was made with a silicone rubber coating, and a spiral wound membrane separator was used to recover ethylene and butene monomers emitted during the production of polyethylene. In the pilot test with Jihua Company, good results were achieved [66]. On this basis, Sinopec Zhongyuan Petrochemical also adopted a commercial silicone rubber-polyamide polyether copolymer composite membrane, and used a roll-type membrane separation component for the tail gas recovery process in polyethylene

Table 6.1 Industrialized membrane method for recovering VOCs development manufacturers [65]

Country	Company	VOCs
Germany	Sulzer Chemtech	Gasoline Vapor
New Zealand	Pervatech BV	Ethyl Acetate
United States	MTR	CFC_2, Ethylene Oxide
Switzerland	Cm-Celfa Membranetrenntechnik AG	Acetonitrile, 1-Butanol
Japan	Nitto Denko	Toluene
China	Jiusi Hi-Tech	Hexane

Fig. 6.12 Membrane method VOC recovery device

production. By coupling membrane separation with compression condensation technology to recover normal butene and isopentane, a profit of 4.8 million yuan/year was created [67].

The Membrane Science and Technology Research Institute of Nanjing Tech University, in conjunction with Nanjing Jiusi High-tech Co., Ltd. (Jiusi High-tech), launched a PDMS composite membrane based on ceramics and related membrane components in 2014. Through the optimization of membrane components and process improvements, a pilot device for permeation vaporization and steam permeation of polymer/ceramic composite membranes was developed, as shown in Fig. 6.12. This device is established by connecting several membrane components in series or in parallel, and integrating the feed end and collection end with the control system. It is used to evaluate the relationship between the amount of waste gas to be treated in the test process and the required membrane area. This device can not only maximize the analytical selectivity of the membrane by controlling the membrane area, but also evaluate and calculate the minimum membrane area according to different processing requirements, saving investment costs [68].

Taking the pharmaceutical industry as an example, a pharmaceutical factory in China has about 800 m^3/h of waste gas rich in hexane. The original method of using adsorption technology to recover hexane from waste gas not only had low recovery efficiency but also had relatively low purity of the recovered hexane. In response to this problem, the Membrane Science and Technology Research Institute of Nanjing Tech University and Jiusi Gaoke jointly developed a hexane recovery coupling process with PDMS/ceramic composite membrane as the core. By setting a − 30 °C condenser at the air inlet, the high-concentration hexane waste gas is pretreated to recover some hexane, and the remaining low-concentration hexane vapor is passed into the membrane assembly. Through the enrichment function of the membrane, the hexane on the permeation side is condensed, the remaining gas can achieve standard discharge, and the unqualified gas flow on the retention side is recycled to the membrane assembly for further recovery until the discharge standard is reached. This process can recover more than 90% of the hexane vapor in the waste gas, recovering more than 2.5 ton of hexane per day, which is much more efficient than the original adsorption process.

In addition, in the petrochemical, coal chemical, storage and other industries, a large amount of volatile oil and gas is produced by tank feeding and loading and unloading. In these situations, condensation/absorption pretreatment, membrane process deep enrichment and coupling pressure swing adsorption or other end treatment technologies can be used to finally stabilize the non-methane total hydrocarbon concentration in the exhaust chimney at 80 mg/m^3 or less. The related application units involve large and medium-sized national and local enterprises such as Sinopec, Shaanxi Coal Group and Jiangsu Overseas Group. So far, more than thirty sets of membrane method industrial application coupling devices have been promoted [68].

6.2 Membrane-Based Absorption Desulfurization

6.2.1 Types and Characteristics of Membrane-Based Absorption Desulfurization

Membrane Gas Absorption (MGA) [69–71] is a separation technology that couples membrane separation technology with absorption technology, fully combining the compact structure of membrane separation and the high-efficiency selectivity of chemical absorption. It is a promising new membrane separation technology, as shown in Fig. 6.13.

The mixed gas and absorption liquid entering the membrane contactor flow on both sides of the tubular membrane. The membrane has no separation effect on the gas components, but only separates the gas and liquid phases, preventing the gas and liquid phases from mixing and dispersing, and the gas and liquid phases contact on a

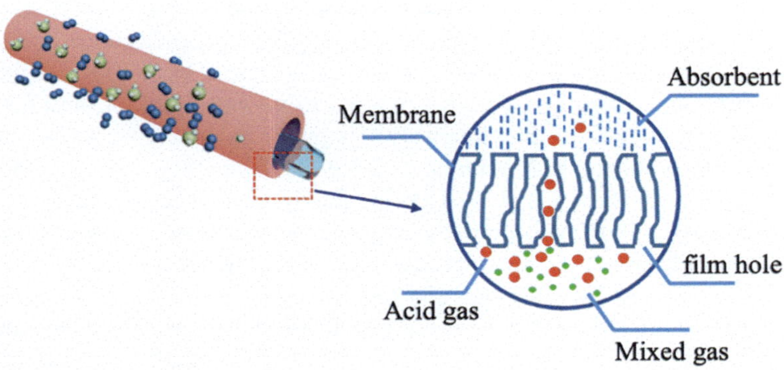

Fig. 6.13 Schematic diagram of membrane gas absorption technology

fixed interface. The acidic gas in the mixed gas is driven by the concentration gradient, passes through the micropores of the membrane material from the gas phase body to the gas–liquid contact interface, and is absorbed, thereby achieving the separation of the mixed gas.

Compared with traditional tower desulfurization equipment, membrane gas absorption technology has the following advantages [72, 73].

(1) Flexible operation: The gas and liquid phases flow on both sides of the membrane and do not come into direct contact. The flow rates of the gas and liquid phase fluids can be controlled, avoiding the problems of leakage, liquid entrainment, mist entrainment, and channel flow in traditional absorption towers.
(2) Compact structure: The membrane contactor has a high packing density and a large specific surface area. The required equipment volume is smaller for processing the same volume of gas.
(3) Performance prediction: Since the gas and liquid phases mass transfer at a fixed interface, the effective contact area is fixed, and its performance can be predicted under stable operating conditions.
(4) Scalable: The modular design not only allows its performance to be predicted but also allows it to be scaled up or down according to the processing volume by series and parallel configurations.

However, membrane-based gas absorption technology also increases the mass transfer resistance of the membrane phase due to the presence of the membrane, thereby increasing the total mass transfer resistance, which is not conducive to absorption. Therefore, reducing mass transfer resistance, preventing membrane contamination, and improving service life are directions for the improvement of membrane-based gas absorption technology.

In recent years, membrane-based gas absorption technology has been increasingly widely used, and its application in the removal of acidic gases (SO_2, H_2S, NO_2), CO_2 capture, natural gas cleaning, ammonia recovery, hydrogen recovery and other fields has attracted much attention [74]. MGA research mainly focuses on the selection of membrane materials, the selection of absorbents, the analysis of the absorption mass transfer process, and the optimization of the design of the membrane contactor.

6.2.2 Membrane Preparation Methods for Membrane-Based Absorption Desulfurization

Polypropylene (PP) [75–77], polyvinylidene fluoride (PVDF) [78–80], polytetrafluoroethylene (PTFE) [81–83], polyether ether ketone (PEEK) [84], polyethylene (PE) [85] and polyethersulfone (PES) [86], polysulfone (PS) [87] are common hollow fiber membrane materials in organic membranes. Qi and Cussler et al. [88, 89] used PP hollow fiber membrane contactor to absorb gases such as SO_2, NH_3, H_2S, and CO_2 using different absorbents. Sun et al. [90] used seawater, deionized water, NaOH solution as absorbents, and used PP hollow fiber membrane assembly to absorb SO_2.

Jeon et al. [91] used water, NaOH, Na_2SO_3 solution to remove SO_2 under PVDF and PTFE hollow fiber membrane contactors, and found that PTFE membrane performance is superior to PVDF membrane. Khaisri et al. [92] compared the absorption performance of PTFE, PP, and PVDF membrane contactors for CO_2. The experimental results indicate that the CO_2 absorption performance follows the order: PTFE > PVDF > PP. Among the three most common organic membrane materials in MGA, PP is cheap but has poor chemical stability; PVDF is more expensive, but can withstand higher temperatures, and is resistant to chemical corrosion; PTFE is also costly but has the most stable chemical properties.

Most organic membranes will experience the phenomenon of absorbing liquid into the membrane pores causing the pores to swell during long-term operation, leading to unstable operation, decreased mass transfer performance, and shortened membrane life [93–97]. Therefore, some researchers have modified organic membranes to improve the stability of membrane materials [98, 99]. Lv et al. [100] used surface modification methods to form a loose porous coating on the surface of the PP membrane to increase the surface roughness and contact angle of the membrane, thereby improving its anti-wetting performance.

Compared with organic membranes, ceramic membranes have advantages such as high temperature resistance, acid and alkali resistance, high mechanical strength, narrow pore size distribution, easy cleaning, and long service life [101]. In recent years, there has been a growing body of research focusing on ceramic membranes in MGA. Yu et al. [102] used monoethanolamine (MEA) as the absorbent, and used hydrophobic modified zirconia ceramic membrane and PP membrane to capture CO_2, and found that the cost per unit area of the ceramic membrane is 12.5 times that of the PP membrane, but at the same CO_2 absorption flux, the cost of the ceramic membrane is lower than that of the PP membrane, and the ceramic membrane has higher stability and anti-pollution ability than the PP membrane. Magnone et al. [103] used fluorinated alkyl silane-modified Al_2O_3 hollow fiber membrane to absorb CO_2, and found that the absorption flux of CO_2 is higher than that of traditional polymer membranes. Luis et al. [104–106] used N, N-dimethylaniline as the absorbent, used Al_2O_3 hollow fiber membrane to remove SO_2, and established a model of the desulfurization mass transfer process. Han et al. [107] used pure water as the absorbent, and used hydrophobic modified zirconia ceramic membrane contactor to absorb SO_2, and found that compared with traditional packed towers, ceramic membranes have a smaller mass transfer unit height.

6.2.3 *Membrane Performance Parameters in Membrane-Based Absorption Desulfurization*

Generally, the total mass transfer coefficient (K_G) and SO_2 removal rate (η) are the two main indicators for evaluating gas absorption performance. Considering the SO_2 in the gas phase. The proportion is very low ($\leq 0.1\%$, volume fraction), so the difference in gas flow into and out of the membrane contactor can be ignored. Therefore, the SO_2 removal rate (η) can be calculated by Eq. (6.10). Where $y_{g,in}$ and $y_{g,out}$ represent the concentration of SO_2 in the inlet and outlet gases, respectively.

$$\eta = \frac{y_{g,in} - y_{g,out}}{y_{g,in}} \times 100\% \tag{6.10}$$

KG can be calculated by Eq. (6.11), where $x_{1,out}$ can be obtained from the conservation of mass.

$$K_G = \frac{Q_g}{A} \frac{y_{g,in} - y_{g,out}}{\frac{\left(y_{g,in} - y^*_{g,out}\right) - \left(y_{g,out} - g^*_{g,in}\right)}{\text{In}\,\frac{y_{y,in} - y^*_{g,in}}{y_{g,out} - y^*_{g,in}}}} \tag{6.11}$$

$$Q_g\left(y_{g,in} - y_{g,out}\right) = Q_1\left(x_{1,out} - x_{1,in}\right) \tag{6.12}$$

When using water absorption, it can be assumed that $x_{1,in} = 0$, and the liquid flow rate at the outlet can be calculated using the gas phase equilibrium formula (6.13).

$$y^*_{g,in} = mx_{1,out} \ldots y^*_{g,out} = mx_{1,in} = 0 \tag{6.13}$$

In the above equations, K_G is the total mass transfer coefficient based on the gas phase, m/s; Q_g is the gas flow rate, m^3/s; Q_l is the liquid flow rate, m^3/s; A is the gas-liquid contact area, m^2; $x_{1,in}$ and $x_{1,out}$ are the molar fractions of SO_2 at the liquid phase inlet and outlet respectively; $y^*_{g,in}$ and $y^*_{g,out}$ are the molar fractions of SO_2 in the gas phase in equilibrium with $x_{1,in}$ and $x_{1,out}$ respectively; m is the phase equilibrium constant.

The mass transfer flux of SO_2 can be calculated by the following formula:

$$F = \frac{\eta Q_g C_g, 0}{A} \tag{6.14}$$

where F is the SO_2 mass transfer flux of the membrane contactor, mol/(m^2 h); $c_{g,0}$ is the inlet SO_2 concentration, mol/m^3.

Furthermore, for hydrophobic desulfurization membranes, the evaluation of their surface hydrophobicity is very important. Usually, measuring the contact angle of the membrane surface is the most direct and effective method. Figure 6.14a shows the water contact angle of the ceramic membrane before and after modification measured by a contact angle analyzer. It is obvious that the droplet will quickly spread on the unmodified membrane surface. In contrast, the contact angle of the modified membrane is stable at 134°, indicating that the membrane has good hydrophobicity.

For desulfurization membranes, the size of the pure water flux and nitrogen permeation flux reflects whether the pore channel is conducive to the passage of liquid and gas. As shown in Fig. 6.14b, the flux of hydrophilic desulfurization membrane almost linearly increases with the increase of transmembrane pressure difference, and the

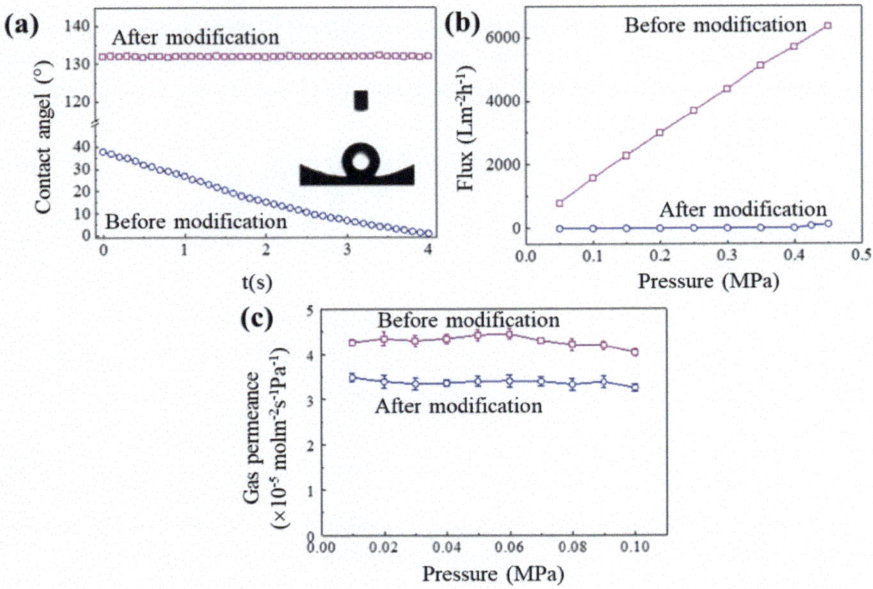

Fig. 6.14 Comparison of the performance of the modified membrane and the original membrane

pure water permeability is about 1800 L/(m^2 h bar). However, when the transmembrane pressure difference is less than 0.40 MPa, the hydrophobic desulfurization membrane shows great hydrophobicity and there will be no water permeation at all. Figure 6.14c shows the relationship between N_2 permeability and transmembrane pressure difference. Compared with hydrophilic membrane, the N_2 permeability of HDTMS modified ceramic membrane slightly decreases. The decrease in N_2 permeability is caused by the reduction of pore size caused by the adhesion of HDTMS molecular layer on the inner surface of the pore channel. Since the HDTMS modification layer is very thin and no more than 3 nm, the decrease in permeability is not significant [102, 108], and after surface hydrophobic modification, the modified membrane still has good air permeability.

6.2.4 Mass Transfer Models of Membrane-Based Absorption Desulfurization

6.2.4.1 Mass Transfer Model of Hydrophobic Desulfurization Membranes

The absorption process of a typical hydrophobic membrane contactor can be divided into several steps: diffusion from the gas phase bulk to the membrane surface, gas

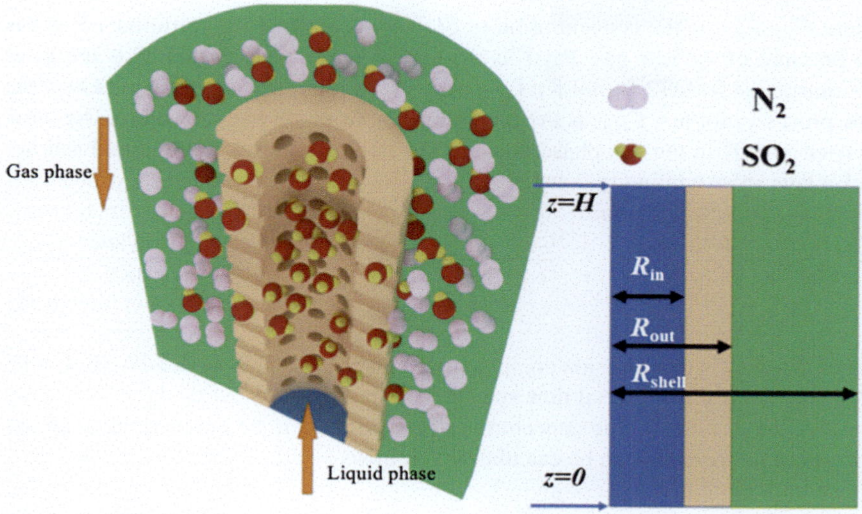

Fig. 6.15 Cross-section of a hydrophobic desulfurization membrane contactor

molecule diffusion through the membrane pores, and transfer from the membrane liquid interface to the liquid phase bulk, followed by liquid phase diffusion and chemical reaction at the interface and within the liquid phase. Therefore, the establishment of the corresponding desulfurization mass transfer model needs to consider the convective and diffusive contributions of SO_2 as well as the reaction between SO_2 and water. During the absorption process, the gas phase and liquid phase counterflow through the membrane contactor.

The mass transfer of SO_2 occurs in three different regions: the shell side, the membrane pores, and the tube side (as shown in Fig. 6.15).

The mass conservation equation for SO_2 in the entire model can be expressed as:

$$\frac{\partial c}{\partial t} = R + D\nabla^2 c - \nabla_C U_Z \tag{6.15}$$

where D is the diffusion coefficient of SO_2 in water or gas, m^2/s; U_z is the axial flow speed of gas or liquid, m/s; R is the absorption reaction rate of SO_2, mol/(m^3 s); t is time, s; c is the molar concentration of SO_2 in the gas or liquid phase, mol/m^3, which can be calculated from the molar fraction of SO_2 in the gas phase (y) or the molar fraction of SO_2 in the liquid phase (x).

$$C_{SO_2-shell} = \frac{yP_{gas}}{RT} \tag{6.16}$$

$$C_{SO_2-tube} = \frac{xc_{H_2O}}{1-x} \tag{6.17}$$

where $C_{SO_2-shell}$ is the concentration of SO_2 in the shell process, mol/m^3; P gas is the pressure of the raw gas, Pa; T is the temperature of the gas, K; R is the ideal gas constant, 8.314472 J/(mol K); C_{SO_2-tube} is the molar concentration of SO_2 in the tube process, mol/m^3; C_{H_2O} is the molar concentration of water, mol/m^3. The mass transfer of SO_2 in the gas phase does not involve a chemical reaction and follows Fick's law, so its continuity equation is:

$$D_{SO_2-gas}\left[\frac{\partial^2 C_{SO_2-shell}}{\partial r^2}+\frac{1}{r}\frac{\partial C_{SO_2-shell}}{\partial r}+\frac{\partial C_{SO_2-shell}}{\partial Z^2}\right]-U_{\text{shell}}\frac{\partial C_{SO_2-shell}}{\partial Z}=0 \tag{6.18}$$

where D_{SO_2-gas} is the diffusion coefficient of SO_2 in the SO_2/N_2 gas mixture, 1.26×0^{-5} m^2/s; U shell is the axial flow speed of the gas, m/s.

Since the gas flow in the concentric tube is assumed to be laminar, the axial gas flow speed distribution can be calculated as follows [109]:

$$U_{shell}=-\frac{8U_{ave-shell}}{(r_{shell}-r_{out})^2}(r-r_{out})(r-r_{shell}) \tag{6.19}$$

where r_{out} is the outer radius of the desulfurization tubular membrane, m; r_{shell} is the inner radius of the membrane component, m, $U_{\text{ave-shell}}$ is the average flow speed of the shell gas, m/s, and can be calculated by the following equation:

$$U_{ave-shell}=\frac{Q_G}{\pi\left(r_{shell}^2-r_{out}^2\right)} \tag{6.20}$$

where Q_{G} is the gas flow rate, m^3/s.

The boundary conditions of the mass transfer equation are as follows:

$$r=r_{shell},\quad \frac{\partial C_{SO_2-shell}}{\partial r}=0 \quad (\text{Symmetric . membrane}) \tag{6.21}$$

$$r=r_{out},\ C_{SO_2-shell}=C_{SO_2-memebrane} \tag{6.22}$$

$$Z=L,\ C_{SO_2-shell}=C_0 \tag{6.23}$$

where c_0 is the concentration of SO_2 at the inlet, mol/m^3; $C_{SO_2-memebrane}$ is the concentration of SO_2 in the membrane pores, mol/m^3; L is the length of the membrane contactor, m.

Typically, an asymmetric membrane is composed of a support layer and a membrane layer. These two layers have different pore sizes and thicknesses. In the membrane layer, SO_2 mainly diffuses through Knudsen diffusion, while in the support layer, SO_2 mainly diffuses through Fick diffusion, and the two layers have different diffusion coefficients. It can be considered that the mass transfer of

SO_2 in the unwetted membrane pores only involves diffusion and does not involve convection and reaction, and its continuity equation is as follows:

$$D_{SO_2-support}\left[\frac{\partial^2 C_{SO_2-support}}{\partial r^2}+\frac{1}{r}\frac{\partial C_{SO_2-support}}{\partial r}+\frac{\partial^2 C_{SO_2-support}}{\partial Z^2}\right]=0 \quad (6.24)$$

$$D_{SO_2-membrane}\left[\frac{\partial^2 C_{SO_2-membrane}}{\partial r^2}+\frac{1}{r}\frac{\partial C_{SO_2-membrane}}{\partial r}+\frac{\partial^2 C_{SO_2-membrane}}{\partial Z^2}\right]=0 \quad (6.25)$$

where $D_{SO_2-support}$ is the diffusion coefficient of SO_2 in the support, m^2/s; is the diffusion coefficient of SO_2 in the membrane layer, m^2/s.

When a molecular gas passes through a porous medium, there are three mechanisms: molecular diffusion, Knudsen diffusion, and transition flow. The transport of the gas depends on the size of the Knudsen diffusion coefficient D_{Kn}, which is defined as follows:

$$D_{K_n}=\frac{\lambda}{d_p} \quad (6.26)$$

$$\lambda=\frac{k_b T}{\pi\sqrt{2}P\sigma^2} \quad (6.27)$$

where λ is the average free path of gas molecules, while dp is the diameter of the pore. The average free path of the gas used in the experiment (T = 293.15 K, P = 1.01 × 10^5 Pa, = 3.67 × 10^{-10}) is about 67 nm, which is of the same order of magnitude as the pore size. Since there is no mechanical pressure difference on the tube wall, viscous transmission through the tube wall is ruled out. The effective diffusion coefficient can be a combination of molecular diffusion rate and Knudsen diffusion rate:

$$\frac{1}{D_{G,eff}}=\frac{1}{D_M}+\frac{1}{D_{Kn}} \quad (6.28)$$

where D_M and D_{Kn} are the molecular diffusion coefficient and Knudsen diffusion coefficient, respectively. At 293.15 K, 1 atm (10^5 Pa), the D_M of SO_2 is 1.23 × 10^{-5} m^2/s; D_{Kn} can be calculated by the following formula:

$$D_{Kn}=48.5d_p\sqrt{\frac{T}{M_A}} \quad (6.29)$$

where d_p is the diameter of the pore, m; M_A is the molecular weight of SO_2, kg/kmol; D_{Kn} is the Knudsen diffusion coefficient, m^2/s.

For hydrophobic membranes, the mass transfer coefficient of the support and membrane layer depends on the pore structure of the membrane, including the porosity ε of the membrane and the tortuosity factor τ as well as diffusion in the membrane pores. It can be calculated by the following formula:

$$\frac{1}{k_m} = \frac{\tau\delta}{\varepsilon D_{G,eff}} \tag{6.30}$$

where k_{m} is the mass transfer coefficient of the membrane layer, m/s, δ is the thickness of the membrane layer, m; τ is the tortuosity factor; ε is the porosity; $D_{\mathrm{G,eff}}$ is the gas diffusion coefficient, $\mathrm{m^2/s}$.

The boundary conditions of the mass transfer equation in the membrane layer are as follows:

$$\mathrm{z} = 0\,\mathrm{or}\,\mathrm{L} \quad \frac{\partial C_{SO_2-support}}{\partial Z} = 0 \tag{6.31}$$

$$\mathrm{z} = 0\,\mathrm{or}\,\mathrm{L} \quad \frac{\partial C_{SO_2-membrane}}{\partial Z} = 0 \tag{6.32}$$

$$r = r_{in}, \quad C_{SO_2-membrane} = C_{SO_2-tube}/m \tag{6.33}$$

$$r = r_{in} + \delta_{membrane}, \quad C_{SO_2-membrane} = C_{SO_2-support} \tag{6.34}$$

$$r = r_{out}, \quad C_{SO_2-support} = C_{SO_2-shell} \tag{6.35}$$

In the formula, m is the solubility of SO_2 in water, which is 21.91 at 293.15 K [110]; $\delta_{\mathrm{membrane}}$ is the thickness of the membrane layer, m; r_{in} is the inner radius of the desulfurization tubular membrane, m. When simulating the transmission of SO_2 under boundary conditions, the upper and lower ends of the tubular membrane are considered to be closed, that is, SO_2 cannot diffuse from the two ends of the membrane.

The diffusion process of SO_2 in water also follows Fick's law. In addition, SO_2 undergoes a reversible chemical reaction with water. Therefore, the continuity equation for the transmission of SO_2 in the tube under steady state can be expressed as:

$$R_{SO_2} + D_{SO_2-water}\left[\frac{\partial C_{SO_2-tube}}{\partial r^2} + \frac{1}{r}\frac{\partial c_{SO_2-tube}}{\partial r} + \frac{\partial^2 C_{SO_2-tube}}{\partial Z^2}\right] - U_{tube}\frac{\partial C_{SO_2-tube}}{\partial Z} = 0 \tag{6.36}$$

where, $D_{SO_2-water}$ is the diffusion coefficient of SO_2 in water, $2 \times 10^{-9}\ \mathrm{m^2/s}$ [111].

The flow in the tube conforms to laminar flow, and the axial flow velocity distribution in the liquid phase is [112]:

$$U_{tube} = 2U_{ave-tube}\left[1 - \left(\frac{r}{r_{in}}\right)^2\right] \tag{6.37}$$

$U_{\text{ave-tube}}$ is the average flow velocity of the liquid in the tube, m/s, which can be calculated by the following formula:

$$U_{ave-tube} = \frac{Q_L}{\pi r_{in}^2} \tag{6.38}$$

where Q_L is the volume flow rate of the liquid, m^3/s.

The boundary conditions of the mass transfer equation are as follows:

$$r = 0, \quad \frac{\partial C_{SO_2-tube}}{\partial r} = 0\ (\text{Symmetric membrane}) \tag{6.39}$$

$$r = r_{in}, \quad C_{SO_2-tube} = mc_{SO_2-membrane} \tag{6.40}$$

$$Z = 0, \ C_{SO_2-tube} = 0 \tag{6.41}$$

The process of SO_2 being absorbed into water involves two reversible reactions, as shown below [113, 114]:

$$SO_2 + 2H_2O \overset{k_1}{\Leftrightarrow} H_3O^+ + HSO_3^- \quad k_1 = 0.014\ \text{mol/m}^3 \tag{6.42}$$

$$HSO_3^- + H_2O \overset{k_2}{\Leftrightarrow} H_3O^+ + SO_3^{2-} \quad k_2 = 6.24 \times 10^{-8}\ \text{mol/m}^3 \tag{6.43}$$

The concentration of SO_3^{2-} can be ignored when the pH value is below 5 [115]. Therefore, the concentrations of HSO_3^- and H_3O^+ can be considered equal under electrically neutral conditions, and the reaction rate of SO_2 can be calculated as follows:

$$-R_{SO_2} = k_3 C_{SO_2} - \frac{k_3}{k_1}\left(C_{HSO_3^-}\right)^2 \tag{6.44}$$

The ionization rate of SO_2, denoted as k_3, is $3.17 \times 10^{-2}\ s^{-1}$.

6.2.4.2 Mass Transfer Models of Hydrophilic Desulfurization Membranes

When using high concentration absorbents to absorb low concentration gases, hydrophilic membrane contactors have better absorption effects than hydrophobic membrane contactors, so it is necessary to establish a mass transfer model for

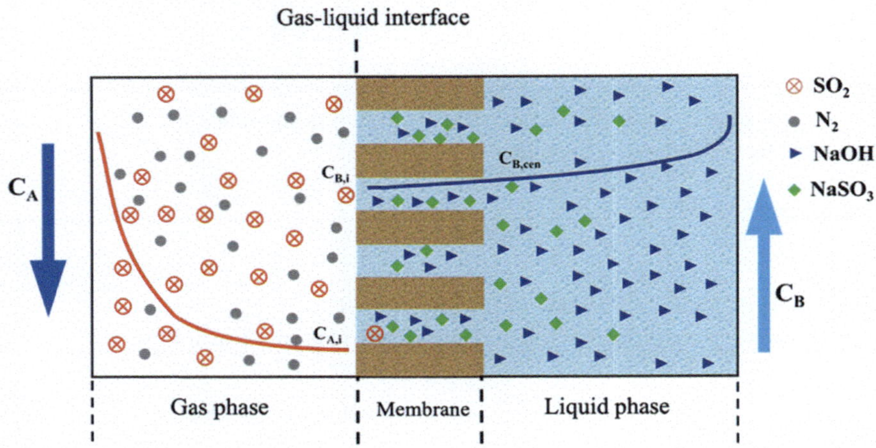

Fig. 6.16 Schematic diagram of the mass transfer process in a hydrophilic desulfurization membrane contactor

hydrophilic membrane contactors to optimize their structure. Taking the absorption of low concentration SO_2 by NaOH as an example, the absorption of SO_2 by NaOH using a hydrophilic membrane can be divided into the following four steps:

(1) SO_2 diffuses from the main body of the gas phase on the feed side to the gas film interface;
(2) SO_2 reacts at the gas–liquid interface to form SO_3^{2-};
(3) SO_3^{2-} diffuses within the membrane pores;
(4) SO_3^{2-} diffuses into the liquid phase and is carried out of the membrane contactor.

At the same time, the diffusion process of NaOH is opposite to that of SO_2 and stops at the liquid–gas interface. The gas phase on the shell side and the liquid phase on the tube side flow in counter-current operation. The hydrophilic membrane provides an effective fixed liquid layer. The mass transfer process is shown in the schematic diagram of Fig. 6.16.

In the entire mass transfer model, the steady-state mass conservation equation of SO_2 can be represented as [116]:

$$R' + D\nabla^2 c - \nabla c U_Z = 0 \tag{6.45}$$

where D is the diffusion coefficient of SO_2 in gas or solution, m^2/s; U_Z is the axial flow velocity of gas or liquid, m/s; R' is the absorption reaction rate of SO_2, mol/(m^3 s); c is the molar concentration of SO_2 in the gas phase, liquid phase, or membrane phase, mol/m^3.

Compared with the hydrophobic membrane mass transfer model, the gas phase mass transfer process is similar to the liquid phase mass transfer process in the hydrophilic membrane mass transfer model, with the difference being the boundary conditions of the gas phase side and the membrane layer:

$$r = r_{shell}, \quad \frac{\partial c_{So_2-shell}}{\partial r} = 0 \tag{6.46}$$

$$r = r_{\text{out}}, \ C_{SO_2-shell} = C_{SO_2-membrane}/m \tag{6.47}$$

$$Z = L, \ C_{SO_2-shell} = C_0 \tag{6.48}$$

Boundary conditions of the liquid phase side and the membrane layer:

$$\text{z} = 0 \, \text{or} \, \text{L}, \quad \frac{\partial C_{SO_2-membrane}}{\partial Z} = 0, \quad \frac{\partial C_{OH^--membrane}}{\partial Z} = 0 \tag{6.49}$$

$$r = r_{in}, C_{SO_2-membrane} = C_{SO_2-tube} \tag{6.50}$$

$$C_{OH^+-membrane} = C_{OH^+-tube} \qquad r = r_{out}, C_{SO_2-membrane} = mc_{SO_2-shell} \tag{6.51}$$

In addition, unlike the hydrophobic desulfurization membrane, the pores of the hydrophilic desulfurization membrane are filled with NaOH solution [117]. Therefore, the concentration at the gas phase interface of the hydrophilic membrane is different from that of the hydrophobic membrane. Because the gas phase is laminar, the flow state of the gas will affect the concentration of SO_2 at the gas–liquid interface. The liquid phase reacting with SO_2 is the NaOH solution in the membrane pores, and the concentration of NaOH and SO_2 will significantly affect the mass transfer and reaction rate.

The interfacial concentration of chemical absorption is often much more complex than physical absorption. Typically, enhancement factors can be introduced to correct the impact of these parameters on the mass transfer model. The desulfurization membrane provides a stable gas–liquid interface during hydrophilic desulfurization. As shown in Fig. 6.16, the concentration of NaOH on the interface ($c_{\text{B,i}}$) is much greater than the concentration of SO_2 ($c_{\text{A,i}}$). The concentration of the liquid away from the interface is basically undisturbed, and in the liquid phase at the concentration centerline, the concentration of NaOH ($c_{\text{B,cen}}$) is similar to the concentration of NaOH in the liquid entering the membrane contactor ($c_{\text{B,in}}$). In this case, the approximate solution of the enhancement factor can use the dimensionless Hatta number [118]. Therefore, establishing a model requires regression of the enhancement factor of the hydrophilic desulfurization membrane contactor, and the mass transfer model suitable for this contactor can be solved through this enhancement factor.

$$E = \frac{J_{chem}}{J_{phy}} = Ha^* = \sqrt{\frac{K_E D_{\text{SO}_2} - membranelC^{AC^{1-A}_{OH+}}_{\text{SO}_2-interface} - interface}{K_L}} \tag{6.52}$$

where J_{chem} and J_{phy} are the fluxes of chemical absorption and physical absorption, respectively; k_{L} is the liquid phase mass transfer coefficient; $D_{\text{SO}_2-membrane}$ is the diffusion coefficient of SO_2 in the membrane; $c_{SO_2-interface}$ and $c_{OH-interface}$ is the concentration of SO_2 and OH^- at the gas–liquid interface. The reaction equation and reaction rate constant of SO_2 and OH^- are as follows:

$$SO_2 + 2OH^- = SO_3^{2-} + H_2O \quad K_E = 1.0 \times 10^6\,m^6/(mol^2\,s) \tag{6.53}$$

where k_E is the reaction rate constant. Since OH^- is in excess, the consumption of SO_2 is half of OH^-. The OH^- concentration at the interface can be calculated by the following formula:

$$N_{OH^-} = K_L(C_{OH^- -liquid} - C_{OH^- -interface}) \tag{6.54}$$

where N_{OH^-} is the diffusion flux, mol/(m^2 s). Since the membrane pores are filled with liquid, there is no Knudsen diffusion in the membrane pores. The membrane layer is much thicker than the liquid boundary layer, so k_L can be represented as:

$$K_L = \frac{D_{OH^- -liquid}\,\varphi}{\tau l_m} \tag{6.55}$$

where $D_{OH^- -liquid}$ is the diffusion coefficient of OH^- in the liquid, m^2/s; φ is the porosity; τ is the tortuosity factor of the membrane pores; lm is the membrane thickness, m.

Similarly, the SO_2 concentration at the interface can be calculated using the following formula:

$$N_{SO_2} = -D_{SO_2}\frac{C_{SO_2-gas} - C_{SO_2-interface}}{\delta_{gas}} = k_g(C_{SO_2-gas} - C_{SO_2-interface}) \tag{6.56}$$

where k_g is the gas phase mass transfer coefficient, m/s; δ_{gas} is the thickness of the gas phase boundary layer, m. In the membrane contactor, the Sherwood number (Sh) is usually related to the Reynolds number (Re) and the Schmidt number (Sc). When gas and liquid flow in a circular straight tube in laminar flow ($Re < 1000$), the Sherwood number can be represented as [119, 120].

$$Sh = \frac{k_g L}{D_{SO_2-gas}} = BRe^c Sc^{0.33} \tag{6.57}$$

where L is the length of the membrane contactor, m; D_{SO_2-gas} is the diffusion coefficient of SO_2 in N_2, m^2/s; B and C are constants [105]. The gas phase mass transfer coefficient depends on the configuration of the membrane contactor and the speed of the gas.

The concentration at the gas–liquid interface is significantly different due to the addition of the membrane compared to direct contact. Due to the lack of research on the use of NaOH solution to absorb SO_2 in hydrophilic membrane contactors, the above unknown constants can be regressed through experiments to establish a complete mass transfer equation.

6.2.5 Comparison of Membrane Absorption and Traditional Techniques

Wet desulfurization is a desulfurization method with high desulfurization rate and large processing capacity. The International Maritime Organization (IMO) suggests in its recommendations on ship exhaust that ships can only use high-sulfur marine fuel when the exhaust is combined with waste gas treatment (such as SO_2 scrubbers) and meets emission requirements. Scrubbers have been used in power plants for many years and have only started to be used in ship exhaust treatment in recent years. Wet scrubbers allow exhaust gas to pass through a liquid medium, and the SO_2 reacts with the active ingredients of the detergent to achieve desulfurization. The most common absorbents are untreated seawater or chemically treated freshwater. Seawater scrubbers are usually open-loop, with seawater directly pumped from the sea and discharged into the sea after treatment, with the seawater only passing through the equipment once. In closed-loop scrubbers, the treated fresh water is continuously circulated back to the scrubber in a closed loop. In the closed-loop system, particles and water are treated to maintain their pH value, making them suitable for repeated use in the scrubber.

The exhaust gas cleaning system (EGCS) is a simple and effective technology that has been used in industrial applications for many years. A wet SO_x scrubber generally includes the following units:

(1) Gas–liquid scrubbing unit of the scrubber. The scrubbing liquid closely contacts the exhaust gas from one or more combustion units. This unit is usually installed at a higher position on the ship;
(2) Wastewater treatment unit. The scrubbing wastewater is treated before being discharged from the ship;
(3) Sludge residue treatment equipment. It treats the sludge residue separated from the scrubber;
(4) Scrubber control and emission monitoring system.

These components are interconnected through pipes with various pumps, coolers, and water tanks, depending on the type of scrubber. A set of sludge residue treatment equipment and a wastewater treatment unit can serve multiple scrubbers. A monitoring and control system determines whether these treatment devices are dedicated to a single scrubber or shared by multiple scrubbers. Marine SO_2 wet scrubbers can mainly be divided into closed-loop scrubbers, open-loop scrubbers, hybrid scrubbers, and membrane contactors.

Membrane absorption is a rapidly developing new technology. The driving force of this technology is the concentration gradient of gas–liquid phase substances. In the membrane absorption process, the membrane is non-selective to gases and only

Table 6.2 Comparison between spray tower and membrane contactor [122]

Parameter	Spray tower	Membrane contactor
Sulfur Dioxide Concentration (μL/L)	1000	1000
Gas Velocity (m/s)	0.65	0.65
Sulfur Dioxide Removal Rate (%)	80	79.87
Liquid to Gas Ratio (*L/G*)	4	4
Height (m)	0.755	0.6
Cross-sectional Area (m^2)	0.138	0.0011
Mass Transfer Coefficient, $K_g a$ ($\times 10^6$ kmol/(m^3 s Pa))	1.734	1094.9

serves as a barrier between the gas flow and the liquid. The selectivity of absorption is provided by the absorbent. The gas mixture and the liquid flow on both sides of the fixed gas–liquid interface in the membrane contactor. The gas diffuses from the gas mixture to the gas–liquid interface through the membrane pores without high pressure, and then contacts the liquid absorbent on the other side. Afterwards, the solution enriched with the absorbed gas can be sent to the second membrane contactor to desorb the absorbed gas. At the same time, the dilute solution is recovered to the absorbent storage tank. The cleaning and capture process of the gas can be completed using the membrane contactor [121].

Compared with the experimental values of the traditional spray tower, the results are shown in Table 6.2. The hydrophilic desulfurization membrane contactor can operate under conditions with smaller liquid–gas ratios. This is because the gas and liquid phases flow on both sides of the membrane without interfering with each other, with small mass transfer resistance and no flooding. Its mass transfer coefficient is much higher than that of the spray tower, which indicates that the hydrophilic desulfurization membrane contactor has strong application potential. In addition to the process of absorbing SO_2 with NaOH, by adjusting the thickness of the membrane or accelerating the rate of absorption reaction to enhance mass transfer, the hydrophilic desulfurization membrane of different absorbed gases and absorbents can achieve a higher mass transfer coefficient than the traditional hydrophobic membrane.

References

1. Committee NSM. Limit of harmful substances in architectural wall coatings: GB 18582—2020. Beijing: China Standards Publishing House; 2020.
2. Zhang XY. The harm and recovery and treatment technology of VOC. Green Environ Protection Build Mater. 2017;9:261–2.
3. Zhao L, Zhang YF, Li RH, et al. The harm and recovery and treatment technology of VOC. Chem Educ. 2015;36:1–6.
4. Wang YF, Liu CX, Cheng J, et al. Comparative analysis of economic measures and engineering technology emission reduction of industrial VOCs. Environ Sci. 2015;36(4):1507–12.
5. Skov H, Hansen AB, Lorenzen G, et al. Benzene exposure and the effect of traffic pollution in Copenhagen, Denmark. Atmos Environ. 2001;35(14):2463–71.

6. Lin L, Lu J, Ma YG, et al. The history of VOCs emission management control at home and abroad. Environ Monit Manage Technol. 2011;5:16–20.
7. Wang HL, Nie L, Li J, et al. Emission characteristics and assessment analysis of volatile organic compounds in key industries. Sci Bull. 2012;19(57):1739–46.
8. Drioli E, Stankiewicz AI, Macedonio F. Membrane engineering in process intensification: an overview. J Membr Sci. 2011;380(1):1–8.
9. Li J, Qiang N. Research progress on the adsorption and recovery of volatile organic compounds by activated carbon. Chem Ind Environ Protection. 2008;28(1):24–8.
10. Feng YY, Xu S, Liu DB, et al. Optimal design of organic solvent recovery by condensation method. Chem Eng. 2012;40(1):35–42.
11. Wang JH, Chen JQ, Cao JS. Analysis of membrane separation technology for VOCs recovery in gas stations. Membr Sci Technol. 2009;3:97–102.
12. Huang G, Li S, Liu L, et al. Ti_3C_2 MXene-modified Bi2WO6 nanoplates for efficient photodegradation of volatile organic compounds. Appl Surf Sci. 2020;503: 144183.
13. Li X, Zhang L, Yang Z, et al. Adsorption materials for volatile organic compounds (VOCs) and the key factors for VOCs adsorption process: a review. Sep Purif Technol. 2020;235: 116213.
14. He X, Hägg M. Membranes for environmentally friendly energy processes. Membranes. 2012;2(4):706–26.
15. Liu CP. Selection and process optimization of VOCs treatment technology in petrochemical intermediate tank area. Chem Ind Environ Protection. 2019;39(3):273–7.
16. Ambrosi A, Cardozo, Nilo Sérgio M, et al. Membrane separation processes for the beer industry: a review and state of the art. Food Bioprocess Technol. 2014;7(4):921–36.
17. Yang W, Zhou H, Zong C, et al. Study on membrane performance in vapor permeation of VOC/N_2 mixtures via modified constant volume/variable pressure method. Sep Purif Technol. 2018;200:273–83.
18. Zhou H, Tao F, Liu Q, et al. Microporous polyamide membranes for molecular sieving of nitrogen from volatile organic compounds. Angew Chem Int Ed. 2017;56(21):5755–9.
19. Liu G, Hou D, Wei W, et al. Pervaporation separation of butanol-water mixtures using polydimethylsiloxane/ceramic composite membrane. Chin J Chem Eng. 2011;19(1):40–4.
20. Wang H, Dai Y, Ruan X, et al. Highly efficient tetrafluoroethylene recovery for batch polymerization system: membrane preparation and process development. J Membr Sci. 2018;549(1):403–10.
21. Ruan X, Wang H, Dai Y, et al. Polyimide membrane system for tetrafluoroethylene recovery: industrial plant, optimal operation and economic analysis. Sep Purif Technol. 2017;188:468–75.
22. Xue J, Wang S, Han X, et al. Chitosan-functionalized graphene oxide for enhanced permeability and antifouling of ultrafiltration membranes. Chem Eng Technol. 2018;41(2):270–7.
23. He X, Wang T, Li Y, et al. Fabrication and characterization of micro-patterned PDMS composite membranes for enhanced ethanol recovery. J Membr Sci. 2018;563:447–59.
24. Li X, Cai W, Wang T, et al. AF2400/PTFE composite membrane for hexane recovery during vegetable oil production. Sep Purif Technol. 2017;181:223–9.
25. Thomas S, Pinnau I, Du N, et al. Hydrocarbon/hydrogen mixed-gas permeation properties of PIM-1, an amorphous microporous spirobisindane polymer. J Membr Sci. 2009;338(1–2):1–4.
26. Barrer RM. Zeolites and their synthesis. Zeolites. 1981;1(3):130–40.
27. Zhou YX, Wang PF, Zhang YM, et al. Preparation of 13X molecular sieve membrane and its application in low carbon olefin purification. Fine Chem. 2018;35(12):1993–8.
28. Li W. Metal-organic framework membranes: production, modification, and applications. Prog Mater Sci. 2019;100:21–63.
29. Li J, Sculley J, Zhou H. Metal-organic frameworks for separations. Chem Rev. 2011;112(2):869–932.
30. Wu Q, Xu R, Li J, et al. Preparation of ZIF-69 membranes for gasoline vapor recovery. J Porous Mater. 2015;22(5):1195–203.

31. Ghanem BS, Msayib KJ, Mckeown NB, et al. A triptycene-based polymer of intrinsic microporosity that displays enhanced surface area and hydrogen adsorption. Chem Commun. 2007;1:67–9.
32. Barrer RM, Chio HT. Solution and diffusion of gases and vapors in silicone rubber membranes. J Polym Sci Polym Symp. 2007;10(1):111–38.
33. Huang DL. Application of membrane technology for separation and recovery of volatile organic compounds. Dalian: Dalian University of Technology; 2006.
34. Birnhack L, Keller O, Tang SCN, et al. A membrane-based recycling process for minimizing environmental effects inflicted by ion-exchange softening applications. Sep Purif Technol. 2019;223:24–30.
35. Nie F, He GH, Zhao W, et al. Preparation and gas separation performance of hydrophobic SiO_2/PTFPMS hybrid composite membrane. J Chem Ind. 2014;65(8):3019–25.
36. Yuan JW, Li QQ, Shen J, et al. Hydrophobic-functionalized ZIF-8 nanoparticles incorporated PDMS membranes for high-selective separation of propane/nitrogen. Asia Pacific J Chem Eng. 2017;12(1):110–20.
37. Feng SC, Ren JZ, Li H, et al. Preparation and separation performance of polyether copolyamide membrane. Membr Sci Technol. 2013;33(4):53–8.
38. Liu L, Chakma A, Feng X, et al. Separation of VOCs from N_2 using poly(ether block amide) membranes. Can J Chem Eng. 2009;87:456–65.
39. Djebbar MK, Nguyen QT, Clément R, et al. Pervaporation of aqueous ester solutions through hydrophobic poly(ether-block-amide) copolymer membranes. J Membr Sci. 1998;146(1):125–33.
40. Wang WQ, Huang WQ, Shen YT, et al. Application of polyether block amide membrane in organic vapor recovery. New Chem Mater. 2012;2:34–6.
41. Baker RW. Membrane technology and applications. New Jersey: Wiley; 2004.
42. Han Q, Zhou HL, Jin WQ, et al. Study on the performance of PDMS/PVDF composite membrane for VOC/N_2 separation. Membr Sci Technol. 2015;1:75–81.
43. Li S, Li P, Si Z, et al. An efficient method allowing for continuous preparation of PDMS/PVDF composite membrane. AIChE J. 2019;65(10): e16710.
44. Si Z, Li J, Ma L, et al. The ultrafast and continuous fabrication of a polydimethylsiloxane membrane by ultraviolet-induced polymerization. Angew Chem. 2019;58(48):17175–9.
45. Li Y, Shen J, Guan K, et al. PEBA/ceramic hollow fiber composite membrane for high-efficiency recovery of bio-butanol via pervaporation. J Membr Sci. 2016;510:338–47.
46. Ma Y, Zhang F, Yang S, et al. Evidence for entropic diffusion selection of xylene isomers in carbon molecular sieve membranes. J Membr Sci. 2018;564:404–14.
47. Liu Y, Zhang B, Liu D, et al. Fabrication and molecular transport studies of highly c-oriented AFI membranes. J Membr Sci. 2017;528:46–54.
48. Aminabhavi TM, Khinnavar RS, Harogoppad SB, et al. Pervaporation separation of organic-aqueous and organic-organic binary mixtures. Polym Rev. 1994;2(34):139–204.
49. Schofield RW, Fane AG, Fell CJD. Gas and vapour transport through microporous membranes. II. Membrane distillation. J Membr Sci. 1990;53(1–2):185.
50. Zong CX, Zhou HL, Jin WQ, et al. Progress in VOCs gas separation technology by membrane method. Membr Sci Technol. 2020;40(1):284–93.
51. Zito PF, Caravella A, Brunetti A, et al. Knudsen and surface diffusion competing for gas permeation inside silicalite membranes. J Membr Sci. 2017;523:456–69.
52. Way JD, Roberts DL. Hollow fiber inorganic membranes for gas separations. Sep Sci Technol. 1992;27(1):29–41.
53. Vallieres C, Favre E, Arnold X, et al. Separation of binary mixtures by dense membrane processes: influence of inert gas entrance under variable downstream pressure conditions. Chem Eng Sci. 2003;58(13):2767–75.
54. Davis EM, Minelli M, Giacinti Baschetti M, et al. Non-fickian diffusion of water in polylactide. Ind Eng Chem Res. 2014;52(26):8664–73.
55. Berens AR. Transport of plasticizing penetrants in glassy-polymers. ACS Symposium Series; 1990.

56. Visser T, Wessling M. When do sorption-induced relaxations in glassy polymers set in. Macromolecules. 2007;40(14):4992–5000.
57. Bohning M, Springer J. Sorptive dilation and relaxational processes in glassypolymer gas systems: I. Poly(sulfone) and poly(ether sulfone). Polymer. 1998;39(21):5183–95.
58. Pandey P, Chauhan RS. Membranes for gas separation. Prog Polym Sci. 2001;26(6):853–93.
59. Ruffolo D, Pianpanit T, Matthaeus WH, et al. Random ballistic interpretation of nonlinear guiding center theory. Astrophys J Lett. 2012;747(2):L34.
60. Zong CX, Zhou HL, Jin WQ, et al. Separation performance of polydimethylsiloxane-ceramic composite membrane for n-hexane-N_2. J Nanjing Tech Univ (Natural Science Edition). 2019;41(5):649–55.
61. Li Q, Liu G, Huang K, et al. Preparation and characterization of Ni2(mal)2(bpy) homochiral MOF membrane. Asia-Pac J Chem Eng. 2016;11(1):60–9.
62. Liu D, Liu G, Meng L, et al. Hollow fiber modules with ceramic-supported PDMS composite membranes for pervaporation recovery of bio-butanol. Sep Purif Technol. 2015;146:24–32.
63. Zhang L, Chen HL, Chai H, et al. Progress in membrane treatment technology for volatile organic waste gas. Chem Ind Environ Protection. 2002;2:75–80.
64. Xing DM, Cao YM, Li H. Application of membrane organic vapor recovery system in industry. Membr Sci Technol. 2000;20(4):43–6.
65. Petrusov Z, Machanov K, Stanovsk P, et al. Separation of organic compounds from gaseous mixtures by vapor permeation. Sep Purif Technol. 2019;217:95–107.
66. Li H, Liu FQ, Cao YM, et al. Process study on membrane separation of organic vapor/nitrogen mixture. Membr Sci Technol. 2000;2:39–42.
67. Yu ZY, Jing XL, Hua KS, et al. Recovery of hydrocarbons from tail gas of gas-phase polyethylene unit using membrane separation technology. Chem Ind Prog. 2007;5:731–4.
68. Jin W, Liu G, Xu N. Organic-inorganic composite membranes for molecular separation. Singapore: World Scientific Publishing; 2017.
69. Mansourizadeh A, Ismail AF. Hollow fiber gas-liquid membrane contactors for acid gas capture: a review. J Hazard Mater. 2009;171(1–3):38–53.
70. Zhao SF, Feron PM, Deng LY, et al. Status and progress of membrane contactors in post-combustion carbon capture: a state-of-the-art review of new developments. J Membr Sci. 2016;511:180–206.
71. Cheng GL, Cheng LH, Zhang L, et al. Progress in gas separation by membrane contactors. Chem Ind Prog. 2006;25:901–6.
72. Drioli E, Criscuoli A, Curcio E. Membrane contactors: fundamentals, applications and potentialities. Amsterdam: Elsevier; 2011.
73. Lv YX. Study on separation of CO_2 by polypropylene hollow fiber membrane contactor. Shanghai: East China University of Science and Technology; 2011.
74. Cui JH, Qi JQ. Research and application progress of membrane absorption technology. Chem Equip Technol. 2005;26:13–7.
75. Demontigny D, Tontiwachwuthikul P, Chakma A. Using polypropylene and polytetrafluoroethylene membranes in a membrane contactor for CO_2 absorption. J Membr Sci. 2006;277(1–2):99–107.
76. Lin S, Tung K, Chen W, et al. Absorption of carbon dioxide by mixed piperazine-alkanolamine absorbent in a plasma-modified polypropylene hollow fiber contactor. J Membr Sci. 2009;333(1–2):30–7.
77. Lin S, Hsieh C, Li M, et al. Determination of mass transfer resistance during absorption of carbon dioxide by mixed absorbents in PVDF and PP membrane contactor. Desalination. 2009;249(2):647–53.
78. Atchariyawut S, Feng C, Wang R, et al. Effect of membrane structure on mass-transfer in the membrane gas-liquid contacting process using microporous PVDF hollow fibers. J Membr Sci. 2006;285(1–2):272–81.
79. Mansourizadeh A, Ismail AF, Matsuura T. Effect of operating conditions on the physical and chemical CO_2 absorption through the PVDF hollow fiber membrane contactor. J Membr Sci. 2010;353(1–2):192–200.

80. Jin PR, Huang C, Li JX, et al. Surface modification of poly(vinylidene fluoride) hollow fibre membranes for biogas cleaning in a gas-liquid membrane contactor system. Royal Soc Open Sci. 2017;4(11):17.
81. Chen S, Li S, Chien R, et al. Effects of shape, porosity, and operating parameters on carbon dioxide recovery in polytetrafluoroethylene membranes. J Hazard Mater. 2010;179(1–3):692–700.
82. Tang HY, Zhang Y, Wang F, et al. Long-term stability of polytetrafluoroethylene (PTFE) hollow fiber membranes for CO_2 capture. Energy Fuels. 2016;30(1):492–503.
83. Wang FS, Kang GD, Liu DD, et al. Enhancing CO_2 absorption efficiency using a novel PTFE hollow fiber membrane contactor at elevated pressure. AIChE J. 2018;64(6):2135–45.
84. Li SG, Pyrzynski TJ, Klinghoffer NB, et al. Scale-up of PEEK hollow fiber membrane contactor for post-combustion CO_2 capture. J Membr Sci. 2017;527:92–101.
85. Mosadegh-Sedghi S, Brisson J, Rodrigue D, et al. Morphological, chemical and thermal stability of microporous LDPE hollow fiber membranes in contact with single and mixed aminebased CO_2 absorbents. Sep Purif Technol. 2012;96:117–23.
86. Li K, Teo WK. Use of permeation and absorption methods for CO_2 removal in hollow fibre membrane modules. Sep Purif Technol. 1998;13(1):79–88.
87. Gomez-Coma L, Garea A, Irabien A. Carbon dioxide capture by [emim][Ac] ionic liquid in a polysulfone hollow fiber membrane contactor. Int J Greenhouse Gas Control. 2016;52:401–9.
88. Qi Z, Cussler EL. Microporous hollow fibers for gas-absorption: 1. Mass-transfer in the liquid. J Membr Sci. 1985;23(3):321–32.
89. Qi Z, Cussler EL. Microporous hollow fibers for gas-absorption: 2. Mass-transfer across the membrane. J Membr Sci. 1985;23(3):333–45.
90. Sun X, Meng F, Yang F. Application of seawater to enhance SO_2 removal from simulated flue gas through hollow fiber membrane contactor. J Membr Sci. 2008;312(1–2):6–14.
91. Jeon H, Ahn H, Song I, et al. Absorption of sulfur dioxide by porous hydrophobic membrane contactor. Desalination. 2008;234(1–3):252–60.
92. Khaisri S, Demontigny D, Tontiwachwuthikul P, et al. Comparing membrane resistance and absorption performance of three different membranes in a gas absorption membrane contactor. Sep Purif Technol. 2009;65(3):290–7.
93. Mosadegh-Sedghi S, Rodrigue D, Brisson J, et al. Wetting phenomenon in membrane contactors—causes and prevention. J Membr Sci. 2014;452:332–53.
94. Zhang HY, Wang R, Liang DT, et al. Theoretical and experimental studies of membrane wetting in the membrane gas-liquid contacting process for CO_2 absorption. J Membr Sci. 2008;308(1–2):162–70.
95. Lv Y, Yu X, Tu ST, et al. Wetting of polypropylene hollow fiber membrane contactors. J Membr Sci. 2010;362(1–2):444–52.
96. Rangwala HA. Absorption of carbon dioxide into aqueous solutions using hollow fiber membrane contactors. J Membr Sci. 1996;112(2):229–40.
97. Wang R, Zhang HY, Feron PHM, et al. Influence of membrane wetting on CO_2 capture in microporous hollow fiber membrane contactors. Sep Purif Technol. 2005;46(1–2):33–40.
98. Zhang Y, Wang R, Zhang LZ, et al. Novel single-step hydrophobic modification of polymeric hollow fiber membranes containing imide groups: Its potential for membrane contactor application. Sep Purif Technol. 2012;101:76–84.
99. Rahbari SM, Ismail AF, Matsuura T, et al. Long-term study of CO_2 absorption by PVDF/ZSM-5 hollow fiber mixed matrix membrane in gas-liquid contacting process. J Appl Polym Sci. 2017;134(14):9.
100. Lv Y, Yu X, Jia J, et al. Fabrication and characterization of superhydrophobic polypropylene hollow fiber membranes for carbon dioxide absorption. Appl Energy. 2012;90(1):167–74.
101. Xu NP, Xing WH, Zhao YJ. Inorganic membrane separation technology and applications. Beijing: Chemical Industry Press; 2003.
102. Yu X, An L, Yang J, et al. CO_2 capture using a superhydrophobic ceramic membrane contactor. J Membr Sci. 2015;496:1–12.

103. Magnone E, Lee HJ, Che JW, et al. High-performance of modified Al_2O_3 hollow fiber membranes for CO_2 absorption at room temperature. J Ind Eng Chem. 2016;42:19–22.
104. Luis P, Garea A, Irabien A. Sulfur dioxide non-dispersive absorption in N, N-dimethylaniline using a ceramic membrane contactor. J Chem Technol Biotechnol. 2008;83(11):1570–7.
105. Luis P, Garea A, Irabien A. Modelling of a hollow fibre ceramic contactor for SO_2 absorption. Sep Purif Technol. 2010;72(2):174–9.
106. Luis P, Garea A, Irabien A. Environmental and economic evaluation of SO_2 recovery in a ceramic hollow fibre membrane contactor. Chem Eng Process. 2012;52:151–4.
107. Han SX, Gao XY, Fu KY, et al. Application of hydrophobic single-tube ceramic membrane contactor in SO_2 absorption. J Chem Ind. 2017;68:2415–22.
108. Li YN, Hao YC, Ye H, et al. Single-sided superhydrophobic fluorinated silica/poly(ether sulfone) membrane for SO_2 absorption. J Membr Sci. 2019;580:190–201.
109. Steimke JL. Natural-convection heat-transfer for a concentric tube thermosiphon. J Heat Transfer-Trans ASME. 1985;107(3):583–8.
110. Hikita HAS, Tsuji T. Absorption of sulfur dioxide into aqueous sodium hydroxide and sodium sulfite solutions. AIChE J. 1977;26:1291.
111. Dutta BK, Pandit A, Ray P. Absorption of sulfur dioxide in citric acid-sodium citrate buffer solutions. Ind Eng Chem Res. 1987;26:1291.
112. Versteeg GF, Van DLAJ, Van SWPM. On the kinetics between CO_2 and alkanolamines both in aqueous and non-aqueous solutions. An overview. Chem Eng Commun. 1996;144:113–58.
113. Karoor S, Sirkar K. Gas absorption studies in microporous hollow fiber membrane modules. Ind Eng Chem Res. 1993;32:674.
114. Roberts DL, Friedlander SK. Sulfur-dioxide transport through aqueous-solutions theory. AIChE J. 1980;26(4):593–602.
115. Shirazian S, Rezakazemi M, Marjani A, et al. Development of a mass transfer model for simulation of sulfur dioxide removal in ceramic membrane contactors. Asia-Pac J Chem Eng. 2012;7(6):828–34.
116. Molina CT, Bouallou C. Carbon dioxide absorption by ammonia intensified with membrane contactors. Clean Technol Environ Policy. 2016;18(7):2133–46.
117. Luis P, Garea A, Irabien A. Zero solvent emission process for sulfur dioxide recovery using a membrane contactor and ionic liquids. J Membr Sci. 2009;330(1–2):80–9.
118. Kumar PS, Hogendorn JA, Feron PHM, et al. Approximate solution to predict the enhancement factor for the reactive absorption of a gas in a liquid flowing through a microporous membrane hollow fiber. J Membr Sci. 2003;213(1–2):231–45.
119. Cote P, Bersillon JL, Huyard A, et al. Bubble-free aeration using membranes—process analysis. J Water Pollut Control Fed. 1988;60(11):1986–92.
120. Ma YH, Ye H, Zhang YZ. Study on the desulfurization performance of polyether sulfone hollow fiber membrane contactor. Membr Sci Technol. 2016;36:60–7.
121. Li JL, Chen BH. Review of CO_2 absorption using chemical solvents in hollow fiber membrane contactors. Sep Purif Technol. 2005;41(2):109–22.
122. Zidar M. Gas-liquid equilibrium-operational diagram: graphical presentation of absorption of SO_2 in the $NaOH$-SO_2-H_2O system taking place within a laboratory absorber. Ind Eng Chem Res. 2000;39(8):3042–50.

Chapter 7
Multifunctional Gas Cleaning Membranes

7.1 Membranes for Nitrogen Oxide Catalytic Degradation

Steel smelting, thermal power generation, automobile exhaust, and urban waste incineration produce a large amount of gaseous pollutants harmful to the ecological environment and human health, such as carbon monoxide, sulfur oxides, nitrogen oxides, etc. [1–3]. In some industrially developed countries, such as Japan and Germany, gas cleaning technology has been industrialized and has achieved significant results. In recent years, with the increase in environmental management requirements, research institutions and industrial production departments in developing countries such as China have begun to conduct research on emission reduction technologies and corresponding materials.

Nitrogen oxides (NO_x) are difficult to control and extremely harmful gaseous pollutants. NO_x can cause photochemical reactions under sunlight, resulting in the formation of photochemical smog, which causes serious air pollution [4, 5]. In addition, problems such as the production of high-concentration nitric acid rain and the reduction of ozone content are also related to NO_x. Therefore, research on nitrogen oxide emission control is increasing.

Nitrogen oxides are divided into stationary sources and mobile sources depending on the source of emissions. Stationary sources refer to waste gases produced in industrial production, such as steel, thermal power generation, cement, and coal-fired boilers, while mobile sources mainly refer to engine exhaust gases produced by various vehicles. However, the production of nitrogen oxides in the actual production process often accompanies a large amount of dust and other fine particulate pollutants, and a single denitration technology can hardly achieve the removal of dust and nitrogen oxides at the same time. The catalytic membrane, as a type of gas cleaning membrane, uses the membrane layer to trap dust, and the catalyst in the membrane material catalyzes the degradation of nitrogen oxides, achieving integrated dust removal and denitration [6, 7]. This section introduces the application of

Z. Zhong and W. Xing, *Membrane Technology for Air Cleaning*,
https://doi.org/10.1007/978-981-96-1144-7_7

catalytic membranes in the removal of dust-containing nitrogen oxides according to different sources of nitrogen oxide emissions.

7.1.1 Membranes for Catalytic Degradation of Stationary Source NO_x

Flue gas denitration technology refers to various physical and chemical processes that reduce and decompose NO_x in flue gas [8]. Over the past few decades, various NO_x control technologies have been developed and applied to stationary-source denitration , such as coal-fired power plants. Existing stationary-source denitration technologies can be divided into two categories: combustion process control technology and flue gas control removal technology. Combustion process control technology reduces the generation of NO_x in the furnace by adjusting the mixing condition of fuel and air, lowering the combustion temperature and initial turbulence. Flue gas cleaning control technology, on the other hand, uses various physical and chemical means to reduce or harmlessly treat NO_x already formed in the furnace. Under normal circumstances, various low-nitrogen combustion technologies can reduce NO_x emissions by up to about 50%. When the NO_x emission requirements of combustion equipment are high, combustion improvement measures alone often cannot meet existing emission standards, so it is necessary to use flue gas control removal technology to further reduce NO_x emissions [9]. In existing high-temperature flue gas treatment systems, dust removal and denitration of flue gas are carried out separately. The flue gas exiting the denitration system needs to be cooled by a heat exchanger before it can enter the dust collector for dust removal. This results in large equipment footprint and energy consumption [9]. If denitration system is placed in front of dust removal system, fine particles in the flue gas will deposit on the surface of the catalyst, causing the catalyst to deactivate quickly and shortening the life of the denitration catalyst. In addition, such multi-step operation processes are long, and the investments in environmental protection facilities, energy consumption, and land occupation are all large, lacking the motivation for sustainable development. The catalytic membrane, by coupling catalysis and membrane separation, can use the small pore size membrane layer to retain dust in industrial flue gas, degrade nitrogen oxides with nano-catalysts in the separation membrane, and achieve clean exhaust gas emission (Fig. 7.1).

The earliest studied catalytic membrane was made by loading precious metal catalysts and vanadium-based catalysts in ceramic membrane supports. Lee et al. [10] prepared a V_2O_5-loaded cordierite catalytic membrane by vacuum impregnation calcination method. The prepared catalytic membrane shows a porosity of 61.6%, a compressive strength of 12.3 MPa, and achieves dust and NO_x removal efficiency of 99.6% and 80%, respectively. In addition, it is found that acid treatment can increase the specific surface area of active components, and then enhance its NO_x removal efficiency to 90%. Nacken et al. [6] prepared a catalytic membrane by immersing

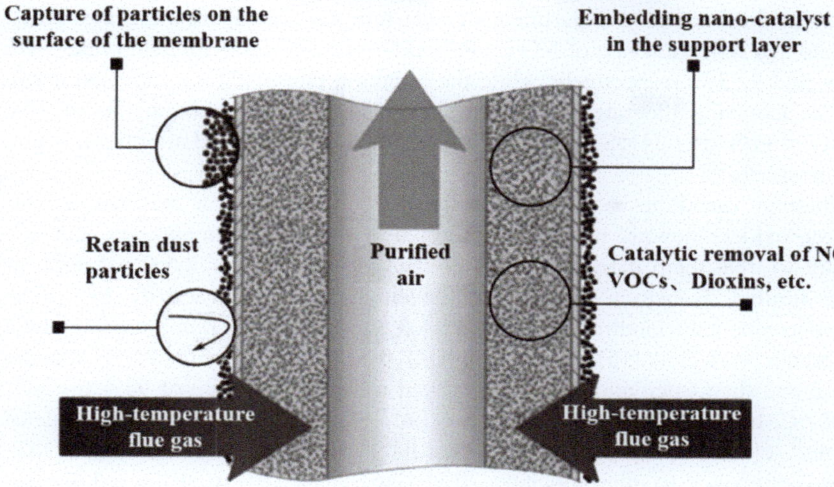

Fig. 7.1 Schematic diagram of catalytic membrane degradation

a SiC–Al_2O_3 porous ceramic membrane in a V_2O_5–WO_3–TiO_2 (V–W–Ti) solution. Under the conditions of a filtration speed of 2 cm/s and 300 °C, the selective catalytic reduction (SCR) conversion rate of NO reached 96%, and the dust retention rate was over 99%. Fino et al. [11] prepared a MnO_x–CeO_2–V_2O_5–WO_3–TiO_2 catalytic membrane by immersing a ceramic membrane in a MnO_x and CeO_2 modified V–W–Ti catalyst, then microwave drying and calcining. The prepared catalytic membrane achieved a conversion rate of over 80% for both NO and benzene, and the dust retention rate remained at 99%. Yuan et al. [12] prepared a V_2O_5–WO_3/TiO_2 honeycomb ceramic catalytic filter and conducted a denitration test in flue gas with a dust concentration of 15 g/m^3. The results showed that the denitration efficiency decreased with the deposition of dust on the catalyst surface (from an initial value of 96.51–80.59% at 140 min). Abubakar et al. [13] prepared a V_2O_5–MoO_3/TiO_2 based catalytic bag filter (as shown Fig. 7.2) by impregnation calcination method, which was used to simultaneously remove particulate matter and NO at a temperature range of 200–250 °C. The increase in V and Mo content in the catalyst can improve the low-temperature activity of the catalyst. When the filtration speed is 0.5 m/min, the NO conversion rate and dust removal efficiency reaches over 80% and 99.9% at 200–250 °C, respectively, proposing that it has good industrial application feasibility. Kim et al. [14] prepared a Pt–V_2O_5–WO_3/TiO_2/SiC catalytic membrane and showed that adding Pt can promote the shift of the NO conversion temperature from 260–340 °C to 160–240 °C. Chen et al. [9] prepared a Pt/SAPO-34@SiC catalytic membrane via sequentially growing a layer of SAPO-34 molecular sieve in the SiC ceramic membrane support and loading Pt nanoparticles to investigate its dust removal and denitration performance (Fig. 7.3). The results showed that growing SAPO-34 molecular sieve can increase the specific surface area of SiC ceramic membrane, which enhance the loading amount of Pt increased by two orders of magnitude, reaching

1.89%. In the integrated experiment of dust removal and denitration, the oxidation rate and reduction rate of NO by Pt/SAPO-34@SiC catalytic membrane were 72.5% and 92.3%, respectively, while the retention rate of $PM_{0.3}$ remained above 99.99%, achieving integrated dust removal and denitration. In addition, the catalytic activity of Pt/SAPO-34@SiC catalytic membrane can be restored after treatment in H_2 atmosphere.

Although vanadium-based catalysts have good degradation performance for nitrogen oxides, they are toxic and pose great harm to the human body, and it is difficult to treat them harmlessly in the environment, so their industrial application is limited. Transition metals, due to their good catalytic activity, low price, and environmental friendliness, are ideal choices for preparing denitration catalytic membranes. Park et al. [15] prepared $CuMnO_x$/PSA catalytic membrane by vacuum suction and then calcination, which achieved a degradation rate of 94% for NO_x at 200 °C, but its dust retention rate was only 90%. In addition, the presence of 150 mg/L of SO_2 would poison $CuMnO_x$ catalyst, resulting in a reduction in NOx removal efficiency to approximately 20%. Chen et al. [16] used the redox reaction between $FeCl_3$ and $KMnO_4$ to prepare MnO_2–Fe_2O_3/PPS catalytic membrane (Fig. 7.4) and found that MnO_2-Fe_2O_3/PPS catalytic membrane has a 99.2% NO_x

Fig. 7.2 Scanning electron microscope images of original filter membrane (**a**, **b**) and V_2O_5-MoO_3/TiO_2 catalytic filter membrane (**c**, **d**) at different magnifications [13]

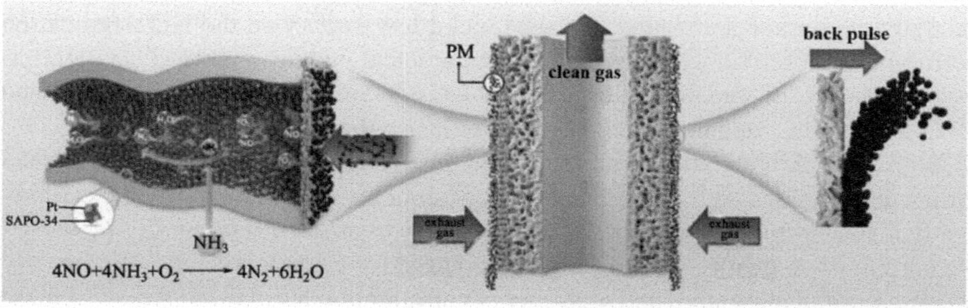

Fig. 7.3 Schematic diagram of dust and nitrogen oxides co-management by Pt/SAPO-34@SiC catalytic membrane [9]

degradation rate at 180 °C and 92% dust retention rate. Furthermore, Fe_2O_3 can inhibit the sulfur poisoning of the MnO_2 catalyst, maintaining a NO_x degradation rate of 85% for the catalytic membrane. Yang et al. [17] prepared a Mn–La–Ce–Ni–O_x/PPS catalytic membrane by vacuum suction and then calcination. This catalytic membrane has a NO_x conversion rate of 85% at 200 °C, a dust retention rate of 90%, and its N_2 selectivity reaches 100%. This is attributed to the Ni-rich surface acidic centers that are conducive to improving the N_2 selectivity of the NH_3-SCR reaction. Yang et al. [18] prepared a Mn–Ce–Nb–O_x/P84 catalytic membrane using a foam coating method and systematically studied the removal performance of the Mn–Ce–Nb–O_x/P84 catalytic membrane for NO and particulate matter under optimized parameters (Fig. 7.5). The results show that the Mn–Ce–Nb–O_x/P84 catalytic membrane prepared by the foam coating method has excellent catalytic and retention performance for NO and particulate matter at 140–220 °C. When the catalyst loading is 450 g/m^2, the NO removal rate of the catalytic membrane can reach 95.3% at 200 °C. In addition, the $PM_{2.5}$ removal rate of the MnCe–Nb–O_x/P84 catalytic membrane is as high as 99.98%. The Mn–Ce–Nb–O_x catalytic membrane also has excellent sulfur resistance. In the presence of 150 μL/L SO_2, the NO removal rate still remains above 85%. Walberer et al. [19] prepared a ceramic catalytic filter using Cu-ZSM as the catalytically active component and carried out synergistic removal of dust, NO_x, and SO_2 produced by power plants on a laboratory scale. By modifying a layer of $Ca(OH)_2$ on the surface of the ceramic membrane, effective adsorption of SO_2 and effective removal of dust can be achieved, and then the catalyst in the support can catalyze the degradation of nitrogen oxides. The results showed that the dust removal rate, NO_x conversion and SO_2 removal efficiency reached 95%, 80% and 86.9% at 200 °C, respectively. Pan et al. [9] prepared a MnO_x/TiO_2/SiC catalytic membrane by a two-step impregnation method for the synergistic removal of dust and nitrogen oxides. Figure 7.6 shows the mechanism of dust and nitrogen oxide synergistic removal by the MnO_x/TiO_2/SiC catalytic membrane. Studies have shown that TiO_2 aerosol is more conducive to enhancing the NO degradation performance of MnO_x than TiO_2 nanoparticles. They systematically investigated the effects of oxygen content, residence time, ammonia nitrogen ratio, dust particle size, dust concentration, and filtration speed on the catalytic and filtration performance of the catalytic membrane and found that dust had a low impact on the NO degradation performance of the catalytic membrane. In the integrated dust removal and denitration test, the NO degradation rate of the $MnO_x/TiO_2/SiC$ catalytic membrane reached 90%, and the dust retention rate for particles larger than 100 nm exceeded 99.97%.

Both ceramic catalytic membranes and fiber catalytic membranes have good dust retention rates and nitrogen oxide degradation rates. Comparatively, ceramic catalytic membranes show higher dust retention rates and thermal stability. However, the current impregnation calcination method used for catalytic membranes not only involves complicated steps and high energy consumption but also leads to a decrease in conversion rate due to catalyst loss during long-term operation, which is not conducive to the promotion and use of catalytic membranes. In addition, there are few studies and industrial application cases of catalytic membranes, and most researchers only focus on the treatment performance of catalytic membranes for one or two

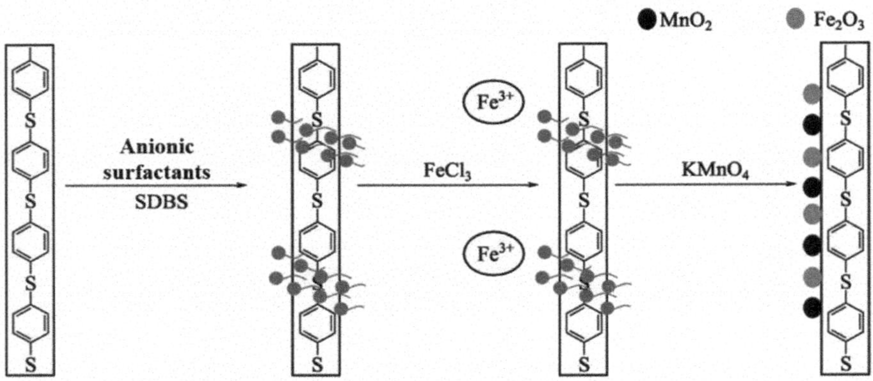

Fig. 7.4 Schematic diagram of the preparation of the MnO_2–Fe_2O_3/PPS catalytic membrane [16]

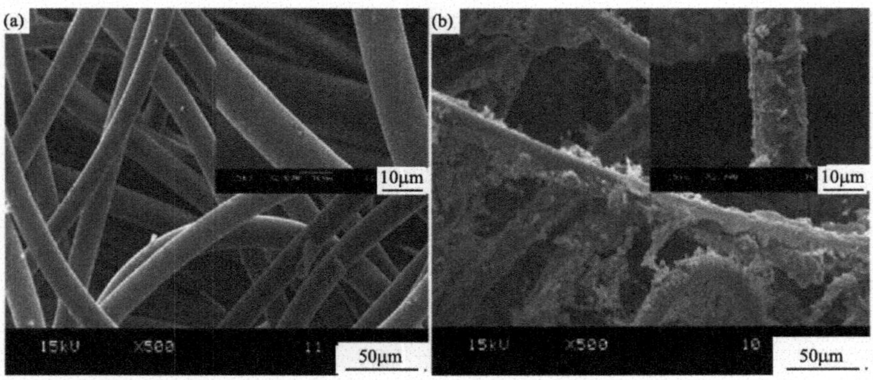

Fig. 7.5 Microscopic morphologies of the P84 fiber membrane (**a**) and Mn-Ce-Nb-O_x/P84 (**b**) fiber catalytic membrane [18]

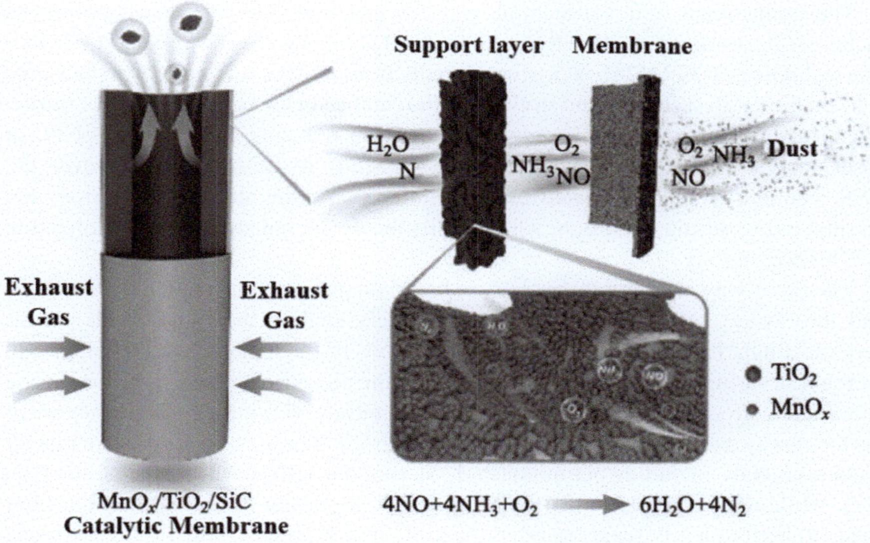

Fig. 7.6 Mechanism of dust and nitrogen oxide synergistic removal by the $MnO_x/TiO_2/SiC$ catalytic membrane [9]

specific gaseous pollutants, neglecting the study of the synergistic treatment performance of catalytic membranes for multiple pollutants. In the future, the development of catalytic membranes should be more focused on the treatment of real systems, exploring the actual synergistic treatment performance of catalytic membranes.

7.1.2 Mobile Source NO_x Catalytic Degradation Membranes

In recent years, the continuous increase in the number of motor vehicles has brought convenience to people, but it has also caused serious environmental pollution. According to the latest annual report on automobile environmental management, annual emission of particulate matter from motor vehicles reached 442,000 tons [20]. Under normal operating conditions, 90% of particulate matter produced in motor vehicle exhaust is composed of dust, heavy metal oxides, and lead oxides, with sizes less than 1 μm ($PM_{1.0}$), which can enter the human body through respiration, endangering people's health [21, 22]. In addition to particulate matter, exhaust pollution produced during the combustion of vehicle fuel is also considered one of the major environmental issues. For example, nitrogen oxides (NO_x) in car exhaust are one of the most harmful gases to the environment and human health, with an annual emission of 5,629,000 tons in China [20]. Therefore, the removal of PM and NO_x from car exhaust is urgent.

The mainstream removal technologies for gasoline vehicle exhaust pollutants include NO_x storage-reduction technology (NSR) [23] and three-way catalytic reduction technology (SCR) [24]. NO_x storage-reduction catalytic technology first adsorbs NO_x on the catalyst surface and stores it on the catalyst in the form of nitrate or nitrite. Then, under brief rich combustion conditions, the reducing agent reduces NO_x to gases such as N_2 and restores the catalyst surface to its initial state. However, the NSR technology is easily deactivated by sulfur poisoning, carbon deposition, and thermal aging during operation, which limits the development and industrialization of NSR technology.

The three-way catalytic technology is based on gasoline particulate filters (GPF) and three-way catalysts (TWCs). By alternately sealing the honeycomb porous ceramic filter, it forces the exhaust gas to flow through the filter channel wall, thereby capturing and filtering particles. Due to the presence of carbon monoxide and unburned hydrocarbons, TWCs can effectively convert NO_x into N_2. The structure of the three-way catalytic converter is shown in Fig. 7.7. The existing TWCs are mainly precious metals, including platinum (Pt) [25], rhodium (Rh) [26], and palladium (Pd) [27], while the high cost limits their large-scale application. In addition, the filter needs to be frequently regenerated in the GPF, thus both GPF and TWCs are prone to irreversible damage due to sudden or accidental temperature increases [28].

In the research of catalytic removal materials for mobile source exhaust, porous metal membrane catalytic materials and porous ceramic membrane catalytic materials (as shown in Fig. 7.8) are mainly used. Tang et al. [20] tried to solve the problem of gasoline vehicle exhaust pollution by preparing multifunctional catalytic membranes through the combination of catalysts and metal fiber membranes (Fig. 7.9). In the environment of automobile exhaust filtration, the exhaust gas temperature can reach up to 600 °C. Under such conditions, organic nanofibers cannot function normally due to their own physical property limitations. Inorganic

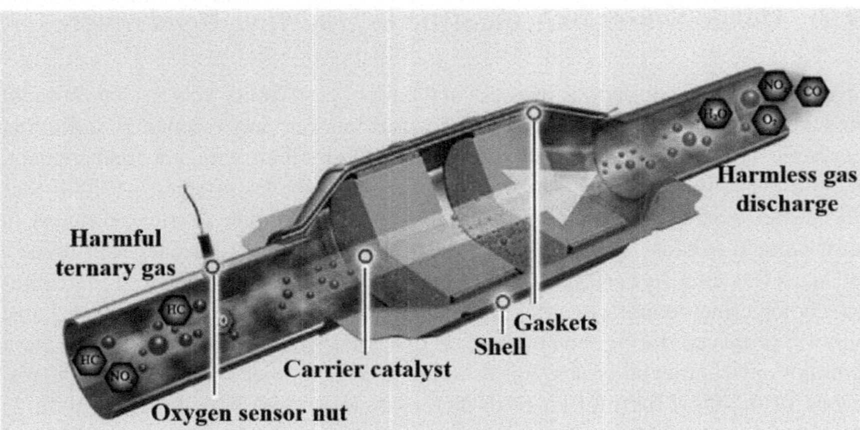

Fig. 7.7 Structure diagram of the three-way catalytic converter

nanofibers, including ceramic fibers and alloy fibers, exhibit excellent performance under these working conditions. Tang et al. [20] prepared large-area (80 mm × 80 mm) Ni nanofibers by electrospinning NiAc/PVP precursor solution, followed by in-situ vacuum heat treatment. These Ni nanofibers have high mechanical strength because the Ni fibers are fused together, not simply stacked. Figure 7.10 shows the schematic diagram and physical picture of the preparation of Ni nanofibers. The prepared nickel nanofibers have a large specific surface area (201 m^2/g), which is beneficial for the adsorption of NO gas. In addition, the large-area Ni nanofibers as filters achieved a filtration efficiency of 99.86% for 300 nm particles at a wind speed of 5.3 cm/s. At the same time, Ni nanofibers can also act as catalysts to reduce NO to N_2 at high temperatures, and can be reused, showing great potential in reducing automobile exhaust pollution. Metal catalytic membranes work well in purifying during cold starts of vehicles, but their development is largely limited due to their complex molding process, high production cost, and less thermal shock resistance compared to honeycomb ceramics.

The most commonly used in porous ceramic catalytic membranes is cordierite ceramic catalytic membrane [29, 30]. The advantages of cordierite catalytic membranes include large specific surface area, smaller gas resistance ratio, higher mechanical strength, thinner catalytic active coating, etc. Moreover, cordierite raw materials are widely utilized due to their ease of access and relatively low cost. Since 60–80% of the harmful substances emitted by cars are due to insufficient fuel combustion within 60 s before cold start, it is particularly important to reduce exhaust emissions within 1–3 min after cold start. In part, this issue has been solved by the development of new catalytic membranes, such as SiC foam ceramic catalytic membranes and fiber porous ceramic catalytic membranes. Porous ceramic fiber

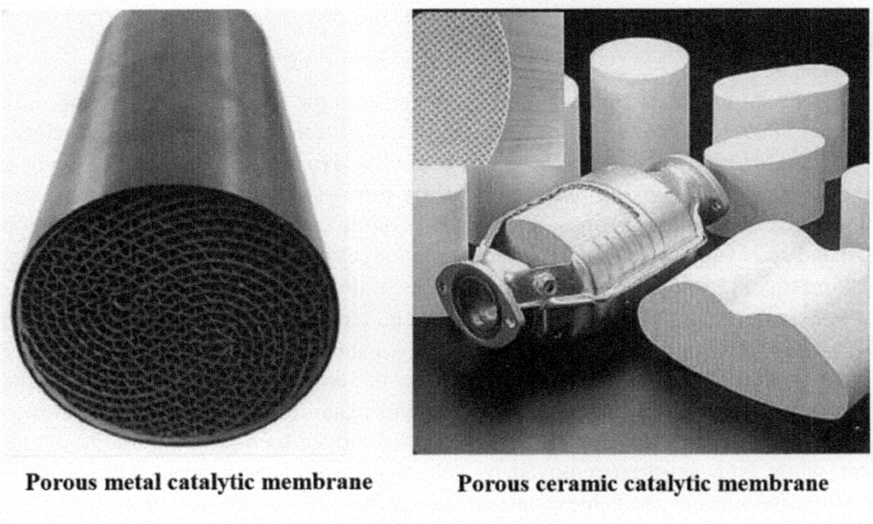

Fig. 7.8 Mobile source nitrogen oxide catalytic membranes

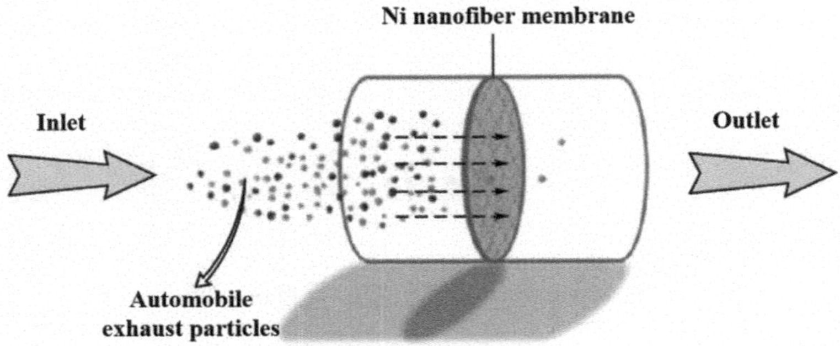

Fig. 7.9 Schematic diagram of Ni nanofiber membranes for simultaneous removal of dust and nitrogen oxides in automobile exhaust

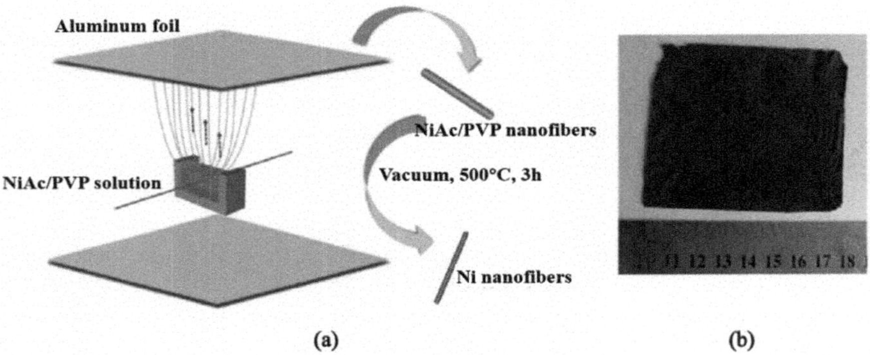

Fig. 7.10 Schematic diagram (**a**) and physical picture (**b**) of Ni nanofiber preparation [20]

catalytic membranes are mainly prepared using ceramic fibers as raw materials, and the commonly used preparation process is vacuum filtration. These ceramic structures are mostly open-pore, with a porosity of over 90%. However, its relatively high cost due to length and weaving properties greatly limits its development. Compared to cordierite catalytic membranes, SiC porous ceramic catalytic membranes have good heat and corrosion resistance, high thermal conductivity, and better impact resistance, making them more suitable for the conditions of drastic temperature change.

The current automobile exhaust treatment technology still mainly relies on three-way catalysis. There is less research on catalytic membranes in automobile exhaust removal due to the lack of a means of removing residual dust from the catalytic membrane. In addition, the components in automobile exhaust are complex. Unburned hydrocarbons and VOCs need to be removed by high-temperature oxidation, while NO_x is often removed by selective catalytic reduction. Although the catalytic membrane can achieve integrated dust removal and denitration, achieving the removal of multiple component gases at the same time is still a challenge and worth exploring.

7.2 Gas–Solid Phase Pollutant Synergistic Cleaning Membranes

With the acceleration of industrialization and the gradual popularization of motor vehicles in recent years, the concentration of inhalable particulate matter, volatile organic compounds (VOCs), mercury vapor, ozone (O_3), and other pollutants in the air has significantly increased. Numerous epidemiological studies have shown that long-term exposure to environments with inhalable particulate matter with a diameter less than 2.5 μm ($PM_{2.5}$) is closely related to cardiovascular diseases. VOCs have strong volatility, lipophilicity, persistence, and toxicity, they can easily enter the human body through the respiratory tract or skin, causing great harm. At the same time, VOCs are also a major factor causing atmospheric environment deterioration (Essential precursors include ozone, organic aerosols, photochemical smog, etc.). The reaction between VOCs and NO_x can enhance the oxidizability of atmosphere, which worsens the air pollution situation, eventually forming "haze" weather after complex reactions. In recent years, the synergistic treatment of $PM_{2.5}$, VOCs and other pollutant gases with membrane materials has been an international research hotspot.

7.2.1 $PM_{2.5}$ and Formaldehyde Synergistic Treatment of Nanofiber Membranes

Formaldehyde and particulate matter (such as $PM_{2.5}$) are the most lethal sources of indoor air pollution. The World Health Organization reports that air pollution leads to about 7 million premature deaths, of which 4.3 million are attributed to indoor air pollution. Indoor air pollution is posing an increasingly serious threat to human health and causing a great negative impact on the socio-economy. Formaldehyde is an important precursor of many home decoration materials and compounds, and the indoor formaldehyde content is usually 10 to 100 times higher than the outdoor level. To control $PM_{2.5}$ and formaldehyde, traditional methods usually use an activated carbon layer along with HEPA filter. However, the activated carbon layer will enlarge the size of the filter unit, and using two independent filter systems will increase the usage cost. Therefore, exploring a new type of membrane material that can simultaneously remove formaldehyde and $PM_{2.5}$ is particularly meaningful [31–33].

Zhao et al. [34] prepared a cellulose nanofiber membrane (CNF). This membrane has multi-level sub-micron channels, which can quickly filter the air and also efficiently retain $PM_{2.5}$. Moreover, the open vacancies of CNF can efficiently adsorb formaldehyde. Experimental research showed that this cellulose nanofiber membrane had a $PM_{2.5}$ retention efficiency of over 95%, and a formaldehyde adsorption capacity of 47.71 mg/g, which can reduce the formaldehyde concentration to below 0.1 mg/m^3 (the international indoor formaldehyde concentration safety standard). In addition, CNF has a thickness of only 5 μm and a gas flow rate of 4.6 cm/s under 200 Pa with a large gas flux. Therefore, this nanofiber membrane has a very large market prospect in areas with frequent smog.

Hu et al. [35] prepared a multilevel pore nanofiber membrane loaded with MnO_2 (MnO_2/PS HPNM in two steps. They first prepared a multilevel pore polystyrene nanofiber membrane (PS HPNM) by electrospinning, then filled the pores of the PS HPNM membrane with layered crystal MnO_2, which increased the specific surface area of the PS HPNM membrane. It effectively filters $PM_{2.5}$ and removes formaldehyde gas under low air flow resistance, with a capture efficiency of up to 99.77% for inhalable particles. Due to the mesopores of the membrane promoting contact between formaldehyde gas and MnO_2, its formaldehyde removal efficiency reached 88.2%. In addition, the large pores formed by the PS HPNM membrane ensure the permeation of clean gas and a lower air flow resistance (82 Pa). The polystyrene multilevel pore nanofiber membrane loaded with MnO_2 provides a new approach for the synergistic treatment of $PM_{2.5}$ and formaldehyde.

Zhu et al. [34] first prepared a silicon dioxide (SiO_2) nanofiber membrane by electrospinning, then grew ZIF-8 nanocrystals on it by diffusion method, successfully developing a multifunctional ZIF-8@SiO_2 composite membrane with a bead-like structure (Fig. 7.11). Thanks to the loading of ZIF-8 nanocrystals, the ZIF-8@SiO_2 composite membrane exhibits a high specific surface area and surface functional group structure, thereby improving the PM removal efficiency and showing strong formaldehyde capture ability. The ZIF-8@SiO_2 composite membrane synthesized for 0.5 h has excellent filtering performance for cigarettes, while the ZIF-8@SiO_2 composite membrane synthesized for 2 h has a higher efficiency for formaldehyde adsorption. The development of the ZIF-8@SiO_2 composite membrane can promote the development of gas cleaning materials for personal protection and environmental management.

7.2.2 $PM_{2.5}$ Synergistic Treatment of Ceramic Catalytic Membranes with Benzene Series

Benzene series (BTEXs) refers to a series of volatile compounds such as benzene, toluene, ethylbenzene, and xylene, which are confirmed by the World Health Organization as highly toxic carcinogens. Long-term inhalation of benzene series substances will damage the nervous system, cause leukemia, obstructive anemia, induce chromosomal mutations and other diseases. Pregnant women stay in an environment containing benzene series substances for a long time, which will lead to congenital defects and deformities in newborns.

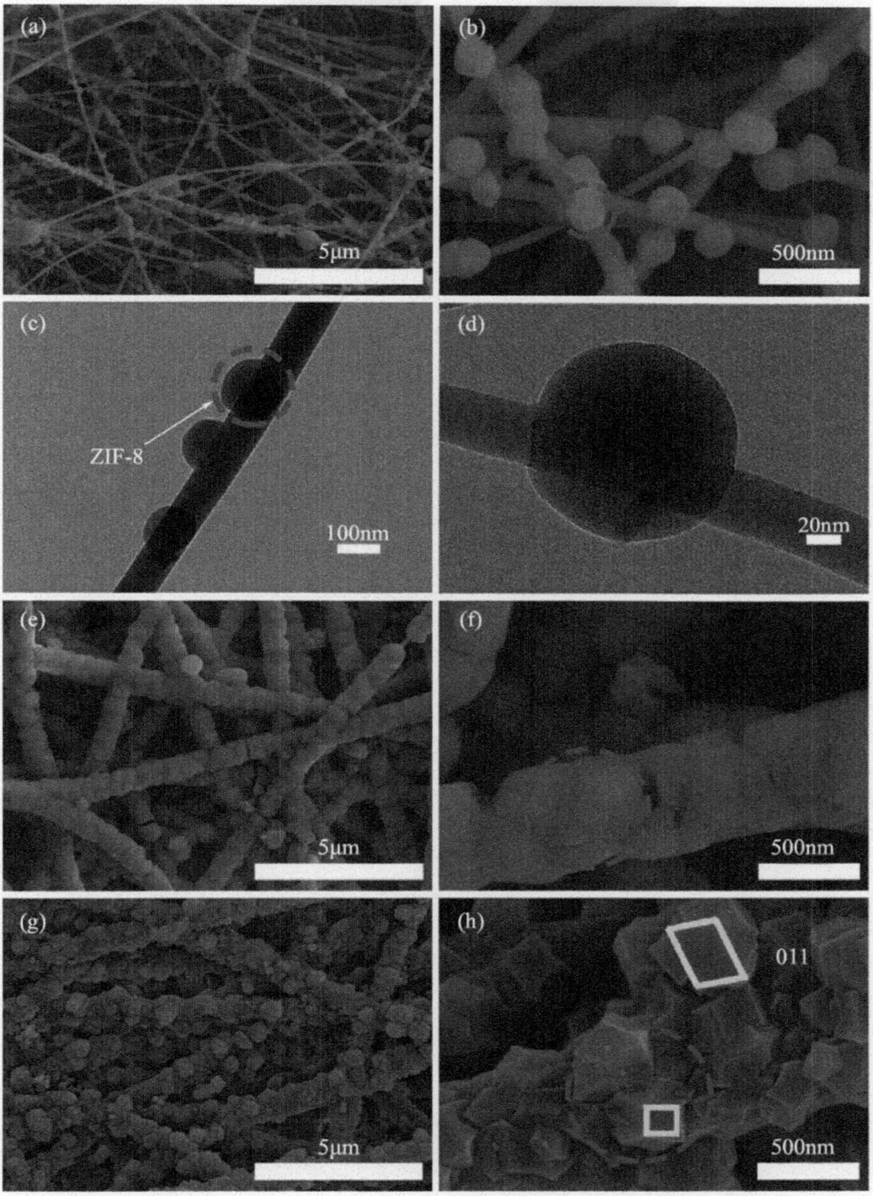

Fig. 7.11 SEM images of ZIF-8@SiO_2 composite membranes synthesized at different times and their corresponding high magnification images: 0.5 h (**a**, **b**), 1 h (**e**, **f**), 2 h (**g**, **h**); (**c**) and (**d**) are ZIF-8@SiO_2 synthesized for 0.5 h. TEM image of the composite membrane (ZIF-8 nanocrystals marked with dashed circles) [34]

Li et al. [36] coated zinc oxide (ZnO) nanoparticles on the silicon carbide ceramic membrane by the sol–gel method to form a secondary carrier and then loaded Pt nanoparticles on the ZnO layer, successfully preparing a new type of Pt/ZnO/SiC catalytic membrane. The microstructure, crystal morphology, composition, and elemental valence of the prepared membrane were characterized by X-ray diffraction, scanning electron microscopy, transmission electron microscopy, and X-ray photoelectron spectroscopy. The microscopic morphology of the membrane is shown in Fig. 7.12. It was found that the ZnO secondary carrier layer improved the dispersion of Pt nanoparticles and significantly enhanced their catalytic performance. Using toluene as a simulated gaseous pollutant, a conversion rate of up to 100% of toluene was achieved at a temperature of 210 °C and a filtration speed of 0.72 m/min on the Pt/ZnO/SiC ceramic catalytic membrane, and the catalytic membrane also has a high filtration efficiency for PM.

Trimethylbenzene mainly comes from the exhaust gas emitted by the petrochemical industry. Trimethylbenzene has three isomers, namely, mesitylene, pseudocumene, and hemimellitene. Trimethylbenzene is a toxic substance that can cause serious pollution to air and water, and it is a flammable and explosive gas, which can easily explode when exposed to open flames in factories. Li et al. [37] coated titanium dioxide (TiO_2) nanoparticles on the silicon carbide ceramic membrane by

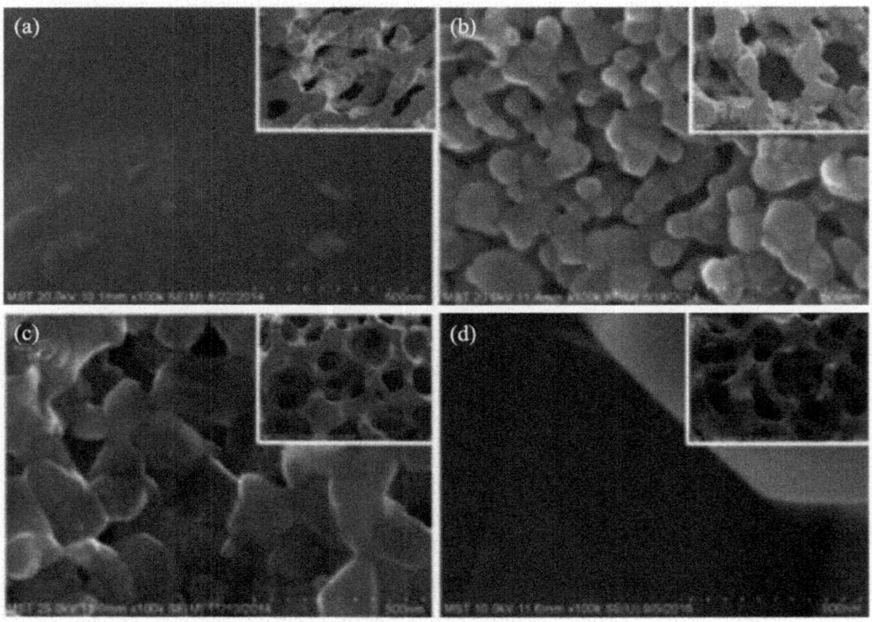

Fig. 7.12 SEM images of SiC ceramic membrane (**a**); ZnO/SiC catalytic membrane (**b**); Pt/SiC catalytic membrane (**c**); Pt/ZnO/SiC ceramic catalytic membrane (**d**) [36]

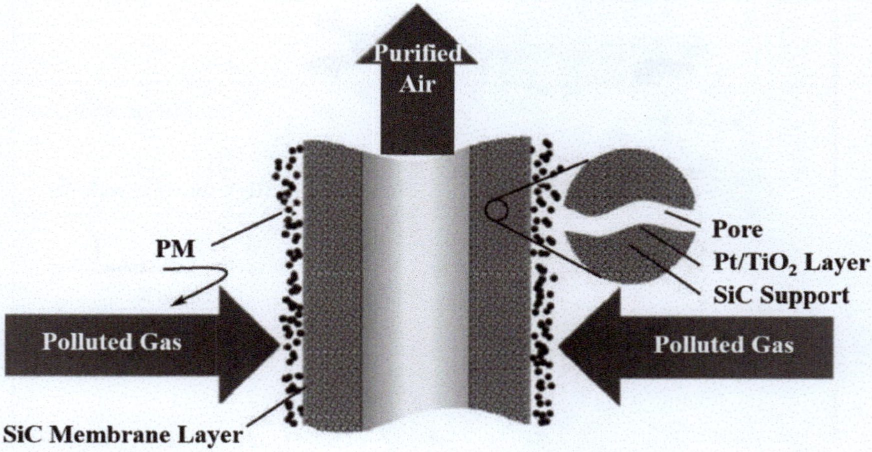

Fig. 7.13 Schematic diagram of the working principle of SiC@TiO_2/Pt catalytic membrane

the sol–gel method to form a secondary carrier, and then impregnated Pt nanoparticles on the TiO_2 layer, successfully preparing a SiC@TiO_2/Pt catalytic membrane. This catalytic membrane showed excellent catalytic filtration performance for both mesitylene and PM (its working mechanism is shown in Fig. 7.13). At a temperature of 262 °C and a filtration speed of 1 m/min, this catalytic membrane can achieve 100% conversion of mesitylene. In addition, when the inlet dust concentration is 240 mg/m^3, the dust removal efficiency is higher than 99.98%.

7.2.3 *Catalytic Membranes for the Cooperative Treatment of $PM_{2.5}$, CO_2/SO_2/O_3 and Other Components*

7.2.3.1 Catalytic Membranes for the Cooperative Treatment of $PM_{2.5}$ and CO_2

Due to the rapid development of industrialization, vehicle exhaust and the combustion of fossil fuels have caused serious air pollution problems. Among them, one of the primary sources of particulate pollutants and toxic pollutant gases is vehicle exhaust emissions. Especially in urban areas, it severely affects the air quality of the city due to high population density and concentrated emissions. Long-term and short-term exposure to high concentrations of $PM_{2.5}$ and CO_2 can lead to stroke, lung infection, asthma and other cardiovascular diseases. Therefore, membrane materials that can cooperatively process $PM_{2.5}$ and CO_2 have gradually become a research hotspot.

Ramachandran et al. [38] prepared ethylenediamine functionalized magnesium tetraphenylporphyrin (MgTPP), and loaded it onto polyetherimide (PEI) nanofibers

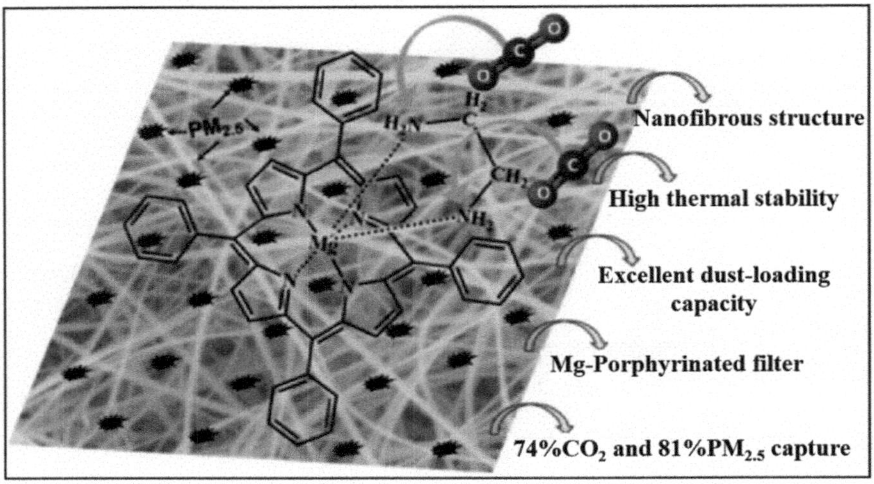

Fig. 7.14 Multifunctional $PM_{2.5}$ and CO_2 capture nano-fiber membrane [38]

using electrospinning. The developed nano-fiber membrane's ability to capture CO_2 and $PM_{2.5}$ was evaluated (Fig. 7.14). The chelating agent ethylenediamine can be verified to be loaded onto the fiber structure through infrared spectroscopy, and the introduced amino chain can effectively capture CO_2 molecules. The amino-modified MgTPP/PEI nanofibers have an average diameter of 1.2–1.4 μm, good thermal stability, and a maximum temperature resistance of 411 °C. At the same time, the prepared nano-fiber membrane has strong hydrophilicity, and the capture efficiency for CO_2 and $PM_{2.5}$ is 74% and 81% respectively.

7.2.3.2 MOF Catalytic Membranes for the Cooperative Treatment of $PM_{2.5}$ and SO_2/O_3

Particulate matter (PM) and toxic gases like ozone (O_3) and sulfur dioxide (SO_2) are the primary pollutants that cause air pollution. They can lead to lung and cerebrovascular diseases when their concentration exceeds the safe range. PM, especially ultrafine particles, can deposit in the brain, thereby increasing the risk of neuroinflammation and Alzheimer's disease. SO_2 and O_3 cause strong irritation to the respiratory tract and lungs. There is ample evidence to support the claim that SO_2 and O_3 at high concentrations exacerbate asthma and lung inflammation by inducing bronchoconstriction and eosinophil proliferation. According to the World Health Organization (WHO) air quality standards, the concentrations of SO_2, O_3, and $PM_{2.5}$ should be kept below 125 μg/m^3, 160 μg/m^3, and 25 μg/m^3 respectively. Metal–organic frameworks (MOFs) have attracted much attention due to their ultra-high porosity, excellent gas storage and separation capabilities, and functional surfaces. However, it is difficult to prepare MOFs thin membranes due to their unstable crystal structure, limiting their potential applications. Currently, there are reports of adding MOFs to polymer spinning solutions to prepare MOF membranes with both PM filtration and pollutant gas adsorption capabilities. However, the preparation process is difficult, and the adsorption capacity of MOFs is reduced because they are encapsulated by polymers. Despite many studies and significant progress in preparing toxic gas selective adsorption membrane materials, selective adsorption of low-concentration toxic gases remains a research bottleneck. In order to prepare a MOF-based nanofiber membrane, Wang et al. [39] synthesized MIL-53(Al)-NH_2 on the surface of polyacrylonitrile (PAN) electrospun membrane through a hydrothermal process. Figure 7.15 shows the microscopic morphologies of the MOF-based nanofiber membrane. Based on the weak acidity of SO_2 and the strong oxidizing ability of O_3, 2-aminoterephthalic acid was chosen as the organic ligand, which has alkaline groups and reductive substrates and can achieve selective adsorption through chemical action. The MIL-53(Al)-NH_2 grown on the PAN fiber membrane has excellent adsorption capabilities for PM, SO_2, and O_3. The nanofiber membrane, after growing MOF has a PM removal efficiency of 99.99%. More importantly, under lower partial pressure and complex gas composition conditions, the MOF nanofiber membrane can selectively adsorb SO_2, reducing the concentration of SO_2 from 7300 nL/L to 40 nL/L. In addition, the MOF nanofiber membrane has a higher removal ability for O_3 than for SO_2, and can rapidly reduce the concentration of O_3 from 3000 nL/L to 7 nL/L. The fiber membrane with MOF grown on the surface can selectively adsorb toxic atmospheric gases without being affected by other gases (such as CO_2 and O_2).

7.3 Antimicrobial Gas Cleaning Membranes

Indoor air may contain bioaerosols, which are composed of fungi, bacteria, viruses, fungal spores, allergenic pollen, and their fragments (including various antigens).

7.3.1 Classification of Antimicrobial Agents in Gas Cleaning Membranes

The most common method currently is to prepare or modify antimicrobial gas cleaning membrane materials by adding antimicrobial agents. Antimicrobial agents are mainly divided into inorganic antimicrobial agents, organic antimicrobial agents, natural antimicrobial agents [40]. According to the classification characteristics of antimicrobial agents, the types of gas cleaning membranes are also different.

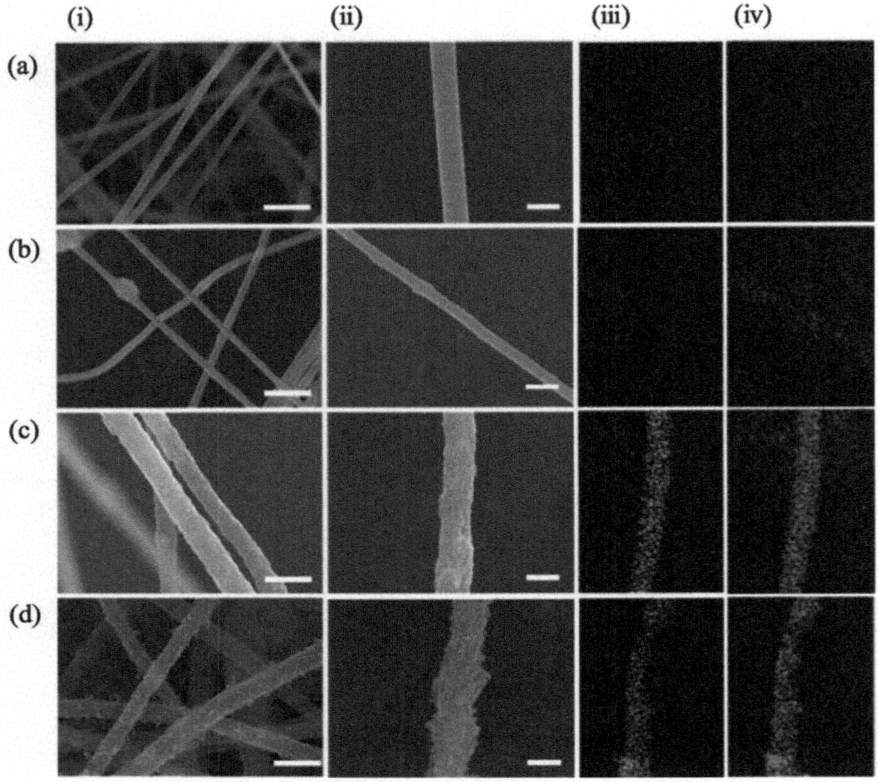

Fig. 7.15 Characterizations of the nanofiber membranes: PAN (**a**), 30% MIL-53(Al)-NH_2 @PAN (**b**), 60% MIL-53(Al)NH_2 @SEM images of PAN (**c**), MGP (**d**). The scale bars represent 2 μm (i) and 500 nm (ii) respectively. X-ray EDS is used to detect Al^{3+} (iii) and O (iv) [39]

7.3.1.1 Inorganic Antimicrobial Agents

Inorganic antimicrobial agents refer to antimicrobial agents made using the antibacterial or bactericidal ability of metals such as silver, titanium, zinc, copper, and their ions. They are currently the most widely used type of antimicrobial agents, mainly including silver series, titanium series, and zinc series. Inorganic antimicrobial agents have advantages such as good heat resistance and processability, high environmental protection, and high safety. Although the antibacterial duration of inorganic antimicrobial agents is not good, life health, long-term effectiveness and safety are the goals people pursue.

Feng et al. [41] wrapped ZnO nanorods on PTFE nanofibers, then loaded with nano-silver particles, to prepare a multifunctional gas cleaning membrane with high-efficiency antibacterial, particulate removal, and formaldehyde degradation. The membrane showed excellent dynamic antibacterial performance, a dust retention rate of over 99.99%, and a formaldehyde degradation rate of 60%. In addition, the author's research group [41] also developed a new type of Ag@MWCNTs/Al_2O_3 ceramic membrane with a graded structure. This multifunctional membrane can simultaneously remove fine particles, microorganisms, and VOCs, with formaldehyde degradation rates of 82.24% at room temperature and 99.99% at 55 °C, and it can completely intercept and kill microorganisms in indoor air.

7.3.1.2 Organic Antimicrobial Agents

Organic antimicrobial agents are antimicrobial agents with organic substances such as phenols, benzimidazoles, quaternary ammonium salts, organic acids. as the active antibacterial components. Organic antimicrobial agents have advantages such as good instant effect, easy operation, not easy to discolor, and relatively cheap price, but while inhibiting harmful microorganisms, they also have certain toxicity to biological cells and systems, and their heat resistance and chemical stability are relatively poor.

Taylor et al. [42] obtained HEPA filters with antibacterial properties via coating quaternary ammonium groups on polyurethane that coated on HEPA filters. They tested 8 strains of bacteria that are prone to cause infections, and the results showed that the HEPA filter samples with polyurethane coating showed high bactericidal activity against all 8 strains, confirming that the quaternary ammonium in the modified polyurethane coating played a good antibacterial role. The research of Ungur et al. [43] showed that polyurethane solutions were modified with copper oxide (CuO) microparticles of different particle sizes (700 nm-1 μm and 50 nm), and polyurethane nanofiber membranes were prepared through electrospinning. The efficiency and stability of these two types of polymer nanofiber filters were tested under simulated filtration conditions. The results confirmed that both micron and nanometer HEPA filter coatings have good antibacterial efficiency. Moreover, from a technical and economic perspective, CuO microparticles are more suitable for antibacterial modification of polyurethane filters than nanoparticles.

7.3.1.3 Natural Antibacterial Agents

There are many types of natural antibacterial agents, most of which are extracted from natural plants. While they demonstrate advantages such as safety, absence of side effects, and high antibacterial rates, these materials also have drawbacks including challenging extraction processes and limited heat resistance. Common antibacterial extracts include peptides, plant oils, and fragrances. According to research reports, natural extracts have high antimicrobial activity, lower toxicity than inorganic antibacterial agents, low cost, environmentally friendly, and abundant in nature.

Choi et al. [44] added Sophora flavescens extract to a polyvinylpyrrolidone (PVP) solution and obtained a nanofiber membrane through electrospinning. Its particle filtration efficiency was 99.99%, and its antibacterial activity against Staphylococcus epidermidis was 99.98%.

The stability of antibacterial agents made from natural product extracts is poor, and further research is needed on their durability, especially long-term testing in real environments, such as the possible effects of temperature and natural oxidation processes [45]. Combining natural extracts with inorganic materials can solve the problem of durability. By electrospinning a soy protein isolate (SPI)/polyvinyl alcohol (PVA) system, Fang et al. [46] prepared a biodegradable multifunctional gas cleaning membrane. This nanofiber membrane had a filtration efficiency of up to 99.99% and showed antibacterial activity against E. coli. SPI/PVA nanofiber membranes with appropriate material composition and microstructure can be used as new high-performance, environmentally friendly filter materials. Lv et al. [47] used konjac glucomannan (KGM) as a raw material, mixed with water-soluble and biodegradable polyvinyl alcohol, and used citric acid as a thermal crosslinking agent to prepare a KGM-based nanofiber membrane loaded with ZnO nanoparticles by electrospinning (Fig. 7.16). This fiber membrane has not only efficient air filtration performance but also has good photocatalytic activity and antibacterial activity and has a good inhibitory effect on both Gram-negative and Gram-positive bacteria.

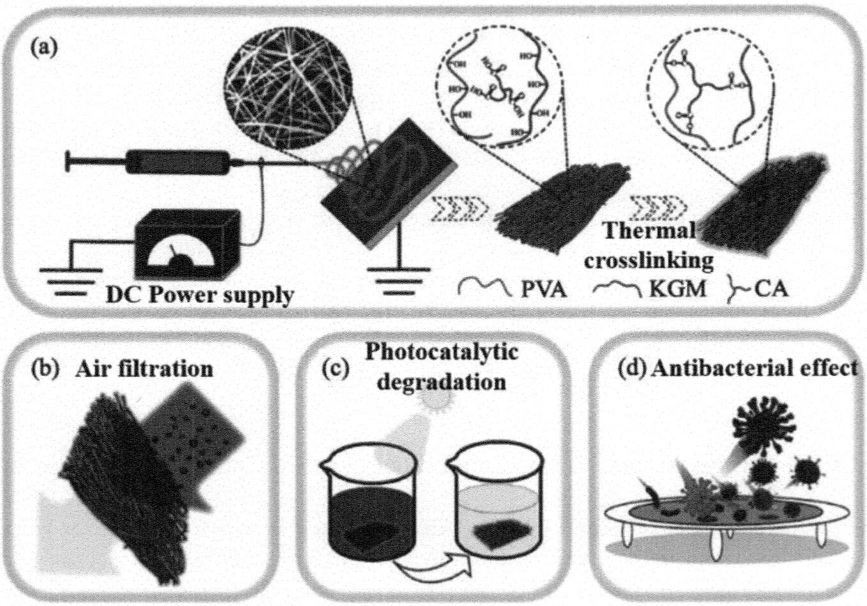

Fig. 7.16 Multifunctional ZnO@PVA/KGM electrospun nanofiber membrane [47]

7.3.2 *Preparation of Antibacterial Gas Cleaning Membranes*

The preparation methods of antibacterial gas cleaning membranes are mainly divided into two types, as shown in Fig. 7.17. One is to modify the membrane material by deposition or coating the antibacterial agent on the surface of the membrane material. Another direct treatment method is to mix the antibacterial agent with the membrane-forming material during the preparation process of the gas cleaning membrane. The antibacterial agents used in the latter are mostly inorganic nanoparticles or natural product extracts.

The deposition method mainly includes immersion coating, atomization process, spraying, electrostatic spraying (Fig. 7.17) and other construction methods [41, 42,

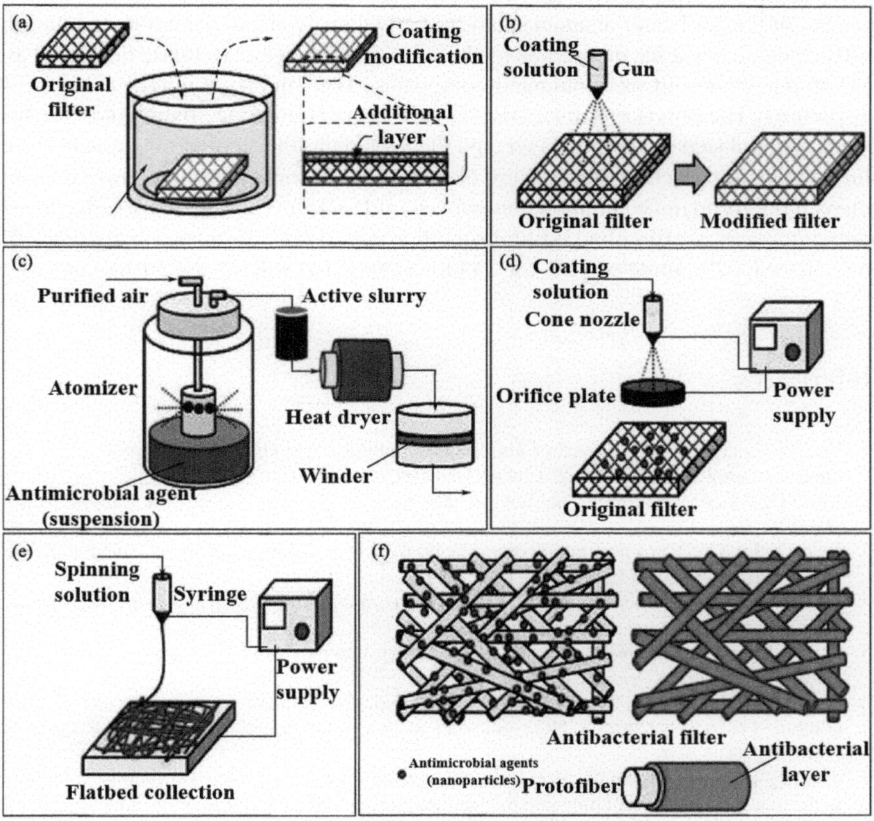

Fig. 7.17 Antibacterial gas cleaning membrane preparation methods [41, 42, 45]: **a** Immersion coating; **b** spraying; **c** atomization process; **d** electrostatic spraying; **e** electrospinning; and **f** deposition and coating of antibacterial agents

45]. Hwang et al. [48] used atomization or aerosol processes to prepare antimicrobial filters. The antibacterial filter is made of natural Euscaphis japonica nanoparticles through atomization extraction (it should be noted that although the toxicity of Euscaphis japonica extract is lower than that of carcinogens, further research is needed to ensure its harmlessness to human health). Han et al. [49] compared the antibacterial activity of grapefruit seed extract (GSE) and propolis, two natural products deposited on PET air filters by electrostatic spraying, and found that the antibacterial effect of GSE is better than propolis when the deposition amount is 5000–8000 $\mu g/cm^2$.

In order to prevent the increase of mass transfer resistance leading to the rise of filter pressure drop, the direct treatment method is usually preferred to effectively combine the nanofiber membrane with the antibacterial agent. The antibacterial agent can be deposited or coated to form a nanofiber membrane during the electrospinning process, or the antibacterial agent can be directly dissolved into the polymer solution and then made into a flat membrane by the solution casting method [50]. Selvam et al. [51] prepared a multifunctional nano-composite gas cleaning membrane using electrospinning. The polyacrylonitrile (PAN) nanofibers containing silver nanoparticles (Ag NP) are placed on the first layer, and the PAN nanofibers containing magnesium aluminate nanoparticles (MA NP) are placed on the second layer, which are located between polypropylene spunbond nonwovens (PP SBN). The study showed that the nano-composite air filter had a filtration efficiency of 99% for bacteria and PM2.5, and a detoxification rate of 95% for 2-chloroethyl ethyl sulfide (2-CEES).

References

1. Wei W. Analysis of the causes of atmospheric pollution and environmental monitoring and control technology. Sci Mass. 2019;4(4):109–109.
2. Liu XH. Based on the harm and control measures of nitrogen oxides. Low Carbon World. 2017;9:8–9.
3. Wang Y, Liu GX. Current situation of waste gas treatment in petroleum and chemical industry. China Petroleum Chem Econ Anal. 2017;7:26–7.
4. Wang YS, Zhang L, Chen JH, et al. The harm and control of nitrogen oxides in the atmosphere. Sci Technol Innov Appl. 2019;7:137–8.
5. Zou GZ. The promotion of ceramic kiln flue gas treatment and low nitrogen combustion technology. Foshan Ceram. 2017;27(5):28–9.
6. Nacken M, Heidenreich S, Hackel M, et al. Catalytic activation of ceramic filter elements for combined particle separation, NO_x removal and VOC total oxidation. Appl Catal B-Environ. 2007;70(1–4):370–6.
7. Zhao R. Analysis of the application of desulfurization, dust removal, and denitration technology in the treatment of glass melting kiln flue gas. Resource Conserv Environ Protection. 2015;7:24.
8. Lin HF, Abubakar A, Li CM, et al. Development of red mud coated catalytic filter for NOx removal in the high temperature range of 300–450 degrees. Catal Lett. 2020;150(3):702–12.
9. Chen J, Pan B, Wang B, et al. Hydrothermal synthesis of a Pt/SAPO-34@SiC catalytic membrane for the simultaneous removal of NO and particulate matter. Ind Eng Chem Res. 2020;59(10):4302–12.
10. Lee JS, Shin KH, Shin MC, et al. Properties of catalytic filter for the hot gas cleaning. Zurich-Uetikon: Trans Tech Publications Ltd; 2004. p. 1181–4.

11. Fino D, Russo N, Saracco G, et al. A multifunctional filter for the simultaneous removal of fly-ash and NO_x from incinerator flue gases. Chem Eng Sci. 2004;59(22–23):5329–36.
12. Yuan GH, Wu HY, Yang GH, et al. Integrated dust removal and denitration using a dual-layer granular bed filter with SCR catalyst. Water Air Soil Pollut. 2020;231(2):11.
13. Abubakar A, Li CM, Lin HF, et al. Simultaneous removal of particulates and NO by the catalytic bag filter containing V_2O_5-MoO_3/TiO_2. Korean J Chem Eng. 2020;37(4):633–40.
14. Kim YA, Choi JH, Scott J, et al. Preparation of high porous Pt-V_2O_5-WO_3/TiO_2/SiC filter for simultaneous removal of NO and particulates. Powder Technol. 2008;180(1–2):79–85.
15. Park YO, Lee KW, Rhee YW. Removal characteristics of nitrogen oxide of high temperature catalytic filters for simultaneous removal of fine particulate and NO_x. J Ind Eng Chem. 2009;15(1):36–9.
16. Chen XH, Zheng YY, Zhang YB. MnO_2-Fe_2O_3 catalysts supported on polyphenylene sulfide filter felt by a redox method for the low-temperature NO reduction with NH_3. Catal Commun. 2018;105:16–9.
17. Yang B, Zheng DH, Shen YS, et al. Influencing factors on low-temperature NO_x performance of Mn-La-Ce-Ni-O_x/PPS catalytic filters applied for cement kiln. J Ind Eng Chem. 2015;24:148–52.
18. Yang B, Huang Q, Chen MD, et al. Mn-Ce-Nb-O_x/P84 catalytic filters prepared by a novel method for simultaneous removal of particulates and NO. J Rare Earths. 2019;37(3):273–81.
19. Walberer J, Giovanny M, Meiller M, et al. Development and tests of a combined filter for NO_x, particulates, and SO_2 reduction. Chem Eng Technol. 2018;41(11):2150–8.
20. Tang X, Wang X, Yang L, et al. Multifunctional nickel nanofiber for effective air purification: PM removal and NO reduction from automobile exhaust. J Mater Sci. 2020;55(14):6161–71.
21. Zhou S, Zhou J, Zhu Y. Chemical composition and size distribution of particulate matters from marine diesel engines with different fuel oils. Fuel. 2019;235:972–83.
22. Yu JS, Lee JC, Park S, et al. Preparation and characterization of catalyst-containing SiC fiber filter media. Key Eng Mater. 2005;287:44–50.
23. Wang JQ, Wang Y, Liu SX, et al. Research progress on NO_x storage reduction technology in lean burn engines. Modern Chem Ind. 2011; 31(4):28–33, 44.
24. Yu NN, Lu J, Guo YP, et al. Mechanism and technological progress of three-way catalysts. Contemp Chem Res. 2011;8(11):9–14.
25. Fan J, Chen Y, Jiang X, et al. A simple and effective method to synthesize Pt/CeO_2 three-way catalysts with high activity and hydrothermal stability. J Environ Chem Eng. 2020;8(5): 104236.
26. Mcatee, Mccullough, Sellick, et al. Characterisation and modelling of the reactions in a three-way pdrh catalyst in the exhaust gas from an ethanol-fuelled spark-ignition engine. Proc. Inst. Mech. Eng. Part D: J. Autom. Eng. 2019;233(12):3222–34.
27. Vedyagin AA, Kenzhin RM, Tashlanov MY, et al. Effect of La addition on the performance of three-way catalysts containing palladium and rhodium. Top Catal. 2020;63:1–2.
28. Yamamoto K, Kondo S, Suzuki K. Filtration and regeneration performances of SiC fiber potentially applied to gasoline particulates. Fuel. 2019;243(62):28–33.
29. Nascimento LF, Serra OA. Wash coating of cordierite honeycomb with ceria-copper mixed oxides for catalytic diesel soot combustion. Process Saf Environ Prot. 2016;101:134–43.
30. Begum S, Misran H, Aminuddin A. Developing of honeycomb shaped cordierite ceramic from indigenous raw materials and its characterization. Adv Mater Res. 2011;264–265:597–601.
31. Bari MA, Kindzierski WB, Wheeler AJ, et al. Source apportionment of indoor and outdoor volatile organic compounds at homes in Edmonton, Canada. Build Environ. 2015;90(8):114–24.
32. Escobedo LE, Champion WM, LiN, et al. Indoor air quality in Latino homes in Boulder, Colorado. Atmos Environ. 2014;92:69–75.
33. Kim J, Kim H, Lim D, et al. Effects of indoor air pollutants on atopic dermatitis. Int J Environ Res Public Health. 2016;13(12):1220.
34. Zhao X, Chen L, Guo Y, et al. Porous cellulose nanofiber stringed HKUST-1 polyhedron membrane for air purification. Appl Mater Today. 2019;14:96–101.

35. Hu M, Yin L, Zhou H, et al. Manganese dioxide-filled hierarchical porous nanofiber membrane for indoor air cleaning at room temperature. J Membr Sci. 2020;605.
36. Li L, Zhang F, Zhong Z, et al. Novel synthesis of a high performance Pt/ZnO/SiC filter for the oxidation of toluene. Ind Eng Chem Res. 2017;56(46):13857–65.
37. Li C, Zhang F, Feng S, et al. SiC@TiO_2/Pt catalytic membrane for collaborative removal of VOCs and nanoparticles. Ind Eng Chem Res. 2018;57(31):10564–71.
38. Ramachandran S, Rajiv S. Ethylenediamine functionalized metalloporphyrin loaded nanofibrous membrane: a new strategic approach to air filtration. J Inorg Organomet Polym Mater. 2020;30(6):2142–51.
39. Wang X, Xu W, Gu J, et al. MOF-based fibrous membranes adsorb PM efficiently and capture toxic gases selectively. Nanoscale. 2019;11(38):17782.
40. Nakashima H, Ooshima T. Analysis of inorganic antimicrobial agents in antimicrobial products: evaluation of a screening method by X-ray fluorescence spectrometry and the measurement of metals by inductively coupled plasma atomic emission spectroscopy. J Healthence. 2007;53(4):423–9.
41. Feng S, Li D, Low ZX, et al. ALD-seeded hydrothermally-grown Ag/ZnO nanorod PTFE membrane as efficient indoor air filter. J Membr Sci. 2017;531:86–93.
42. Taylor M, Mccollister B, Park D. Highly bactericidal polyurethane effective against both normal and drug-resistant bacteria: potential use as an air filter coating. Appl Biochem Biotechnol. 2016;178(5):1053–67.
43. Ungur G, Hruza J. Modified polyurethane nanofibers as antibacterial filters for air and water purification. RSC Adv. 2017;7(78):49177–87.
44. Choi J, Yang BJ, Bae GN, et al. Herbal extract incorporated nanofiber fabricated by an electrospinning technique and its application to antimicrobial air filtration. ACS Appl Mater Interfaces. 2015;7(45):25313–20.
45. Chong ES, Hwang GB, Nho CW, et al. Antimicrobial durability of air filters coated with airborne Sophora flavescens nanoparticles. Sci Total Environ. 2013;444:110–4.
46. Fang Q, Zhu M, Yu S, et al. Studies on soy protein isolate/polyvinyl alcohol hybrid nanofiber membranes as multi-functional eco-friendly filtration materials. Mater Sci Eng B Solid State Mater Adv Technol. 2016;214:1–10.
47. Lv D, Wang R, Tang G, et al. Eco-friendly electrospun membranes loaded with visible-light response nano-particles for multifunctional usages: High-efficient air filtration, dye scavenger and bactericide. ACS Appl Mater Interfaces. 2019;11:12880–9.
48. Hwang GB, Heo KJ, Yun JH, et al. Antimicrobial air filters using natural euscaphis japonica nanoparticles. PLoS ONE. 2015;10(5): e0126481.
49. Han B, Kang JS, Kim HJ, et al. Investigation of antimicrobial activity of grapefruit seed extract and its application to air filters with comparison to propolis and shiitake. Aerosol Air Qual Res. 2015;15(3):1035–44.
50. Tamayo L, Azócar M, Kogan M, et al. Copper-polymer nanocomposites: an excellent and cost-effective biocide for use on antibacterial surfaces. Mater Sci Eng C. 2016;69(69):1391–409.
51. Selvam AK, Baskar D, Nallathambi G. Layer by layer nanocomposite filter for ABC filtration. Chem Eng Commun. 2019;208(6):1–7.

Chapter 8
Gas Cleaning Membrane Separation Equipment

8.1 Factors Affecting Membrane Filtration

8.1.1 Properties of Flue Gas Particles

The properties of dust in flue gas affect and determine the design of the membrane dust collector, so it is often necessary to scientifically design the membrane dust collector based on accumulated design experience according to some special properties of dust.

8.1.1.1 Adhesion and Agglomeration

When dust with adhesion and agglomeration enters the membrane dust collector, the dust tends to agglomerate and enlarge. The dust accumulated on the surface of the membrane bag can also continue to agglomerate during the process of being shaken off. The dust removal efficiency and the amount of dust passing through the filter material are related to the adhesion and agglomeration of the dust. The stronger the adhesion and agglomeration of the dust, the easier it is to accumulate on the surface of the membrane, and the back-blowing effect will be greatly reduced. Therefore, it is necessary to select the type of membrane material based on actual parameters for the dust with remarkable adhesion and agglomeration. Membrane materials with more hydrophobic surfaces or good anti-pollution performance can be selected, and the dust collector must also be designed with different measures according to the characteristics of the dust.

Z. Zhong and W. Xing, *Membrane Technology for Air Cleaning*,
https://doi.org/10.1007/978-981-96-1144-7_8

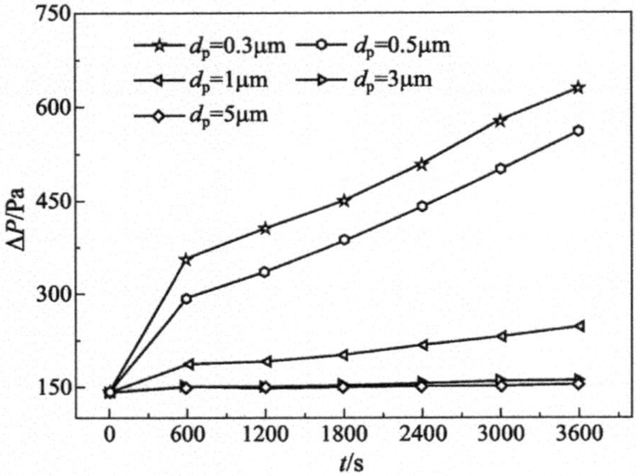

Fig. 8.1 Impact of particle size on filter pressure drop

8.1.1.2 Dust Particle Size

The size and distribution of dust particles mainly affect the resistance during the operation of the membrane dust collector and the wear of the dust collector. Fine dust has a greater impact on pressure loss, and coarse dust has a more obvious impact on wear. When the inlet dust concentration is high, and the hardness is large, the wear of the dust collector will increase. Usually, the effect of particle size on membrane dust removal performance is mainly reflected in the dust retention rate and filtration resistance. The larger the dust particle size, the slower the rate of pressure drop increase and the higher the retention rate; the smaller the dust particle size, the faster the pressure drop increases, and the lower the retention rate. This is because pore blockage occurs in the initial stage of filtration, and then particles deposit on the surface of the filter material. Tang et al. [1] investigated the impact of particle size on filter pressure drop. As observed in Fig. 8.1, the particle size increases, the filter pressure drop gradually decreases. When the particle size is 0.3 μm and 0.5 μm, respectively, the pressure drop increases significantly over time, and the rate of increase is relatively fast. This is because the smaller the particle size, the easier it is for particles to enter the membrane pores and cause pore blockage, thereby increasing the pressure drop.

8.1.1.3 Particle Shape

According to the principle of gas cleaning membrane separation, the separation process mainly relies on surface sieving. However, before entering the membrane surface for filtration, the gas is disturbed by the airflow in the dust collector, and dust will also be separated accordingly. This is not only related to the size of the dust particles but also to the shape of the dust particles. The shape of a particle refers to the image formed by all points on the outline or surface of a particle. The shapes of dust in industrial flue gas vary greatly, and the drag coefficient of particles closer to spherical is smaller. The smaller the sphericity, the lower the separation efficiency. For larger particles (> 10 μm) and smaller particles (< 1 μm), the effect of sphericity on separation efficiency is weak. Therefore, when dust enters the dust collector, large particles and particles with low sphericity are easily settled or separated first under the action of gravity and centrifugal force. The dust particles finally trapped on the surface of the gas cleaning membrane are closer to spherical in shape, have relatively high sphericity, and are also finer.

8.1.1.4 Particle Concentration

The impact of particle concentration on gas filtration is mainly reflected in the aspects of filter pressure drop, filtration efficiency, wear of membrane materials, and dust removal cycle. The impact of particle concentration on filter pressure drop is relatively single. Within a single back-blow cycle, the filter pressure drop basically increases linearly with time; the filter pressure drop increases with the increase of inlet concentration. The greater the particle concentration, the thicker the filter cake deposited on the membrane surface per unit time, and the faster the pressure drop increases (Fig. 8.2). Figure 8.3 shows the impact of inlet dust concentration on the retention rate. When the inlet concentration is 600 mg/m^3, the retention rate is 99.995%. As the inlet concentration increases, the retention rates are 99.993% and 99.991% at concentrations of 700 and 800 mg/m^3, respectively. Although as the inlet dust concentration increases, there are more small dust particles, and the possibility of passing through the membrane increases, from actual filtration experiments, using membrane materials to remove dust can achieve a dust retention rate of 99.99% [2].

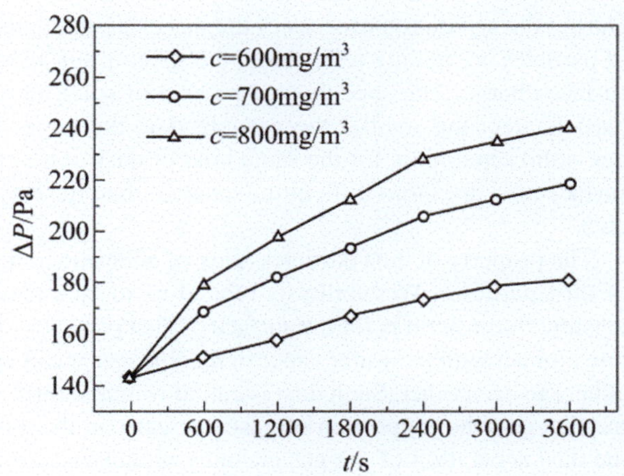

Fig. 8.2 Impact of inlet dust concentration on filter pressure drop

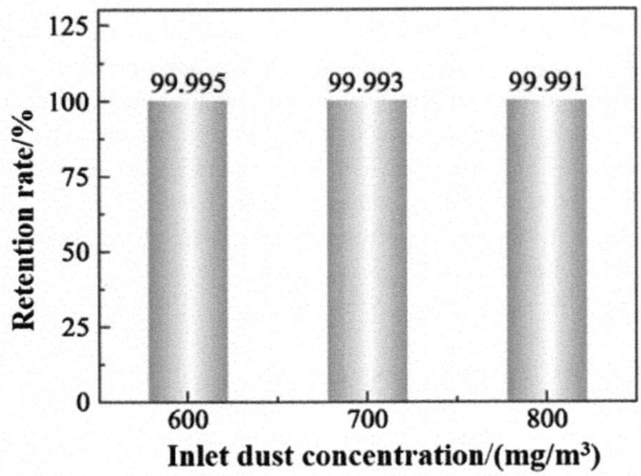

Fig. 8.3 Impact of inlet dust concentration on retention rate

8.1.1.5 Particle Packing Density

The accumulation density of dust is related to particle size, agglomeration, adhesion, as well as the resistance loss and filtering area of the membrane dust collector. The smaller the accumulation density, the more difficult it is to clean the dust, so a lower filtering wind speed should be chosen during design. Moreover, the accumulation density of dust is crucial for selecting the dust collector's ash hopper and ash discharge device.

8.1.1.6 Particle Surface Properties

During the separation process of the gas cleaning membrane, the surface properties of particles are mainly reflected in the specific surface area, surface wettability, and surface charge. The specific surface area of solid particles is greatly related to the particle size; the smaller the particle size, the larger the specific surface area. In gas–solid separation, for the same kind of dust, solid particles with a larger specific surface area are more difficult to separate than dust with a smaller specific surface area.

The property of dust particles' ease of adhesion to liquids is called the wettability of dust particles. Wettability is related to surface tension; the smaller the surface tension, the easier it is for the surface of dust particles to be wetted. Once dust comes into contact with the water vapor in the flue gas, it can expand the wetted surface and adhere to each other. Such dust is called wettable dust. When wettable dust exists in the flue gas, the proportion of gaseous water in the gas has a significant impact on the dust separation of the gas cleaning membrane, so it is necessary to ensure that the flue gas temperature does not fall below the dew point to reduce the appearance of liquid water. Failure to do so may result in the filter bag sticking, causing a rapid increase in resistance on both sides of the membrane, ultimately rendering the dust collector inoperable.

Yuan et al. [3] researched the effect of environmental humidity on the filtration performance of PTFE membranes and conducted a force analysis on the liquid bridge force formed between particles. The results showed that the filtration resistance gradually increased with the decrease in particle size, and the resistance of the filter cake also gradually increased with the increase in humidity. When Tang et al. [2] increased the relative humidity from 60 to 70%, the filtration pressure drop rapidly increased, and the permeability showed a rapid downward trend (Fig. 8.4). This is mainly because agglomeration occurs between particles with the increase of humidity, leading to an increase in particle size, and the particle accumulation is relatively loose, resulting in an increase in filter cake porosity and a decrease in filter cake filtration pressure drop, indicating that the filling state of water between particles is pendulum-like. When the number of liquid bridges stabilizes, an increase in environmental humidity leads to compression of the filter cake. Because water occupies the gaps between particles, leading to a decrease in the porosity of the filter cake, and the growth rate of filtration pressure drop accelerates. The filling state of water between particles changes from pendulum-like to ribbon-like (Fig. 8.5). At this time, the effect of environmental humidity on dust removal performance has changed from enhancement to deterioration. Therefore, there is an optimal value for the effect of environmental humidity on the dust removal performance of the PTFE membrane.

When the flue gas temperature is high, tiny dust particles may collide and agglomerate due to the action of surface charge and Brownian motion. This characteristic plays an indispensable role in the gas–solid separation mechanism. If it occurs in high-concentration dust flue gas, it will have a significant impact on the efficiency

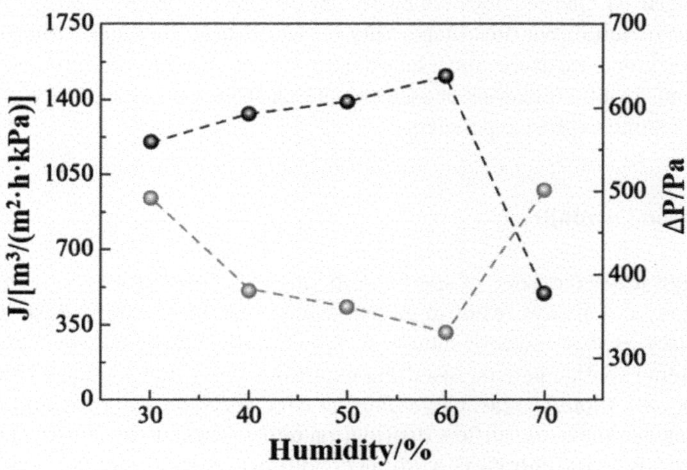

Fig. 8.4 The effect of humidity on pressure drop and permeability

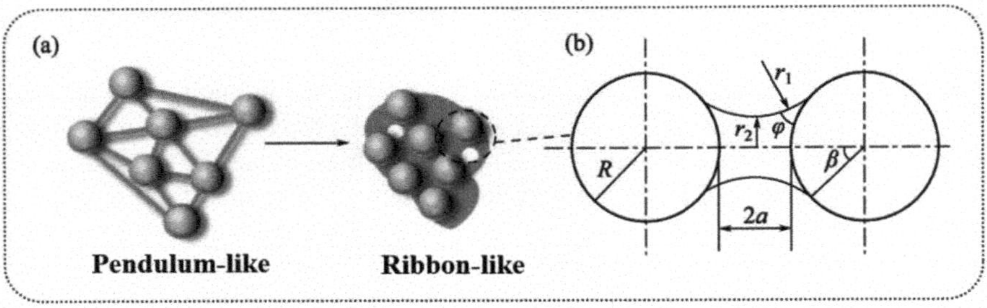

Fig. 8.5 The transformation of the filling state between particles (**a**) and the liquid bridge formed between two spherical particles (**b**)

of gas cleaning membrane separation. Therefore, when agglomeration occurs, it is generally necessary to disintegrate the agglomerates through ultrasonic treatment before entering the dust collector for dust removal.

During the operation process of the membrane dust collector, dust with strong hygroscopicity and deliquescence easily absorbs and solidifies on the surface of the filter material, or it will become a thick substance when encountering water deliquescence, causing difficulties in dust removal, increasing equipment resistance, and affecting the normal operation of the membrane dust collector. For dust containing strongly deliquescent substances such as KCl, $MgCl_2$, NaCl, CaO, etc., necessary technical measures need to be taken.

8.1.1.7 Static Electricity

Dust that is easily charged does not easily fall off once static electricity is generated on the filter material. For dust that is very easy to charge, anti-static filter materials and other technical measures must be used to prevent dust from adhering to the filter material and avoid explosions caused by sparks from static electricity during the filtration of some explosive powders.

8.1.1.8 Combustibility

Even though the occurrence of sparks in the process flow before reaching the dust collector may not always result in an explosion when combustible dust is present, precautionary explosion-proof measures should still be implemented. These may include adding spark catchers, installing explosion-proof doors, and so forth. In the design, not only should a reasonable filtering wind speed be designed, and diversion components that make the airflow distribution uniform should be added, but also the dust particle size, concentration, working conditions, and a reasonable airflow rising speed should be considered.

8.1.2 Flue Gas Properties

The properties of flue gas mainly refer to the temperature of the dust-containing gas, the water content, and the acidity and alkalinity. Generally, when the water content is small, there is no acidic gas, and the temperature is ≤ 130 °C, the membrane material used for air cleaning is commonly 500–550 g/m^2 coated polyester needle-punched felt; when the temperature is between 130 and 260 °C, aramid Nomex needle-punched felt, 800 g/m^2 glass fiber needle-punched felt, 800 g/m^2 weft double glass fiber fabric or Flumex high-temperature coated filter material is selected, and the surface membrane material is made of PTFE material; when the water content is large and the dust concentration is also large, the water-proof characteristics of the membrane material need to be considered, and generally water-proof, oil-proof filter material or coated filter material, such as PTFE coated filter material, will be chosen; when the flue gas temperature is high (> 260 °C) and contains acidic and alkaline gases, the temperature resistance and corrosion resistance of the membrane material need to be considered, and inorganic membrane materials with strong adaptability to working conditions, such as silicon carbide ceramic membrane, Fe–Al metal membrane, oxide ceramic fiber membrane, etc., should be selected.

8.1.3 Membrane Performance Parameters

The performance parameters of the membrane mainly affect the dust collector in two aspects: dust emission concentration and filtration pressure drop. Usually, after selecting the type and parameters of the membrane material, as long as there are no operational errors and sudden changes in flue gas conditions during operation, it will not have a significant impact on emission indicators or operational stability. For flue gas conditions with a high content of water vapor, the membrane material needs to have a certain hydrophobic performance. Otherwise, under long-term conditions of high humidity, the membrane material itself may undergo hydrolysis, causing material damage, exceeding emission standards, and causing the dust collector to be unable to operate. For some complex flue gas conditions, such as gases that not only contain water vapor but also contain oily components such as tar, ordinary membrane materials cannot be used. It is necessary to choose a double sparse PTFE membrane material to effectively prevent the resistance of the dust collector from rising too quickly.

At present, Jiangsu Jiulang High-tech Co., Ltd. in China has successfully developed this series of products and realized industrial applications. Figure 8.6 shows the dual-phobic PTFE membrane product and the hydrophobic and oleophobic test situation. Table 8.1 gives the performance parameters of this product.

Fig. 8.6 Dual-phobic PTFE membrane product

Table 8.1 Dual-phobic PTFE membrane product performance parameter table

Main indicators	Value	Main indicators	Value
Membrane pore size/μm	0.5–15	Oil contact angle/°	≥ 130
Porosity	> 80%	Operating temperature/°C	< 260
Membrane thickness/μm	10–150	Gas permeability/[m/(h kPa)]	> 200
Water contact angle/(°)	≥ 150	Outlet dust concentration/(mg/m^3)	< 5

8.1.4 Equipment Operating Parameters

8.1.4.1 Equipment Resistance

The so-called equipment resistance refers to the total pressure difference from the inlet to the outlet of the dust collector under operating conditions. The pressure loss of the bag filter is usually between 1000 and 2000 Pa, and the pressure loss of the membrane dust collector is usually less than 1500 Pa. Equipment resistance is the main basis for fan selection, and a certain range of pressure loss is allowed during equipment operation. The fan selection should be determined through the allowance of equipment resistance change margin during design. Improper selection of filtration wind speed or uneven distribution of compartments will affect the life of the filter bag. Similarly, improper selection of air rise speed or uneven air rise speed in the compartment will also affect the life of the filter bag.

8.1.4.2 Equipment Pressure Resistance

The pressure resistance of the membrane dust collector is determined according to the process requirements and the static pressure of the fan. The design pressure resistance of the equipment must be determined according to the normal use pressure of the dust collector. For general-purpose dust collectors, the equipment pressure resistance is 4000–5000 Pa; for long bag pulse dust collectors, the equipment pressure resistance is generally 6000–8000 Pa; for negative pressure air conveying devices powered by Roots blowers, the design pressure resistance of the dust collector is 15–50 kPa. In addition, for dry pulse dust collectors of blast furnace gas, the design pressure resistance is 0.3 MPa or higher, and the equipment is generally designed to be round to meet the pressure resistance requirements.

8.1.4.3 Back Blow Pressure

The back blow pressure for dust removal is an important parameter in the design of the membrane dust collector. Depending on the pressure of the compressed air used, it is divided into high pressure (0.5–0.7 MPa), medium pressure (0.35–0.5 MPa), low pressure (0.2–0.35 MPa) and ultra-low pressure (below 0.2 MPa), and the pulse valve is divided into high pressure valve (right angle valve) and low pressure valve (submerged valve) according to the pressure inside the air bag. Box-type spraying does not have a spray pipe and belongs to unordered spraying. The dust removal airflow is directly sprayed into the upper box by the pulse valve to increase the pressure, which is then transmitted to each membrane bag in the chamber to achieve dust removal. For high-temperature dust collectors, the back blow air needs to be heated. Therefore, to prevent the tubular membrane from being damaged due to multiple cold and hot cycles, the back blow air pressure is generally set to 0.4–0.6 MPa.

Fan et al. [4] conducted an experimental simulation of the high-temperature back-blow process of metal filter tubes prepared with iron-aluminum alloy. The back blow gas temperature was 227 °C, the spray pressure was 0.4 MPa, and the pulse spray time was 0.1 s. Through the optimization of back blow pressure and nozzle, the arrangement of filter tubes can be better guided. Liu et al. [5] optimized the back blow parameters, and investigated the back blow effect from 0.1 to 0.3 MPa. As shown in Fig. 8.7, the greater the back blow pressure, the more obvious the back blow effect.

8.2 Types of Membrane Separation Elements and Filtration Methods

8.2.1 Types of Membrane Elements

According to material classification, gas cleaning membranes are mainly divided into organic and inorganic matters, which are used in low and medium as well as high temperature flue gas dust removal processes, respectively. The low and medium gas

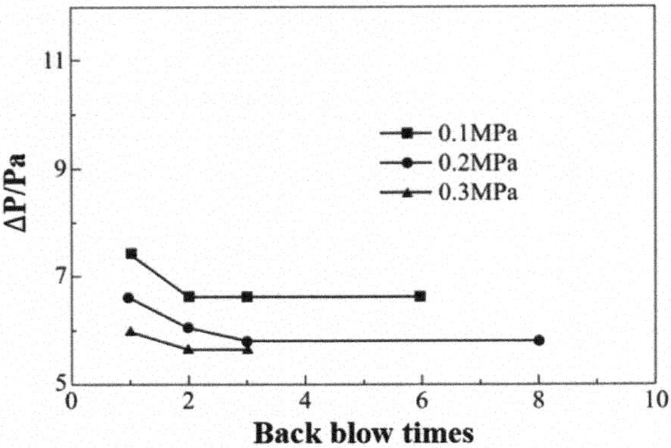

Fig. 8.7 The effect of back blow pressure on pressure drop recovery (back blow time: 2 s)

cleaning membranes mainly include folded filter tube type, folded plate type, and round bag type, etc., among which the folded filter tube type and folded plate type are mainly used in domestic air cleaning or fresh air systems. The round bag type is mainly used in industrial flue gas treatment, with lengths ranging from 1000 to 8000 mm and diameters of 130–160 mm. High-temperature gas cleaning membranes mainly have honeycomb type, hollow fiber tube type, candle type, etc. The honeycomb type configuration is a multi-channel symmetrical structure, loaded with denitrification catalysts for the flue gas denitrification process or loaded with ternary catalysts for diesel engine exhaust filter collectors; the hollow fiber tube type configuration is generally used to load molecular sieve membrane layer, bundled in the membrane component for the separation of VOCs or CO_2 in the gas. The candle-shaped tubular membranes are generally made of metal, silicon carbide or oxide ceramic fiber materials, mainly used in high-temperature flue gas cleaning processes with a usage temperature of 300–800 °C. Several common industrial gas cleaning membrane element configurations are shown in Fig. 8.8.

(a) Bag type (b) Tube type (c) Honeycomb type

Fig. 8.8 Common industrial gas cleaning membrane elements

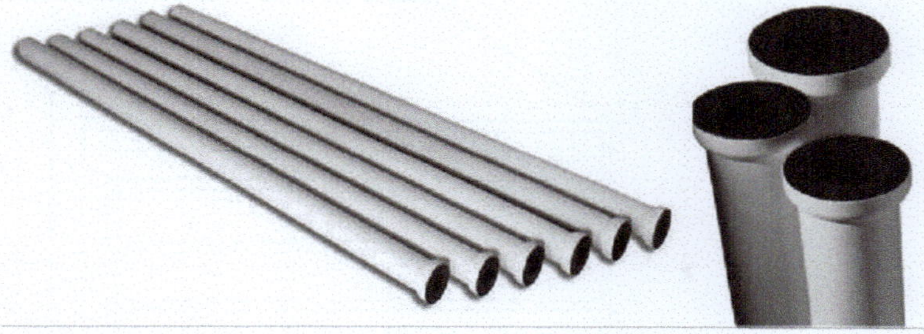

Fig. 8.9 Basic external structure of candle-shaped ceramic filter element

8.2.2 Terminal Filtration Mode

In the terminal filtration mode, also known as straight tube or candle-shaped filtration mode, the standard filter element form is blocked at one end and open at the other. The open section has a flange for fixing on the flower board, which separates the dust-containing gas and the purified gas (Fig. 8.9). Terminal filtration is generally surface filtration. Dust is trapped on the surface of the membrane filter element and forms a filter cake. After a period of time, part of the filter cake layer can be removed by back blowing. The existence of surface filter cake increases some filtration resistance but also improves the dust removal efficiency. In general, the dust concentration of outlet gas in this membrane filtration mode is less than 5 mg/m^3.

8.2.3 Wall Flow Filtration Mode

Whether it is a candle-shaped or tubular ceramic filter element, there are problems, such as large volume, high strength requirements, inconvenient installation, sealing, replacement and maintenance. Because the ceramic filter element has a certain rigidity, it can also be made into a unit module type honeycomb form filter element (Fig. 8.10), in addition to the traditional tubular form.

This structure has many parallel channels. The bottom of the air inlet channel and the top of the air outlet channel are blocked. With this structure, the filtration area per unit volume will greatly increase, the overall strength will increase, and the wall thickness of the channel will also be greatly reduced. The wall thickness is only 1 mm, the filtration resistance is greatly reduced, the material is mainly cordierite, the porosity is 30–50%, the pore diameter of the air inlet channel membrane is 0.2–0.5 μm, and the membrane thickness is less than 50 μm.

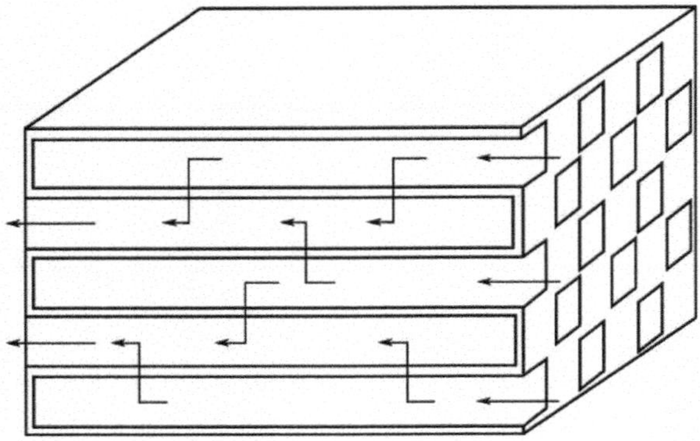

Fig. 8.10 Wall flow filter

8.2.4 Cross-Flow Filtration Mode

Cross-flow filtration has always been the mainstream filtration form in the field of liquid filtration. In the field of high-temperature gas dust removal, cross-flow filtration has not yet entered a large-scale industrial application stage.

Figure 8.11 shows two designs of cross-flow filtration systems. In design (a), the raw material gas enters the inside of the tubular membrane, part of the gas is purified vertically through the tube wall, the rest of the gas enters the cyclone separator along the pipe, ultrafine particles form agglomerates on the membrane surface due to increased concentration, and then are removed by the cyclone separator; in design (b), the raw material gas enters from the outer wall of the membrane, the clean gas comes out from the inside of the tube, the dust-laden gas is separated by the cyclone separator and then mixed with the raw material gas for multiple separations. The gap between the outer wall of the tubular membrane and the shell is small, which can provide a higher cross flow rate [6].

8.3 Design and Structure Type of Membrane Equipment

8.3.1 Material Selection of Membrane Equipment

The material selection for membrane equipment mainly consists of membrane material and membrane equipment selection. The selection of membrane materials is mainly based on the properties of flue gas, particle concentration, investment cost, etc., while the steel selection for membrane equipment mainly depends on whether there are corrosive components in the flue gas, operating pressure, and maximum temperature conditions, etc. When the temperature of the flue gas is below 260 °C, organic composite membranes suitable for low and mediums are generally selected, such as PTFE membranes, P84, and fiberglass; when the temperature exceeds

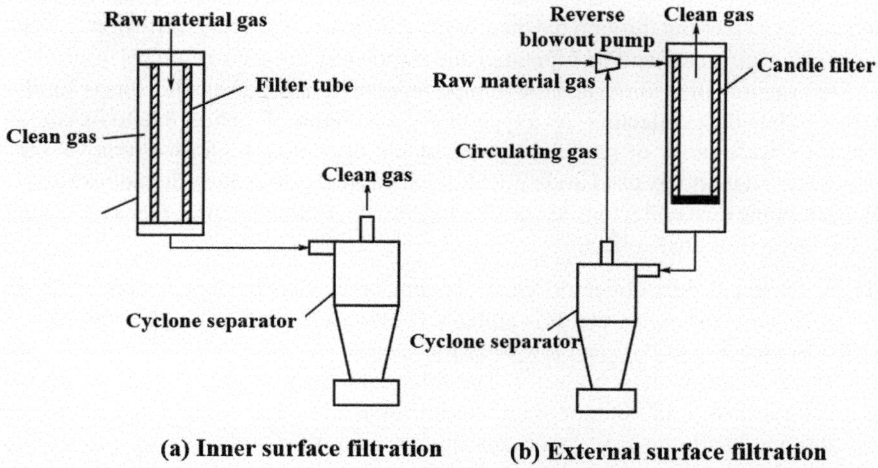

Fig. 8.11 Cross-flow filtration

260 °C, or even higher, high-temperature resistant inorganic composite membranes are needed. According to the actual demand, porous metal membranes or porous ceramic membranes can be chosen. The temperature resistance of metal membranes is lower than that of ceramic membranes, but metal membranes can be welded and installed to improve sealing; ceramic membranes have advantages in terms of filtration accuracy, thermal shock resistance, and corrosion resistance, etc. Under normal temperature conditions without corrosive flue gas, carbon steel is generally chosen as the shell material. If it is high temperature and contains sulfides, the shell is generally made of stainless steel, such as 304 or 316 L. When the flue gas contains Cl^-, the shell also needs to consider pre-surface treatment to prevent the corrosion of chloride ions on the grain boundaries of the steel particles.

8.3.2 *Medium and Low-Temperature Membrane Filtration Equipment*

8.3.2.1 Overview

Medium and low-temperature gas filtration mainly refers to the filtration of flue gas with a temperature below 260 °C. The principle of medium and low-temperature membrane filtration equipment is similar to that of traditional bag dust collectors. It relies on organic composite membranes as filter elements during operation. When the dust-containing gas passes through the outer surface of the membrane, the dust will be trapped on the surface of the membrane, and the filtered gas will then permeate to the inner surface of the membrane, thereby achieving the goal of separating dust and gas. Dust passing through the membrane will generate sieving, diffusion, inertia, static electricity, etc., and be separated and trapped by the membrane [7].

The structure of medium and low-temperature membrane dust collectors is similar to that of bag dust collectors. As a type of dry dust removal device, due to its simple structure, wide range of particle capture, stable operation, high dust removal efficiency, etc., it is widely used in chemical, steel, thermal power and other industries [8, 9]. Membrane dust collectors can be divided into the following three types according to the method of dust collector:

(1) Mechanical dust collector. Dust cleaning is carried out by manual shaking, mechanical shaking, etc. It is suitable for the case of short bag length, uneven dust cleaning of bag, and low filtration speed.
(2) Reverse airflow dust collector. The filter cake layer is made to fall off by the reverse airflow of outdoor or circulating air and the dust-containing airflow. On the one hand, the cake layer can directly peel off. On the other hand, due to the different directions of the airflow, the expansion and contraction vibrations occur in the membrane bag, which can further accelerate the fall of the dust cake.
(3) Pulse dust collector. Compressed air is released and sprayed out through the nozzle on the blowing pipe, causing the dust around the membrane bag to be impacted and fall off. The pulse dust cleaning method has a large dust cleaning intensity, can perform on-line dust cleaning, can use long membrane bags, and can handle a large amount of dust-containing gas.

The pulsed membrane dust collector is the most widely used equipment in membrane dust collectors, and its working principle is as follows: the dust-containing gas enters the dust collector from the inlet, a portion of the larger particles settle in the dust hopper due to gravity, and smaller particles enter the middle body with the airflow. Dust collides, intercepts, and diffuses between the membrane surface fibers and settles on the surface of the filter material, forming a dust layer. The purified gas enters the upper body and is discharged from the outlet. As the filtering time increases, more and more dust accumulates on the surface of the filter material, and the filtering resistance continues to increase. When the filtering resistance reaches a certain value, it needs to be cleaned [10]. Figure 8.12 is a schematic diagram of the general structure of a membrane dust collector. The membrane bag installation structure is shown in Fig. 8.13. The membrane bag is fixed to the flower board with an elastic expansion ring, and then put into a gantry to ensure the verticality of the membrane bag, so that the membrane bag has a certain rigidity during the filtering and back-blowing process.

The application of baghouse dust collectors has a history of more than 100 years. In a large number of industrial applications, researchers and engineers have optimized and improved the dust collector, and its performance has been greatly improved. In recent years, researchers have conducted a lot of research on the intake method, injection pressure, airflow velocity, and pressure distribution of the membrane dust collector based on the baghouse dust collector [11–13]. However, in practical engineering applications, due to the complexity of different working conditions and the

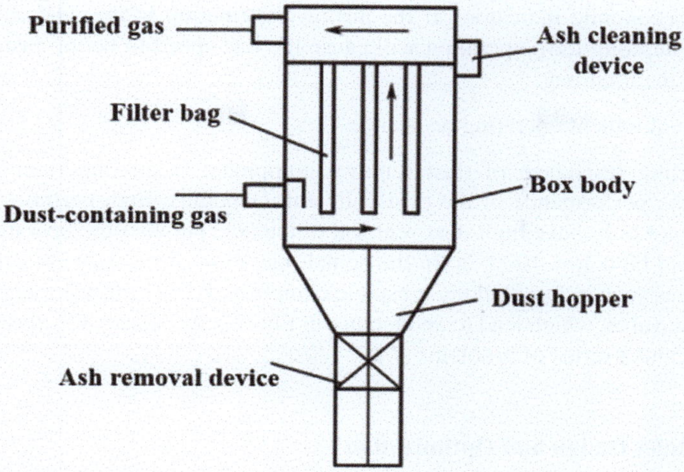

Fig. 8.12 Schematic diagram of baghouse dust collector structure

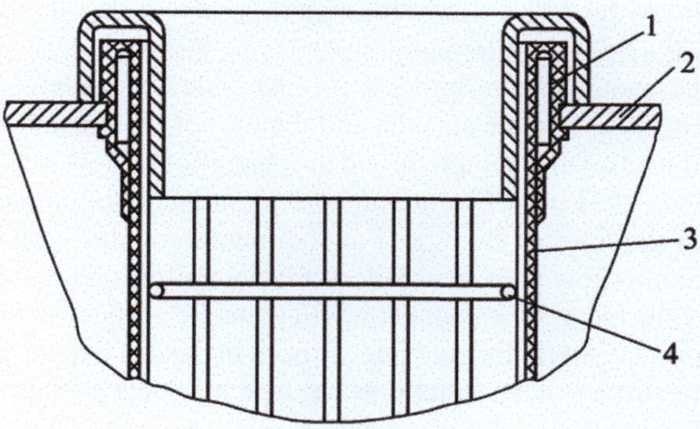

Fig. 8.13 Fixing method of low and medium organic composite membrane bag. 1—Elastic expansion ring; 2—Flower board; 3—Membrane bag; 4—Membrane bag dragon frame

differences in dust-containing gas temperature and humidity, there are still some problems in the operation of the dust collector:

(1) Filter bag issue

The filter bag is the core of the dust collector and determines the performance of the dust collector. The innovation and improvement of filter materials further promote the progress of the dust collector. The main reasons for the damage of the filter bag during the operation of the dust collector are: high temperature burning, corrosion, spark burning, high temperature shrinkage, dust hopper dust heat storage burning filter bag, etc.

(2) Design issue

Although the current baghouse dust design has a certain standardization, the uniformity of the airflow distribution in the dust collector is often ignored due to the impact of various economic indicators. If the airflow distribution in the dust collector is unreasonable, long-term operation will cause the bag speed to be too fast, causing damage to the filter bag.

(3) High equipment operating resistance

The operating resistance of dust removal equipment is an important indicator affecting the performance of the dust collector. Generally, high-performance dust collectors not only have high dust removal efficiency, but also the operating resistance should be below 1000 Pa. In the actual operation of the dust collector, there are problems such as bag sticking, unstable compressed air of pulse jet, and damage of the pulse valve, which lead to an increase in filtering resistance, low dust removal efficiency, and a series of problems [14].

8.3.2.2 Inlet Design and Optimization

In practical industrial applications, membrane dust removal systems have complex structures and large volumes, making it very difficult to obtain certain parameters inside the dust collector through experimental methods. Therefore, many researchers use numerical simulation methods to study the dust collector, and ultimately guide the manufacturing process of dust collector equipment through optimized design. The design of the air inlet and the way of air intake have a significant impact on the impact on the membrane and the change of airflow entering the dust collector. Pereira et al. [15] used Fluent software to simulate the impact of different inlet positions on the internal flow field of the membrane dust collector (Fig. 8.14). The double counterflow inlet dust collector has a uniform mass flow distribution on the surface of the membrane bag, but it will generate vortices in the ash hopper part; the middle air intake will not generate vortices in the ash hopper part, but the mass flow rate on the surface of the membrane bag near the inlet side is too large, which can easily cause the damage of the membrane bag. Wang et al. [16] studied the effects of different air inlet positions on the airflow distribution and pressure loss of the membrane dust collector, and found that compared with the upper and middle air inlets, the lower air inlet has a more uniform flow distribution through each membrane bag; under a certain inlet wind speed, the pressure loss of the membrane dust collector increases with the increase of the inlet position.

8.3.2.3 Design and Optimization of Baffle

In the membrane dust collector, the reasonable arrangement of baffles and barriers at the flue gas inlet and dust removal room entrance will improve the distribution of airflow. Liu et al. [17] concluded that adding baffles can make the airflow distribution in the box more uniform by comparing the velocity cloud maps inside the dust collector with and without baffles. Wang et al. [18] established a small pulse bag type dust collector model, and measured the distribution of the internal flow field

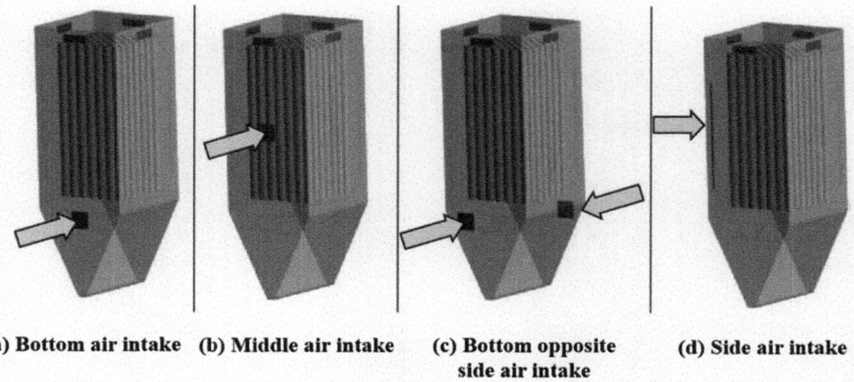

Fig. 8.14 Membrane dust collector with 4 different inlet positions

of the grid baffle, step baffle, and oblique baffle respectively (Fig. 8.15). The results show that the oblique baffle has the best effect, reducing the system resistance by more than 20% compared to the step baffle and grid baffle.

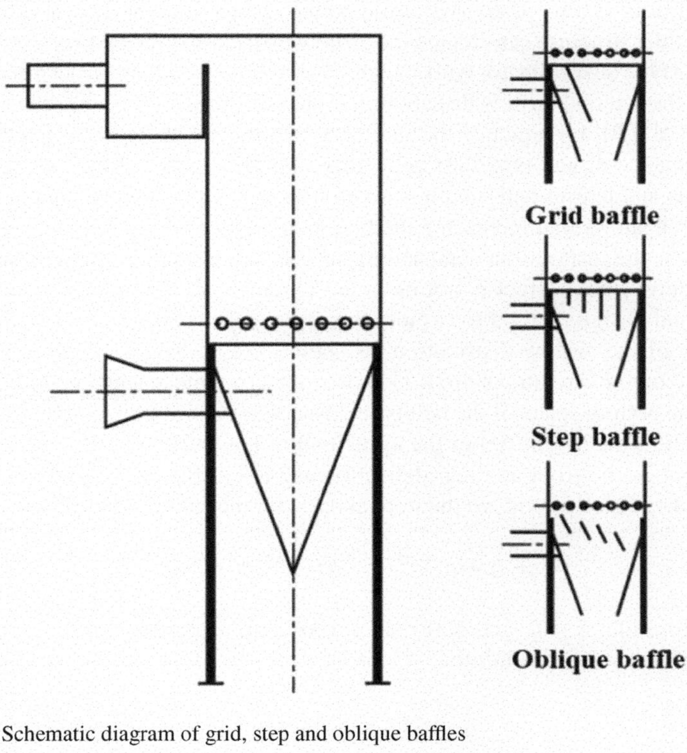

Fig. 8.15 Schematic diagram of grid, step and oblique baffles

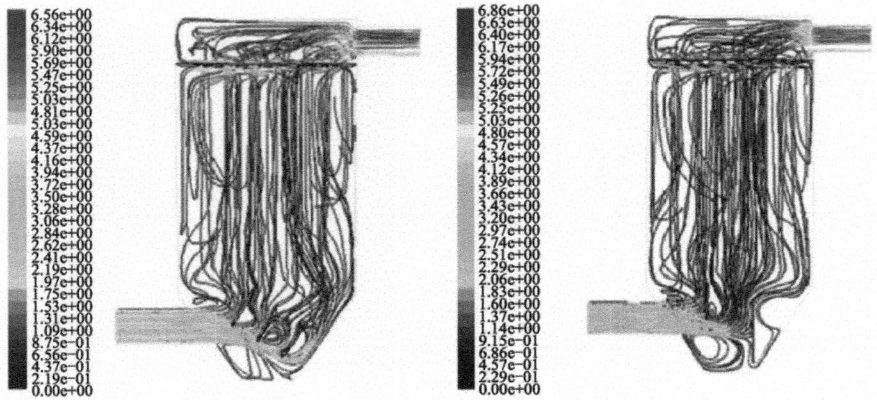

(a) Diagonal inlet airflow diagram **(b) Staggered inlet airflow diagram**

Fig. 8.16 Streamline diagram of two different shape inlet airflows

Gao et al. [19] simulated the internal flow field of the bag type dust collector with two different inlet structures, oblique type and staggered type, under the condition of lower air intake (Fig. 8.16). According to the results of numerical simulation, the change of inlet structure can change the airflow distribution inside the dust collector, and the bag type dust collector with staggered air inlet will reduce the jet phenomenon and make the internal airflow distribution more uniform.

Ning et al. [20] investigated the influence factors of four types of inlet pipes on the performance of bag type dust collectors, and also analyzed the factors causing airflow bias and jetting-bell mouth geometry (taper), airflow speed, and bend form. The research shows that increasing the bell mouth taper or reducing the bend angle can improve the airflow bias and jetting in the dust collector to a certain extent, but the improvement effect is not obvious. Zhang et al. [21] used the orthogonal experimental design method to explore the impact of the shape of the baffle on the uniformity of the airflow distribution, as shown in Fig. 8.17. The research results provide the function relationship (8.1) between the root mean square σ of the airflow speed in the bottom space of the belt type dust collector with the number of plates n, plate length L, the height ΔH of the upper end of the baffle from the top section of the ash hopper, the front and rear deflection angle θ of the baffle, and the filtration wind speed v, which provides a theoretical basis for the setting of the diversion plate.

$$\sigma = 0.97n^{-0.03}L^{0.75}\Delta H^{-0.14}\theta^{-0.12}v^{0.01} \tag{8.1}$$

In the formula, $2 \leq n \leq 4$, $500 \leq L \leq 1000$ mm, $0 \leq \Delta H \leq 150$, $-15^\circ \leq \theta \leq 15^\circ$, 0.7 m/min $\leq v \leq$ 1.3 m/min.

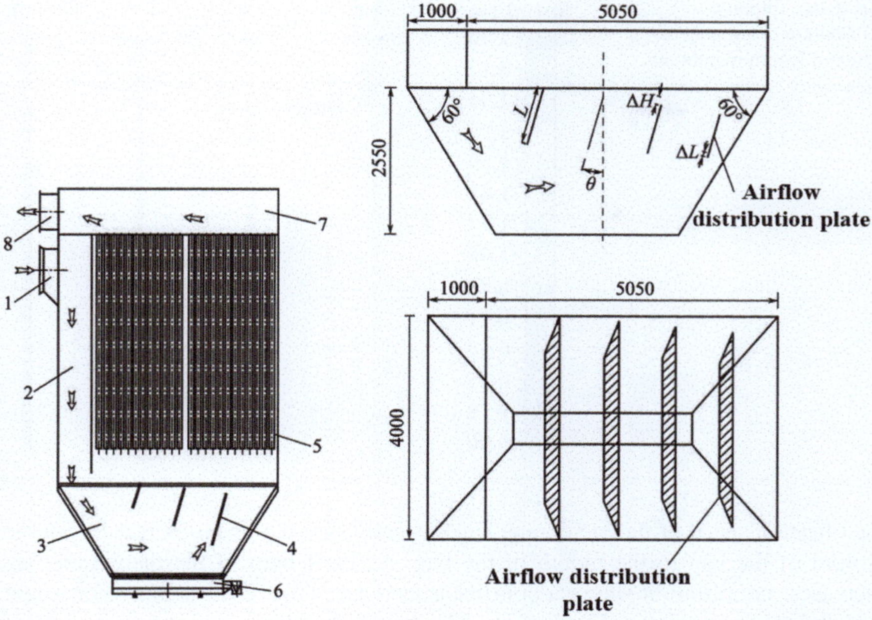

Fig. 8.17 Schematic diagram of dust collector physical model and airflow distribution plate structure. 1—Inlet; 2—N-shaped flue; 3—Ash hopper; 4—Airflow distribution plate; 5—Membrane bag; 6—Fluidized bed; 7—Upper box; 8—Outlet

8.3.2.4 Optimization of Membrane Bag Length

The length of the organic composite membrane commonly used in industry at low and mediums is between 6000 and 8000 mm. The longer the membrane bag, the higher the dust collector equipment. Especially during the dust removal process, this may cause poor back-blowing effect due to the too long membrane bag. Jia et al. [22] addressed this problem in the membrane dust collector, optimized the structure of the dust collector with the help of Computational Fluid Dynamics (CFD) technology, and conducted numerical simulation studies on gas–solid two-phase flow and unsteady pulse jetting process. The study found that at the same inlet speed, the velocity cloud maps of different length membrane bags are different (Fig. 8.18). The surface speed of the bottom area of the membrane bag is low, and the surface speed of the membrane bag near the flower plate position is larger, but the surface speed distribution in this area of different length membrane bags is relatively small. This is because the airflow gathers in the flower plate area after entering the dust collector. Therefore, under the same dust collector structure, when the processing air volume is the same, for different lengths of membrane bags, the difference in the surface speed of the membrane bag is mainly in the middle area of the membrane bag.

Figure 8.19 is the filtration speed of the membrane bag along the height direction [23]. As can be seen from the figure, when the length of the membrane bag is 1 m,

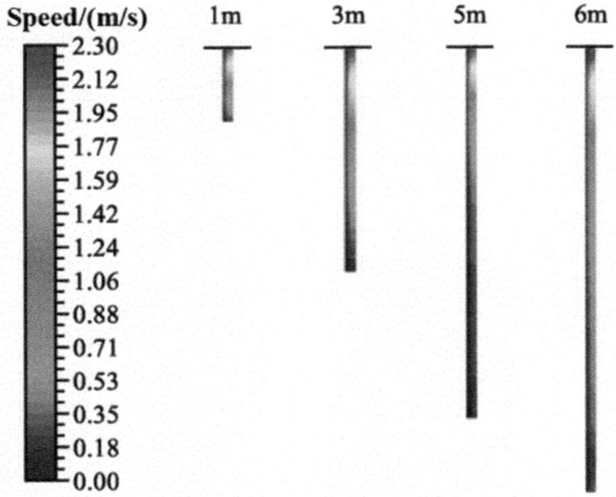

Fig. 8.18 Velocity distribution cloud map of different length membrane bags

the filtration speed of the membrane bag basically shows a linear increase from the bottom of the bag to the mouth of the bag. As the length of the membrane bag increases, the filtration shows a non-linear increase. This result is more consistent with the research results of Park et al. [23], that is, the filtration speed at the bottom of the membrane bag is small, and the filtration speed near the flower plate area is larger. This also shows that the air volume handled by the membrane bag area at the bottom of the flower plate is large. Experimental research has found that the lower the filtration speed, the higher the dust removal efficiency of the coated filter material. Therefore, when designing a membrane bag type dust collector, the dust-containing gas should be settled as much as possible in the middle and lower area of the membrane bag.

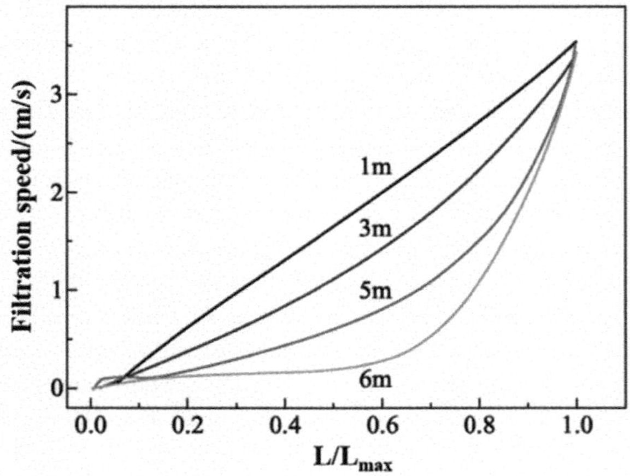

Fig. 8.19 Relationship between membrane bag length and filtration speed along the height direction

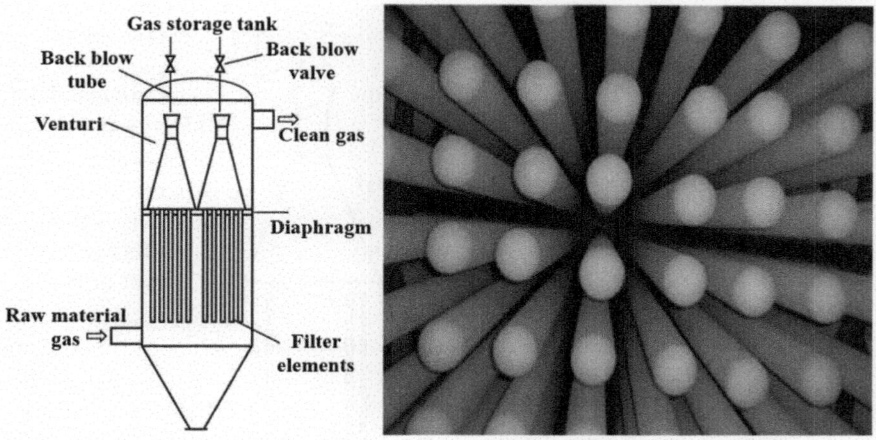

Fig. 8.20 Single-layer diaphragm design principle and tubular membrane arrangement diagram

8.3.3 High-Temperature Membrane Filtration Equipment

Currently, all high-temperature filter designs are based on single-layer or multi-layer diaphragm arrangements, both of which can be used for different filters. The single-layer diaphragm type is the most widely used structure in high-temperature membrane filtration applications. Typically, a candle-shaped membrane filter element with one end open and the other end closed is hung parallel on a diaphragm, separating the raw gas and purified gas through the diaphragm (Fig. 8.20). The filtering gas enters from the outer surface of the filter element, and the clean gas is discharged from the inner surface of the element. Filter elements can be combined into multiple units. The filter unit is fixed on the diaphragm of the shared gas chamber and connected with a sleeve and a high-temperature sealing pad.

8.3.3.1 Venturi Back Blow System

During the pulse cleaning process, the pulse gas is introduced into the filter element from top to bottom, which is opposite to gas filtration. Common back blow systems include Venturi-based jet pulse systems. The gases used include air and nitrogen. Sometimes, to avoid unnecessary condensation, the back blow gas needs to be preheated. The filter housing can be round (for high pressure) or square (for normal pressure). The advantage of the single-layer diaphragm type is that it is convenient to install and maintain, and the main disadvantage is that when the size of the round filter housing is fixed, the number (filling area) of membrane filter elements is limited.

Heidenreich et al. [24] used a Venturi ejector to study the back blow effect of high-temperature dust collectors of different lengths. The device contains 48 tubular membranes, using silicon carbide tubular membranes of 1500 and 2000 mm as the

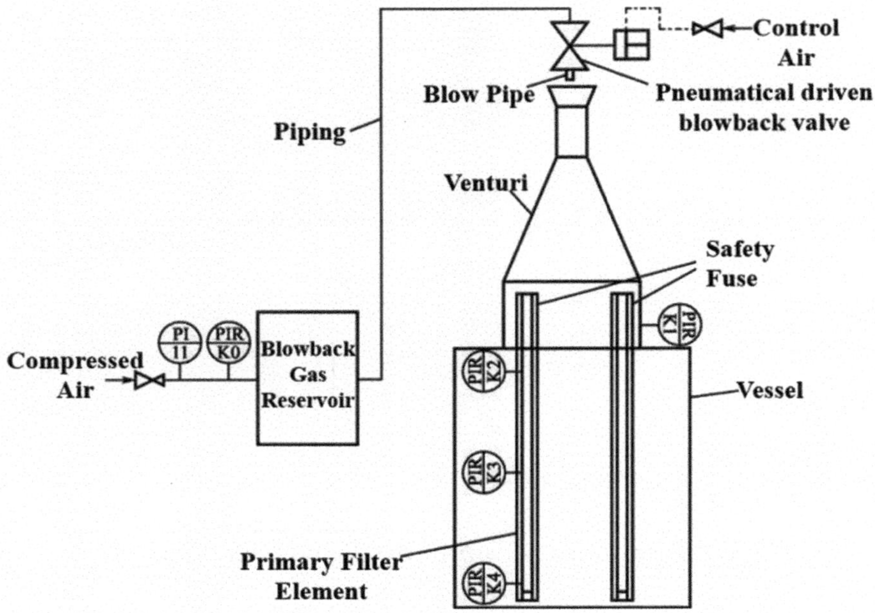

Fig. 8.21 Schematic diagram of the Venturi jet ceramic membrane filtration device process

filter core (Fig. 8.21). The experimental results show that even if the length of the tubular membrane is increased, as long as the design equipped with a Venturi ejector is used, it does not affect the back blow effect, and the extension of the tubular membrane increases the filtration area and reduces the investment cost of the system.

8.3.3.2 Non-pulse Cleaning System

This design uses a series of injectors to generate very high annular (or shear) forces on the surface of the tubular membrane, allowing gas to continuously infiltrate the tubular membrane without affecting normal filtration, thereby controlling the surface speed of the tubular membrane to control the thickness of the filter cake. The shear force generated on the surface of the tubular membrane keeps the dust particles in suspension, reducing the pressure drop of the entire filter. Its equipment structure is shown in Fig. 8.22 [25]. The raw gas enters the dust collector from the side, allowing the gas to flow at an appropriate speed on the surface of each filter element and then penetrate into the inside of the tubular membrane. The clean gas is discharged from the top, and a large amount of concentrated dust-containing gas moves down on the surface of the filter into the cyclone separator chamber to remove large particles. The gas treated by the cyclone separator leaves upwards, mixes with the incoming raw gas, and then filters through the dust collector.

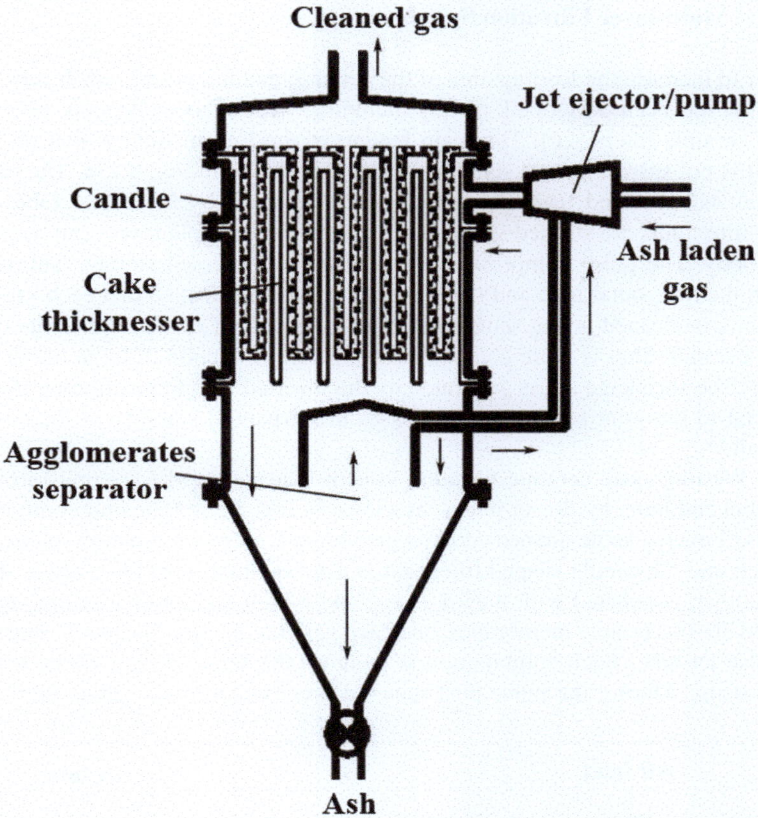

Fig. 8.22 Schematic diagram of a ceramic membrane filtration device without a pulse cleaning system

Comparative experiments have found that under the pulse back-blowing mode, the pressure drop of the dust collector rises much faster than the non-pulse mode. Moreover, under the same dust concentration and filtration speed, both the back-blowing gas consumption and filtration pressure drop are significantly lower than the pulse back-blowing mode. The experiment also found that under the same filtration area, the gas flow rate of the pulse back-blowing system is 12.9–129 L/min, but when using the non-pulse dust removal system, the gas flow rate of the dust collector is 200–245 L/min. This also means that the investment cost of membrane materials can be greatly reduced, and the thickness of the filter cake on the membrane surface is only 0.33 mm.

8.3.3.3 Multi-layer Filtration System

In order to increase the loading area of the ceramic membrane, a multi-layer flower plate filter has been developed, mainly including Westinghouse ceramic filters and Lucchi ceramic filters [26]. The main feature of the Lucchi filter is that the filter element is not suspended but standing upright, as shown in Fig. 8.23a. The tubular filter unit is supported from the bottom and positioned by horizontal tubes. The tubular filter units are stacked vertically in groups, so multiple layers can be placed in the body. The pulse compressed air from the compressed storage tank passes through the horizontal tube and then enters the vertical tube, so that each group of filter units is cleaned at the same time. The arrangement of the filter units of the Lucchi ceramic filter is more compact, making the production of large filters more flexible. The shrinkage stress generated by the inverted weight positioning block is beneficial to the joint sealing and increases the structural stability of the ceramic filter unit.

The Westinghouse ceramic filtration system is made up of filter units arranged in parallel and hung by flower plates, as shown in Fig. 8.23b. The unit combination consists of many candle-shaped ceramic membranes, each combination shares a gas chamber, and the candle-shaped filter unit is fixed on the flower plate of the shared gas chamber, connected with a card sleeve and a high-temperature sealing pad. It can hold 30–60 ceramic membranes, and each gas chamber has its own Venturi tube and pulse jet tube. Each combination is hung on the same support frame to form a filter string. During the pulse dust removal process, the pulse gas is introduced

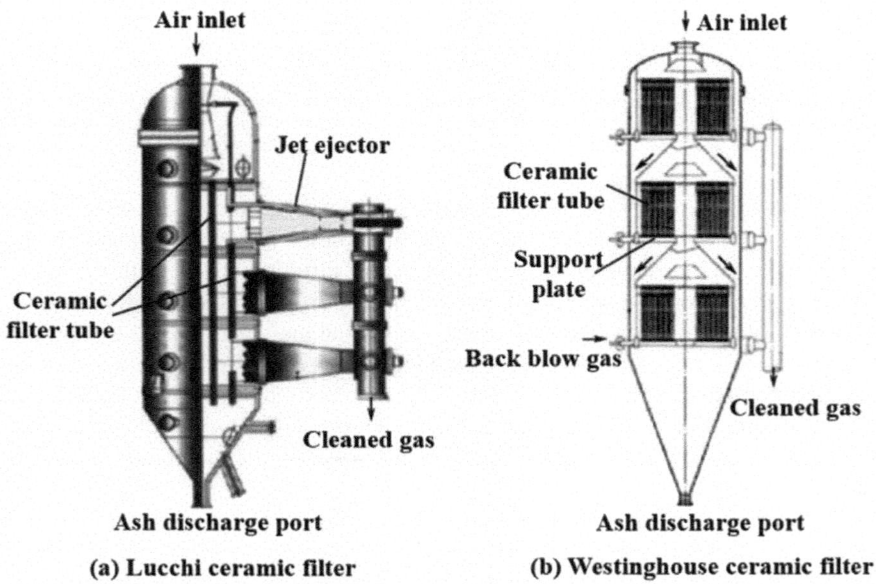

Fig. 8.23 Multi-layer flower plate high-temperature ceramic filtration system

from top to bottom into the shared gas chamber. Depending on the flow rate of the filtered flue gas, the system can be designed to have more ceramic membranes in each gas chamber, or increase the number of gas chambers in each filter string, or accommodate more filter strings [27].

References

1. Tang X, Zhao SF, Feng SS, et al. Exploring the key factors in dusty gas filtration: experimental and modeling studies. Ind Eng Chem Res. 2019;58(42):19633–41.
2. Tang X. Separation mechanism and process simulation of ultrafine dust by PTFE membrane. Nanjign: Nanjign Tech University; 2019.
3. Yuan XL. Study on the influence of environmental humidity on the filtration performance of PTFE membrane filter material. Beijing: Beijing University of Chemical Technology; 2015.
4. Fan LL. Numerical simulation of filter tube in pulse reverse blowing process of high temperature dust collector. J Changchun Inst Technol (Natural Science Edition). 2018;19(1):66–71.
5. Liu W. Preparation and gas-solid separation properties of porous silicon carbide ceramics. Nanjing: Nanjing Tech University; 2014.
6. Sibanda V, Greenwood R, Seville J, et al. Particle separation from gases using cross flow filtration. Powder Technol. 2001;118:193–201.
7. Yao X, Yu YX. Research on bag filter technology and equipment development. Process Equip. 2017;6:135–6.
8. Shaaban S. Numerical optimization and experimental investigation of the aerodynamic performance of a three-stage gas-solid separator. Chem Eng Res Des. 2011;89(1):29–38.
9. Ribeiro D. Applications of CFD techniques in the design of fabric filters. Chem Eng Trans. 2014;39:1369–74.
10. Oliver K, Jörg M, Gerhard K. The contribution of small leaks in a baghouse filter to dust emission in the $PM_{2.5}$ range: a system approach. Particuology. 2017;30:40–52.
11. Joe YH, Shim J, Park HS. Evaluation of the can velocity effect on a bag filter. Powder Technol. 2017;321:454–7.
12. Koch M, Krammer G. Filter performance with non-uniformly distributed concentration of dust-evidence from experiments and models. Powder Technol. 2016;292:149–57.
13. Saleem M, Tahir MS, Krammer G. Measurement and simulation of axial velocity in a filter bag. Chem Eng Technol. 2012;35(12):2161–9.
14. Li WH, Gao GJ. Study on flow field uniformity of straight-through down-inlet bag filter. Coal Mine Mach. 2015;36(6):289–92.
15. Pereira TWC, Marques FB, Pereira FAR, et al. The influence of the fabric filter layout of in a flow mass filtrate. J Clean Prod. 2016;111:117–24.
16. Wang DD, Qian FP, Wu XQ, et al. Measurement and numerical simulation of airflow distribution uniformity in bag filter. J Anhui Univ Technol (Natural Sciences). 2013;30(3):343–9.
17. Liu LR, Wang ZJ, Liu LB, et al. Automatic optimization design of flow field uniformity of bag filter. Mach Des Res. 2017;33(2):171–5.
18. Wang G. Experimental study on inlet guide plate of pulse bag filter. Environ Prot. 2007;20:71–4.
19. Gao GD, He LL. Numerical analysis of flow field of bag filter influenced by flow guide plate. Coal Mine Mach. 2010;31(12):38–40.
20. Ning B, Wang ZJ, Zhang S, et al. Study on influence factors of inlet duct on bag filter performance. J Saf Environ. 2013;13(6):58–63.
21. Zhang XL, Shen HG, Zhou R, et al. Analysis of the influence of the structure parameters of the inlet plate on the airflow distribution of the bag filter. Environ Eng. 2012;30(4):76–9.
22. Jia LX. Structure design and research of membrane bag filter. Nanjing: Nanjing Tech University; 2020.

23. Park S, Joe YH, Shim J, et al. Non-uniform filtration velocity of process gas passing through a long bag filter. J Hazard Mater. 2019;365:440–7.
24. Heidenreich S, Haag W, Salinger M, et al. Next generation of ceramic hot gas filter with safety fuses integrated in venturi ejectors. Fuel. 2013;108:19–23.
25. Sharma SD, Dolan M, Ilyushechkin AY, et al. Recent developments in dry hot syngas cleaning processes. Fuel. 2010;89:817–26.
26. Meng GY, Dong Q, Liu XQ, et al. Recent advances in inorganic porous separation membranes. Membr Sci Technol. 2003;23(4):261–8.
27. Ahmadi G, Smith DH. Gas flow and particle deposition in the hot gas filter vessel at the Tidd 70 MWE PFBC demonstration power plant. Aerosol Sci Technol. 1998;29:206–23.

Chapter 9
Application of Gas Cleaning Membranes in Indoor Air Cleaning

9.1 Overview of Indoor Air Pollution

The indoor air environment is fundamental to people's work and daily life, as individuals spend over 80% of their time indoors. Therefore, the quality of indoor air significantly influences human health. Common indoor air pollutants comprise particulate matter (PMs), microbial aerosols, volatile toxic gases, and more [1]. Particulate matters are the main pollutant in indoor air, which can be divided into PM_{10}, $PM_{2.5}$, $PM_{1.0}$ etc., according to the size of the particle aerodynamic diameter. The smaller the particle size, the greater the harm to the human body and the more difficult to be removed. Among them, particulate matter with the aerodynamic equivalent diameter less than or equal to 2.5 μm is called $PM_{2.5}$, which corresponds to the main particle size distribution of particulate matter pollutants. It can suspend in the air for a long time. This type of the particulate matter has a relatively large surface area, and is easy to adsorb toxic and harmful substances; at the same time, this type of the particulate matter has a small particle size, which can enter the human respiratory tract and then inhale into the lungs through human breathing, thereby causing some respiratory diseases. Even some substances from $PM_{2.5}$ can directly enter the cardiovascular system, causing greater harm to human health [2]. Airborne microorganisms mainly exist in indoor air in the form of aerosols, including floating bacteria, fungi, and spores, with particle sizes generally ranging from 0.002 to 30 μm [3, 4]. Airborne microorganisms originate from various sources in nature, including the Earth's surface, rivers, animals, plants, and humans. Additionally, environments closely associated with human activities, such as wastewater treatment plants, food factories, garbage dumps, and breeding farms, serve as significant sources of airborne microorganisms [5, 6].

Z. Zhong and W. Xing, *Membrane Technology for Air Cleaning*,
https://doi.org/10.1007/978-981-96-1144-7_9

9.1.1 Sources of Indoor Air Pollution

There are mainly two sources of indoor air pollution, one is the pollution brought to the air by indoor materials and human indoor activities, and the other is the pollution caused by outdoor pollutants entering the room [7]. The air pollution caused by indoor materials mainly comes from building materials, decoration materials, etc., which are the leading causes of chemical pollution in indoor air. Indoor air pollution caused by human activities mainly includes cooking fumes, smoke from smoking, harmful substances such as bacteria and microorganisms bred in the environment, etc. [8]. Outdoor pollutants mainly include sulfur oxides, nitrogen oxides, particulate matters, etc., produced by the combustion of fossil fuels, dust generated during construction, and various chemical pollutants emitted from industrial production, etc. These pollutants enter indoors through windows, doors, holes, and gaps, causing indoor air pollution [9].

9.1.1.1 Pollution Caused by Building Materials and Decoration Materials

Building materials and decoration materials are the main causes of chemical pollution in indoor air. During the interior decoration process, a large number of building materials and decorative materials are applied to the indoor environment, which brings serious gas pollution to the indoor air. The chemical pollution caused by building materials and decoration materials mainly includes man-made panels (particle planks, plywood, marble planks, etc.) and chemical adhesives (including paint coatings, waterproofing agents, adhesives, etc.) used in decoration. These materials can produce volatile organic compounds (VOCs), such as formaldehyde, benzene, toluene, etc. In addition, building materials can also release ammonia, radon and other chemical pollutants into the indoor air [10].

9.1.1.2 Pollution Caused by Indoor Household Items

As people's living standards improve, the use of household items during indoor activities becomes inevitable. However, these household items themselves contribute to indoor air pollution. These household items are mainly divided into two categories. One is chemical products used in daily life, such as insecticides, cosmetics, detergents, etc. And the other is home decoration and furniture, such as beds, wardrobes, curtains, etc. These household items will cause chemical pollution in the indoor air [11].

9.1.1.3 Pollution Caused by Human Activities

Human activities can also bring pollution into the indoor air. For example, during indoor cooking, a large amount of cooking fumes will be produced, which contain many harmful chemicals such as hydrocarbons, aldehydes, alcohol, etc. These compounds will also react with other substances in the air to form new harmful substances, thereby causing serious pollution to the indoor air. In addition, the smoke produced by smoking is also a cause of indoor air pollution. The smoke produced after burning contains a large amount of nicotine and tar. These chemicals not only cause air pollution but also harm humans themselves [12].

9.1.1.4 Pollution Caused by Human Metabolism

Human metabolism releases many pollutants through the skin, feces, urine, and respiratory tract. These pollutants will cause indoor air pollution. For example, human metabolism produces substances such as benzene, hydrogen sulfide, chloroform, etc. In addition, various bacteria and viruses brought back from the outside world by the human body will be discharged into the indoor air through sneezing and coughing, thereby causing chemical pollution in the indoor air [13].

9.1.2 The Harm of Indoor Air Pollution

The pollution caused by formaldehyde, benzene, and other volatile organic compounds, which are produced during renovation, can disrupt the body's immune level, affecting the functions of the central nervous system and the digestive system. Mild exposure can cause symptoms such as skin allergies, sore throat, headache, dizziness, chest tightness, fatigue, loss of appetite, and nausea. Severe cases can cause bronchitis, allergic asthma, pneumonia, pulmonary edema, etc. In serious cases, it can damage the liver, causing disorders in the hematopoietic system function [14]. Well-known organizations such as the World Health Organization (WHO) and the National Academy of Sciences/National Research Council (NAS/NRC) of the United States have always emphasized that VOCs are an important class of air pollutants. The World Health Organization has long classified formaldehyde as a class of carcinogens and explicitly warned formaldehyde is strongly associated with cancer, especially childhood leukemia [15–17].

Bacteria, fungi, dust mites, and other microorganisms are extremely easy to breed under conditions of confined spaces, overuse of air conditioning, and very humid indoor environments. Long-term exposure to this environment can easily cause skin allergies, respiratory discomfort, and other reactions, causing allergic asthma, allergic rhinitis, bronchitis, and allergic dermatitis. Furthermore, bacteria and viruses can become airborne through activities like talking, coughing, and sneezing, allowing them to spread to others and potentially causing epidemics of influenza and other diseases [18].

In addition to solid particles, the combustion products from fuel and cooking fumes also contain large amounts of harmful gases, such as sulfur dioxide, nitrogen oxides, carbon monoxide, nitrogen dioxide, and organic hydrocarbons. Carbon monoxide poisoning incidents occur from time to time. The smoke produced by smoking contains a large number of organic carcinogens, and each cigarette can produce 0.6–3.6 mg of nicotine [14]. Nicotine has many adverse effects on human health. First, nicotine can raise blood pressure, accelerate heartbeat, even cause arrhythmia and induce heart disease. Second, it can damage the bronchial mucosa, causing tracheitis. Thirdly, nicotine poisons brain cells, leading to symptoms affecting the central nervous system in smokers. Finally, it can also promote the formation of cancer. In fact, indoor smoking can cause $PM_{2.5}$ values to surge more than 10 times [19].

Research has shown that people who live near roads and are long-term exposed to nitrogen dioxide or $PM_{2.5}$ pollution are more likely to have left and right ventricular hypertrophy. Furthermore, the severity of air pollution is directly linked to the prominence of ventricular hypertrophy. Similar heart changes can also be observed in the early stages of heart failure [20].

9.1.3 *Indoor Air Pollution Control*

In order to improve indoor air quality, make living and working conditions better, and promote physical and mental health, it is necessary to control indoor air pollution. The commonly used techniques for indoor air treatment worldwide are described as follows.

9.1.3.1 Chemical Absorption Method

The principle of the chemical absorption method is mainly to utilize the characteristic that most indoor gas pollutants are easily soluble in water, which is to reduce the concentration of pollutants by solvent absorption. Complexing agents or oxidants are commonly added to absorption liquids to enhance absorption efficiency. These agents alter the chemical structure or disrupt the molecular structure of toxic gases, thereby reducing pollutant concentrations in indoor air [21]. The chemical substances contained in the absorption liquid mainly include organic nitrogen-containing compounds or polymers, inorganic ammonium salts, or sulfites, etc. Taking formaldehyde as an example, because the ammonium salt solution contains ammonium (NH_4^+), it can react with formaldehyde to form hexamethylenetetramine. Another compound, the sulfite ion contained in the solution of sulfite, can easily react with formaldehyde to form precipitates, thereby achieving the purpose of removing formaldehyde [22].

9.1.3.2 Adsorption Method

Physical adsorption technology is one of the more widely used air cleaning technologies in practical applications than absorption. It is used more popularly in various air purifiers. The main principle of adsorption technology is to use the strong adsorption performance of adsorbents to remove formaldehyde and other toxic gaseous pollutants [23]. Common adsorbents include various forms of activated carbon, activated carbon fiber, porous carbon materials, honeycomb activated carbon, and other novel activated carbon. Other adsorbents include porous clay minerals, molecular sieves, zeolites, activated alumina, silica gel etc. Because the adsorption method has a strong enrichment function and high removal efficiency, it has a strong effect on the effective control of low-concentration toxic gases. However, when using ordinary activated carbon for physical adsorption, the adsorption of activated carbon will quickly reach equilibrium.

The characteristic of the physical adsorption of activated carbon is that the molecules of the adsorbed substance move freely on the surface of the adsorbent rather than being fixed on the surface of the adsorbent. When the adsorption process reaches equilibrium, because the adsorption force is mainly intermolecular force that shows poor stability, toxic gas is easy to desorb, and the thermal motion of the adsorbed molecules will cause the adsorbed molecules to leave the surface of the adsorbent, so the adsorption of ordinary activated carbon is easily affected by changes in formaldehyde concentration and temperature. To solve the problem of poor stability of the physical adsorption of activated carbon, researchers often modify porous materials to enhance their adsorption performance. For example, HNO_3, O_3, and H_2O_2 can be used to modify the surface of activated carbon fiber. Dynamic adsorption tests show that the activated carbon fiber modified with H_2O_2 has the best adsorption effect on formaldehyde [24].

9.1.3.3 Photocatalytic Technology

Photocatalytic technology has received widespread attention and research since its emergence in the 1970s. Photocatalytic degradation is based on the oxidation ability of photocatalysts under ultraviolet light to purify pollutants in the air. The photocatalytic technology directly utilize oxygen in the air as an oxidant. It possesses strong redox, low cost and long service life. Photocatalytic technology can decompose organic pollutants in the air into carbon dioxide and water. Because there is no adsorption equilibrium phenomenon during photocatalytic reaction, the photocatalytic air cleaning is thorough and have little secondary pollution [25].

Photoconductive catalysts usually use n-type semiconductor materials (such as ZnO, CdS, WO_3, and TiO_2 etc.) as catalysts. The catalytic reaction mechanism can be elaborated as follows: n-type semiconductor materials possess a distinct band structure characterized by a forbidden band, also known as a bandgap, located between the valence band and the conduction band. When the energy of the photon is higher than the threshold that the semiconductor can absorb, the semiconductor continues to be irradiated with light, and the valence band electrons on the semiconductor transfer conduction band via band-to-band transition. And then, free electrons (e^-) and holes (h^+) are generated. The dissolved oxygen adsorbed on the adsorbate surface will generate superoxide anions with electrons, and the holes will oxidize the water or hydroxyl ions, which are adsorbed on the adsorbate surface into the hydroxyl radicals. These two newly generated substances have strong oxidizing properties and can oxidize most organic substances into carbon dioxide and water.

The studies that combine photocatalytic technology with adsorption technology have gained great attention currently. The material first adsorbs toxic gases and then degrades them with highly efficient photocatalysis, so as to achieve the synergistic benefits of adsorption and photocatalysis. Zhang et al. [25] studied the photodegradation activity of the TiO_2 and activated carbon composite material. The results show that in the early stage of catalyst preparation, the calcination temperature affects the photocatalytic degradation activity. The catalyst obtained by calcination at 500–700 °C shows higher photocatalytic activity, and the photocatalytic degradation rate of formaldehyde can reach more than 95%.

9.1.3.4 Combined Technology

Due to the adsorption equilibrium and adsorption saturation of activated carbon adsorbents, they need to be replaced regularly. Activated carbon has poor absorption performance for small molecule VOCs, but those VOCs have important impacts on human health [26]. Therefore, a combined technology has been developed, combining activated carbon adsorption with photocatalytic oxidation technology. After VOC concentration reduces to a certain level via activated carbon adsorption, the photocatalysts can fully oxidize the residual VOCs, thereby improving the efficiency of catalytic oxidation. Activated carbon can also adsorb intermediate by-products, allowing them to be further catalytically oxidized to achieve thorough purification. In addition, because the pollutants adsorbed by activated carbon participate in the oxidation reaction under the action of photocatalysts, the regeneration of activated carbon is performed by photocatalytic oxidation to remove surface pollutants, extending the recycling rate of activated carbon [27]. The combination of photocatalytic oxidation and ozone can allow the ozone produced by the ozone device to enter the photocatalytic reaction device. The synergistic effect of photocatalytic oxidation and ozone is able to decompose organic pollutants, deodorize, sterilize, etc., which is also one of the important research directions at present [28].

9.1.3.5 Low-Temperature Plasma Technology

Low-temperature plasma technology for exhaust gas treatment uses a dielectric barrier discharge to generate the plasma. When the applied voltage reaches the discharge voltage of the gas, the gas is broken down, producing a mixture including electrons, various ions, atoms, and free radicals. Low-temperature plasma generates high-energy electrons under the acceleration of the electric field. When the energy of the electrons exceeds the chemical bond energy of the gaseous pollutants, the chemical bonds will break, thereby achieving the purpose of eliminating gaseous pollutants [29]. In the 1980s, Professor Oda of the University of Tokyo proposed the high-voltage pulsed corona discharge method as the simplest and most effective method to obtain low-temperature plasma under normal pressure and temperature [30]. The method of removing gaseous organic matter by pulsed corona discharge is to obtain non-equilibrium plasma under normal temperature and pressure through high-voltage pulsed corona discharge, that is, to generate a large number of high-energy electrons and active particles such as O, OH, and oxidize and degrade harmful substance molecules so that pollutants are finally transformed into harmless substances. Yamamoto and other researchers [31] first proposed the technology of treating organic waste gas and nitrogen oxides with pulsed corona discharge and conducted exploratory research. Professor Huang of Zhejiang University, Professor Yan of Shanghai Jiao Tong University, and others have conducted in-depth research in this area [32–34].

9.2 Household Air Purifiers

An air purifier can remove one or more pollutants from the air, which has a desirable effect on indoor air quality control [35]. The air purifier is mainly divided into standalone air purifiers that are used in rooms and module air purifiers used in centralized air conditioning ventilation systems. Air purifiers, also known as "air cleaners" and "air fresheners, refer to products that can capture, adsorb, decompose or convert various air pollutants (generally including $PM_{2.5}$, dust, pollen, odors, formaldehyde, bacteria, and allergens, etc.). Air purifiers can effectively improve the cleanliness of indoor air, and they are popular indoor air cleaning products [36]. Figure 9.1 shows different styles of air purifiers currently on the market.

Fig. 9.1 Different styles of air purifiers

9.2.1 Principle of Household Air Purifiers

Indoor air pollutants are diverse and complex and can be classified as chemical pollutants, physical pollutants, and biological pollutants. Chemical pollutants include not only the various volatile organic compounds (VOCs) but also various harmful inorganic small molecule pollutants, such as sulfur dioxide and nitrogen oxides. Except for the chemical pollutants, indoor physical pollutants are also not be ignored. Physical pollutants mainly include various inhalable particles such as $PM_{2.5}$, PM_{10}, etc. The biological pollutants refer to various air-suspended microorganisms such as bacteria and viruses, etc. [37].

Air purifiers come in various forms, tailored to address different types of air pollutants and the diverse needs of individuals and regions. Based on their operation mode, air purifiers can be categorized into passive, active, and hybrid types. Passive purifiers are adsorption-filtration purifiers that use pressure differences generated from fans as the driving force. They capture air pollutants by air passing through the multi-layered filters. Active purifiers remove air pollutants by releasing purifying factors into the air. They do not rely on the fans to drive the air passing through the filter. Hybrid air purifiers combine the features of both active and passive air purifiers [38].

9.2.1.1 Passive Air Purifier

As shown in Fig. 9.2, passive air purifiers are the most commonly used air cleaning products (such as HEPA air purifiers). They are known as filtration-adsorption purifiers [39]. The air is drawn into the machine by a fan and filtered through the built-in filters. It mainly filters dust, removes odors, adsorbs toxic gases, and kills bacteria [40].

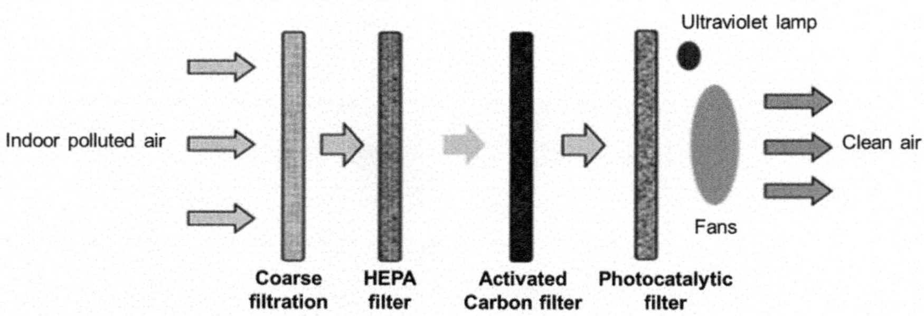

Fig. 9.2 Principle of passive air purifier

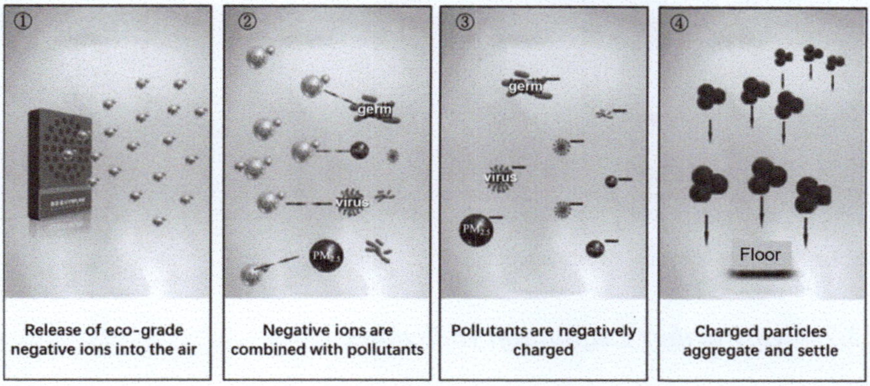

Fig. 9.3 Principle of active air purifier (the type of negative ion) [41]

9.2.1.2 Active Air Purifier

As shown in Fig. 9.3, the removal principle of the active air purifier is distinctly different from the passive air purifier. It is not passively waiting for indoor air to be drawn into the purifier for filtration. Active purifiers are widely used in the applications of chemical and biological pollutant removal because the chemical and biological pollutants are difficult to remove by filtration. Chemical pollutants need to react with chemicals to produce harmless substances, and the removal of biological pollutants needs antibacterial and disinfection materials and methods. The mainstream active purifiers on the market are mainly based on negative ions, electrostatics, and photocatalysis [41].

9.2.1.3 Hybrid Air Purifier

The principle of the hybrid air purifier is shown in Fig. 9.4. The hybrid air purifier combines the features of active and passive air purifiers. First, the polluted indoor air is drawn into the purifier by a fan, and it is purified layer by layer through the filter. After passive removal, purifying factors are released into the air from the hybrid air purifier. The purifying factors can actively remove the indoor air. The hybrid air purifier combines passive and active removal technologies, enhancing the air removal capability [42].

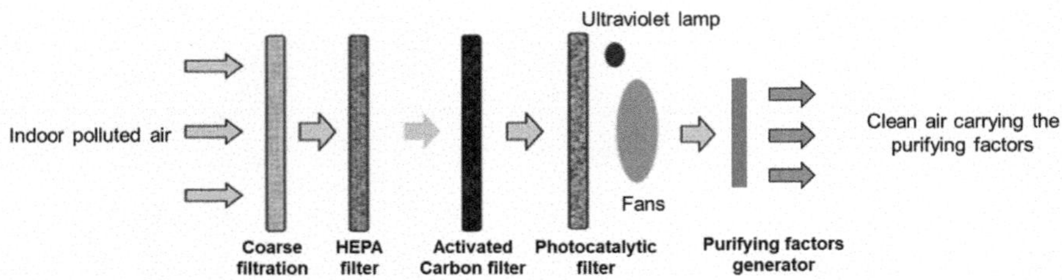

Fig. 9.4 Principle of hybrid air purifier

9.2.2 *Air Purifier Evaluation Indicators*

The pollutants removed by air purifiers are determined by the nature of the purifier's filter materials or removal mechanisms. Currently, indoor air pollutants are generally divided into three major categories: particulate matter, gaseous pollutants, and microorganisms. Particulate matter is further divided into PM_{10} and $PM_{2.5}$ based on the range of "aerodynamic particle size". There are many types of gaseous pollutants, including formaldehyde, toluene, xylene, SO_2, ammonia, ozone, and other VOCs. Microorganisms refer to bacteria that affect human health. The evaluation parameters of purifier technology include [43, 44]: harmful substance release, standby power, clean air delivery rate (CADR), cumulative purification amount (CCM), removal efficiency (η), noise, and microorganism removal efficiency.

9.2.2.1 Clean Air Delivery Rate

The parameter that is generally used to evaluate the removal performance of air purifiers is the clean air delivery rate (CADR). The clean air delivery rate is a parameter that evaluates the ability of the air purifier to remove pollutants. It is calculated after experimental testing in a standard test cabin with constant temperature and humidity. This parameter indicates how efficiently the air purifier provides clean air. The clean air delivery rate is a core indicator of purifier performance, and the actual value should be greater than 90% of the calibrated value. The calculation formula is as follows:

$$Q = 60(k_e - k_n)V \tag{9.1}$$

where Q is the clean air delivery rate, m^3/h; k_e is the total decay constant, min^{-1}; k_n is the natural decay constant, min^{-1}; V is the volume of the test cabin, m^3.

In the above formula, the values of the decay constants k_e and k_n can be obtained from the concentration function of particulate matter over time in the test cabin, which is given by function fitting.

9.2.2.2 Energy Utilization Efficiency of Air Cleaning

Energy utilization efficiency of air cleaning refers to the amount of purified air per unit energy consumption of the air purifiers under the rated operating conditions. Energy utilization efficiency is related to the purifier's CADR and power consumption. It is an indicator of purifier performance and power consumption evaluation. In the evaluation of air purifier performance, the larger the energy utilization efficiency value, the more purified air produced per unit energy consumption, which indicates better energy-saving and environmental performance of the air purifier. The actual measured energy utilization efficiency value of the particulate-matter removal of the air purifier must be $\geq 90\%$ of the calibrated value. The energy utilization efficiency of particulate-matter removal (η) of air purifiers in China has been divided into a qualified level ($2 \leq \eta \leq 5$) and a high-efficiency level ($\eta \geq 5$). The calculation formula is as follows:

$$\eta = \frac{Q}{P} \tag{9.2}$$

where η is the removal efficiency, $m^3/(W\ h)$; Q is clean air volume, m^3/h; P is power measurement, W.

9.2.2.3 Input Power and Standby Power

Generally, the input power of the air purifier is not equivalent to the power measurement. The input power is the power value of the air purifier during the purification work. If other functions of the air purifier consume more power, the extra power from these other functions should be separately discussed. In this case, the purifier's power number is only considered as the input power if the fluctuation of the purifier's power within half an hour is less than 1%. Standby power refers to the power of the air purifier in the standby non-working state. The standby power stipulated by the Chinese national standard cannot be higher than 2.0 W. The smaller the standby power, the less electricity the purifier consumes in a non-working state, and the better the energy-saving effect.

9.2.2.4 Cumulative Purification Volume

Cumulative Purification Volume (CCM) refers to evaluating the air purifier's cumulative purification ability, which is used to characterize the service life of the air purifiers. Cumulative purification volume is tested in a standard condition where it is in a cabin under a constant temperature and humidity environment. In general terms, CCM refers to the cumulative pollutant mass accumulated by an air purifier's filter from the start of its operation to the point of failure. The Chinese national standard stipulates that the cumulative mass of particulate matter is the air purifier's cumulative

Table 9.1 Particulate matter CCM index grading

Grade	CCM/mg	Grade	CCM/mg
P1	$3000 \leq M \leq 5000$	P3	$8000 \leq M \leq 12{,}000$
P2	$5000 \leq M \leq 8000$	P4	$12{,}000 \leq M$

purification volume when the air purifier's CADR decreases by 50%. Therefore, the cumulative purification volume is not the characterization of the filter in air purifiers but the characterization of the other purifier components behind the filter. For filters with the same performance parameters, the higher the purifier's CADR, the larger the particulate matter purified by the purifier when it is determined to fail. Since the purifier's CADR continuously decreases with the use time, if the air purifier has a high CCM, it can also indicate that the purifier can maintain a high CADR for a long time. The main factors affecting the purifier's cumulative purification volume are the working area of the purifier's filter and the materials of the filter. The Chinese national standard has graded the CCM of particulate matter, and the evaluation indicators are shown in Table 9.1.

9.2.3 *Application of Membrane Technology in Household Air Purifiers*

Filtration technology is the main method for household air purifiers to remove indoor inhalable particles and microorganisms. The traditional filtering material is a high-efficiency air filter (HEPA), which has a capture efficiency more than 99.97% for $PM_{0.3}$ [45]. At the end of the twentieth century, PTFE membranes for indoor air cleaning appeared. Michael and others [46] compared the PTFE membranes with HEPA, the removal efficiency of PTFE membranes for 0.3 μm particles can reach 99.9999%, far higher than HEPA's 99.97%, which can meet the standard of ultra-high efficiency air filters (ULPA), and the resistance is only half of that of HEPA [47–49].

Expanded polytetrafluoroethylene (ePTFE) microporous membrane is a composite membrane material with surface filtration performance. The ePTFE membrane is obtained by expanding and stretching polytetrafluoroethylene resin particles, and then compounding with various filter materials such as needle-punched felt, woven fabric, non-woven fabric, or glass fiber [50] (Fig. 9.5). Compared with the symmetrical structure of traditional HEPA filters, the composite membrane structure transforms "deep filtration" into "surface filtration". This revolutionary transformation can not only maintain high filtration efficiency, but also reduce filtration resistance and extend the service life. The porosity of PTFE membrane can exceed 88%, making it suitable for installation as the core filter in air purifiers, replacing traditional HEPA filters. The ePTFE membrane can quickly and effectively intercept sub-micron ultrafine dust, and the dust removal efficiency for 0.3 μm particles can reach more than 99.99%, and the service life is up to 3 years [51].

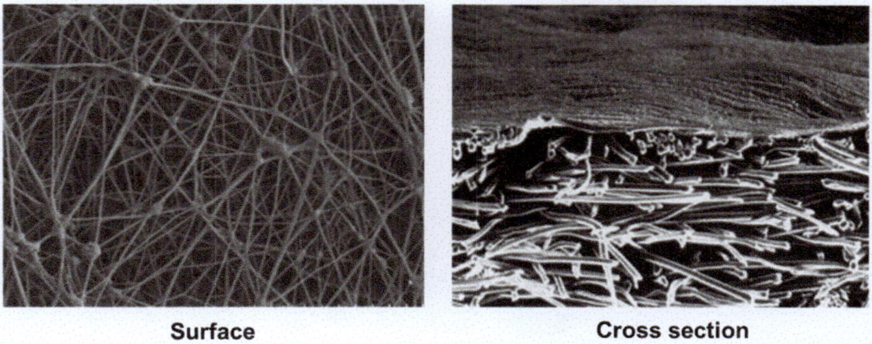

Fig. 9.5 SEM of ePTFE membrane

ePTFE membrane is breathable to air but impermeable to water. It has the characteristics of large air permeability, low resistance, high particle filtration rate, good temperature resistance, strong acid and alkali resistance, good organic solvents and oxidants, good aging resistance, non-stickiness, non-flammability, non-toxicity, biocompatibility, etc. It is currently the most advanced air filter material in the world, and it is the best choice for various vacuum cleaners, air filter elements, air cleaning equipment, high-efficiency air filters, etc. [46]. Its main technical parameters are shown in Table 9.2. From the data in the table, it can be seen that without increasing gas resistance, PTFE membrane material is far superior to existing HEPA filters in terms of filtration efficiency, operating temperature, and service life.

The nanofiber membrane is the core filtering component of the air cleaning membrane. Figure 9.6 shows the multifunctional nanofiber membrane filter, which can be used to simultaneously remove pollutants such as $PM_{2.5}$, bacteria, and formaldehyde. The removal efficiency for $PM_{2.5}$ can reach over 99.999%, and the removal rate for volatile organic chemicals such as formaldehyde is higher than 95%.

In addition to PTFE membranes, nanofiber membranes prepared by electrospinning are also used in household air purifiers, of which fiber diameters are in the nanometer to micrometer scale. The removal efficiency of nanofiber membrane for 0.3 μm particles is over 99.9% [52]. Polyvinylidene fluoride (PVDF) membranes can demonstrate excellent performance in bacterial filtration. Integrating PVDF membranes into a three-stage filter in air purifiers, which includes pre-filtration, coarse filtration, and fine filtration, can achieve a retention of 99.99995% of bacteria in the air [53].

Table 9.2 Technical parameters of ePTFE membrane for household air purifiers

Pore size/μm	Porosity/%	Filtration efficiency (0.3 μm)/%	Wind resistance (5.33 cm/s)/ Pa	Thickness/μm	Maximum use temperature/ °C	Service life/year
0.2–5	80–95	> 99.99	30–50	2–10	260	> 3

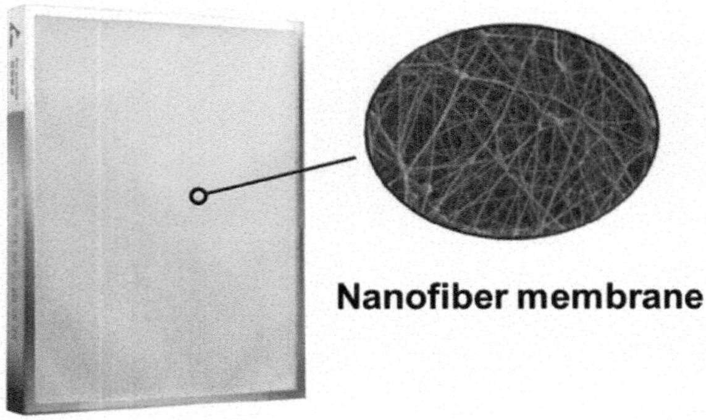

Fig. 9.6 Nanofiber membrane filter

9.3 Household Fresh Air Cleaning System

The fresh air cleaning system is an effective approach to improve indoor air quality. The household fresh air cleaning system expels dirty indoor air by introducing outdoor fresh air to dilute indoor air pollutants. Studies by Liddament et al. [54] have proven that ventilation using a fresh air cleaning system is a relatively effective method to control indoor air pollution. Xie et al. [55] conducted a study on a residence, establishing a mass balance equation to study the removal methods of indoor $PM_{2.5}$. The calculation results show that improving the performance of the fresh air filter can significantly reduce the indoor $PM_{2.5}$ concentration. Ye et al. [56] tested a gym with a G4 + H12 two-stage filter fresh air system and found that the $PM_{2.5}$ value in the gym was less than 10 $\mu g/m^3$. Chao et al. [57] found that in addition to having a good removal efficiency on fine particles, CO_2, TVOC, and other toxic gases, the fresh air system also has some control effect on total volatile organic compounds.

9.3.1 Principle of Household Fresh Air Cleaning System

The indoor fresh air cleaning system is composed of a fan, an air inlet, an air outlet, various pipes, and some joints. The fan installed in the ceiling is connected to a series of air outlets through pipes. When the fan is started, the indoor polluted air is discharged to the outside through the air outlet and the fan; a negative pressure is created inside the room. In order to achieve indoor air cleaning, so that people inside can breathe high-quality fresh air. The fresh air outside then enters the room through the air inlet installed above the window frame (between the window frame and the wall), filtering, sterilizing, oxygenating, and preheating (in winter) the indoor air while supplying air. When the exhaust air passes through the main unit, heat exchange is performed with the fresh air for energy recovery, returning most of the energy to the room through the fresh air [58, 59]. The ventilation of the fresh air cleaning system is not just about discharging polluted air, it also has the functions of deodorization,

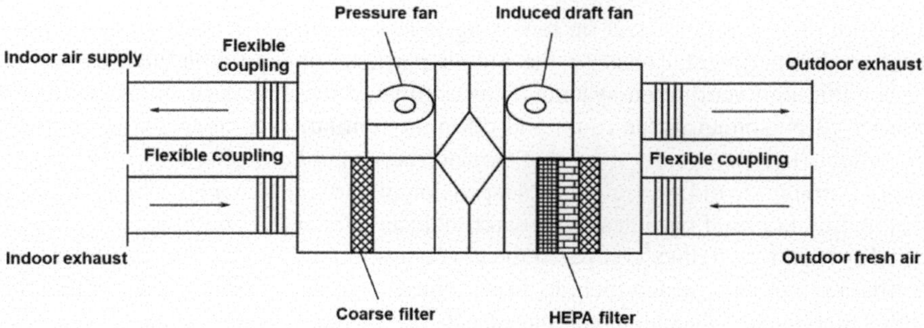

Fig. 9.7 Working principle of the fresh air cleaning system

dust removal, dehumidification, and temperature regulation. Figure 9.7 shows the working principle of the fresh air cleaning system.

Wang [60], Halek [61], Lange [62], and others have studied the indoor particulate matter concentration in rooms with fresh air systems. They found that the cleaning performance of the fresh air system on indoor particulate matter is related to the installation position of the filter, indoor/outdoor particulate matter concentration, filter efficiency, and the duration of fresh air operation. Chen et al. [63] took particulate matter as the research object and tested the concentration of particulate matter at various points in the fresh air laboratory. Next, they analyzed the distribution pattern of fresh air particulate matter. The fresh air volume is the main factor affecting the removal efficiency of the fresh air system. Some scholars have studied the changing rule of indoor pollutant concentration under different fresh air volumes. Li et al. [64], through the simulation studies, found that the change in fresh air volume in air-conditioned rooms has a significant impact on the cleaning effect of the fresh air system on indoor air pollutants. However, when the number of air changes increases to a certain level, the change in fresh air volume has no obvious effect on pollutant removal. Hu et al. [65] studied the distribution of indoor pollutant concentration when the air conditioning system and mechanical ventilation are turned on at the same time and found that the independent fresh air system can act as a supplement to the air conditioning system, provide the required fresh air volume indoors. They pointed out that fresh air volume has a significant impact on the concentration of indoor pollutants. Different ventilation and air supply methods also have different impacts on indoor pollutants.

Wu et al. [66] used tracer gas to study the diffusion of gas pollutants in apartment-style residences through experiments. The results showed that natural ventilation can reduce the risk in the room with an infection source, but it has no obvious effect on other rooms. Yaghoubi [67] and others simulated indoor pollutants and studied the impact of different ventilation methods and indoor pollutant source positions on the distribution of pollutant concentrations. Yuan et al. [68] changed the ventilation strategy of the room and analyzed the cleaning effect of the fresh air system on indoor pollutants. He found that the combination of the fresh air system and the air purifier has a better effect on removing indoor air pollutants. Mundt et al. [69] conducted experimental tests on the indoor particle concentration changes in the air conditioning ventilation system. They compared the effects of the presence or absence of personnel in the room and different ventilation frequencies on particle movement. The test results show that displacement ventilation can effectively expel indoor particles, avoiding secondary suspension of particulate matter. Zhao [70], Lin [71] and others used particles as research objects. Under the condition of constant initial concentration, they compared the concentration of indoor particles under two ventilation methods, which include displacement ventilation and mixed ventilation. They found that displacement ventilation can more effectively expel indoor particles. Zhu and others [72] used Fluent simulation software to simulate the diffusion distribution of indoor particle pollution when changing the particle size and divergence source position under different air supply forms.

9.3.2 Application of Membrane Technology in Household Fresh Air Cleaning Systems

The air filter in the fresh air system is a device for purifying air pollutants. The air filter has a simple structure; it is widely used in various production and living places. The principle of the air filter is to intercept pollutants from the air through physical methods when the particles in the air pass through the filter materials, thereby reducing the concentration of indoor pollutants [73].

The filter can be divided into coarse filters, medium filters, and high-efficiency filters, depending on its filtration efficiency. The materials used for coarse filters mainly include metal-mesh filters, non-woven fabric, etc. Coarse filters are often used for fresh air filtration in air conditioning ventilation systems, and as pre-filters for fresh air cleaning devices, they mainly filter particles larger than 5 μm. Medium filters often use non-woven fabrics with medium and fine pore sizes, glass fibers, etc. Medium filters can generally be used as filters for fresh air and return air, mainly filtering particles larger than 1 μm. High-efficiency filters can be further divided into high-medium efficiency, sub-high efficiency, and ultra-high efficiency filters, which mainly filter particles smaller than 1 μm. The end of various cleaning systems usually chooses high-medium efficiency filters, while the end filters in clean rooms will use sub-high efficiency filters. Sub-high-efficiency filters are also often used for the final filtration of fresh air units. Ultra-high efficiency filters are usually used for air filtration in clean operating rooms and clean factories where the indoor environment requirements are extremely strict [74].

High-Efficiency Particulate Air (HEPA) filters are the core cleaning components of household fresh air cleaning systems. Figure 9.8 is an electron microscope image of the surface and cross-section of the traditional HEPA filter. Since the HEPA filter is a symmetrical filter material, it is difficult to ensure a large amount of fresh air and effectively filter fine particles in actual applications. Some research institutions

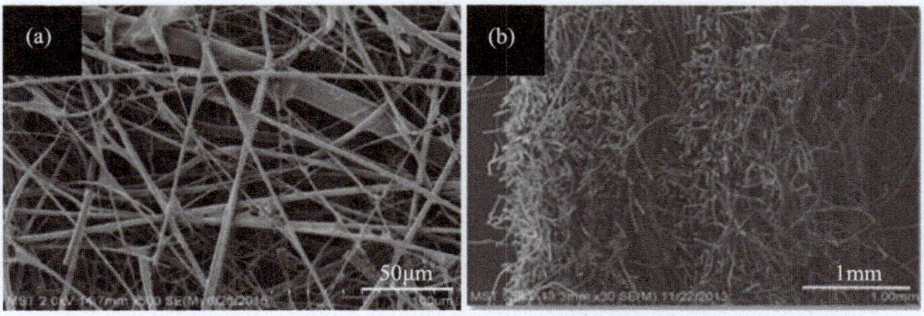

Fig. 9.8 SEM images of the surface (**a**) and cross-section (**b**) of a traditional HEPA filter [75]

have found that due to various conditions, the actual particle removal efficiency of the current commercial air cleaning system is only 50% to 95%, which is far lower than the filtering capacity of the filter material itself. This is mainly because the merchants use larger pore separation materials to increase the fresh air volume of the fresh air cleaning system. In addition, HEPA mainly relies on deep filtration of the filter mesh channel. When the outdoor fresh air is purified by HEPA, it will cause serious blockage of the filter mesh channel. The wind resistance of the filter will rapidly increase, resulting in a hotbed for bacteria to breed. In addition, the service life of the air filter will be rapidly reduced, which means that the HEPA filter needs to be replaced frequently [75].

Membrane separation technology is a new type of high-efficiency separation technology that can effectively remove $PM_{2.5}$ from the air. Incorporating membrane separation technology into fresh air cleaning systems as a replacement for traditional HEPA filters significantly enhances the interception efficiency of PM2.5 particles. This is because the nanofiber membranes have high porosity and specific surface area, resulting in more effective contact between $PM_{2.5}$ particles and nanofibers, so nanofiber membranes have higher filtration efficiency than HEPA filters. On the other hand, the scale of nanofibers is comparable to the mean free path of gas molecules (about 66 nm), and the "slip effect" can reduce the filtration resistance [76].

The membrane materials used in fresh air cleaning systems are mainly biaxially stretched PTFE membranes and electrospun nanofiber membranes. Figure 9.9 shows the SEM images of these two types of membranes. The porosity of the PTFE membrane used in the fresh air cleaning system can reach more than 88%. Using a PTFE membrane as the main filter element can greatly enhance the cleaning effect of the fresh air cleaning system by combining photocatalysis, plasma technology, and other removal methods [77]. The fiber diameter of the electrospun nanofiber membrane can be flexibly adjusted within a range of a few nanometers to a few micrometers. Table 9.3 shows the parameters of the nanofiber membrane used in the fresh air cleaning system. Using a nanofiber membrane to make a filter for the fresh air cleaning system has the advantages of high efficiency, low resistance, and long service life. The removal efficiency for 0.3 μm particles can reach more than 99.99% [78, 79].

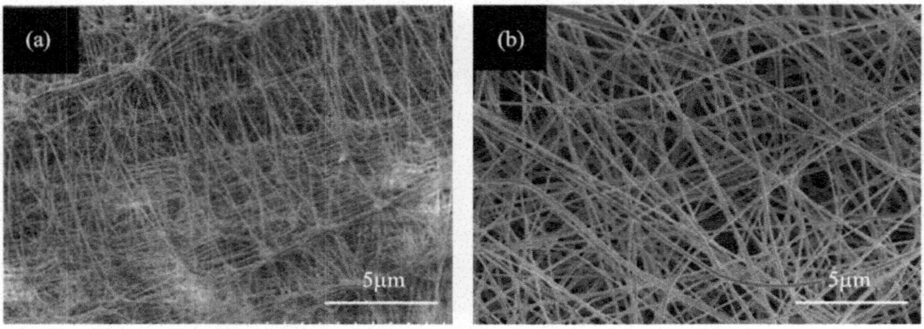

Fig. 9.9 Membrane materials used in fresh air purification systems. **a** Biaxially stretched PTFE membrane; **b** electrospun nanofiber membrane

Table 9.3 Parameters of the nanofiber membrane used in the fresh air cleaning system [78, 79]

Pore size/μm	Porosity/%	Filtration efficiency (0.3 μm)/%	Air resistance/Pa	Thickness/μm	Maximum operating temperature/°C	Service life/year
0.1–3	85–90	> 99.99	25–40	2–10	120	> 2

9.4 Domestic Anti-smog Window Screens

Anti-smog window screens, while ensuring ventilation and light transmission, can effectively block outdoor dust from entering the room. As a new type of civilian anti-smog product, anti-smog window screens are gradually being recognized and applied by society [80]. The core of the anti-smog window screen is gas cleaning technology, and the gas cleaning technology currently used for domestic anti-smog window screens mainly includes electrostatic and filtration technology.

The electrostatic anti-smog window screen fixes an electrostatic generator on the window frame. The input end of the electrostatic generator is connected to the low-voltage direct current. One wire of the output end is grounded, and the other wire is connected to the metal mesh. Then, the metal mesh is covered on the window screen made of glass fiber, which will carry static electricity. In this way, the dust will be adsorbed by static electricity when the polluted air passes through the window screen. Regularly cleaning the dust on the window screen can achieve the purpose of purifying indoor air [81–83]. Zhang et al. [84] designed a set of new anti-smog window screen devices. An electrostatic dust removal device is installed on the outer layer of the ordinary window screen to efficiently remove smog from the air that is going to enter the room from window screen outside. A humidity detection probe is installed at the outer end of the window screen and connected to an indoor humidity detection control device. This setup enables control over the working humidity of the electrostatic dust removal device. Field tests found that when using ordinary window screens, the indoor $PM_{2.5}$ value was 401 μg/m^3, and after using the anti-smog window screen, the indoor $PM_{2.5}$ value increased to 20 μg/m^3.

Another type of electrostatic anti-smog window screen uses the principle of a capacitor to adsorb and remove fine particles. It is composed of two window screens, each of which is woven with an individual wire. One end of the wire from the window screens is connected to the positive pole of the high-voltage control system through the current-limiting resistor in the high-voltage control system, and the other end is sealed with high-voltage resistant insulating glue, which does not come into contact with the air. The wire from the other window screen is connected to the negative pole of the high-voltage control system, and the other end of the wire is sealed with high-voltage-resistant insulating glue and does not come into contact with the air. These two window screens are placed vertically side by side to form a capacitor, as shown in Fig. 9.10 [85].

The anti-smog window screen uses filtration technology and applies filtering material as the core component to intercept fine particles. The fine particles outside the room are intercepted by the filtering material, and the air can freely pass through the window screen, realizing the ventilation function of the window screen [86, 87]. As the filter material for anti-smog screens, in addition to efficiently intercepting outdoor particulates and ensuring ventilation, it also needs to have good light transmission. The material is often made by a composite of high permeability membrane materials with substrate materials, as shown in Fig. 9.11.

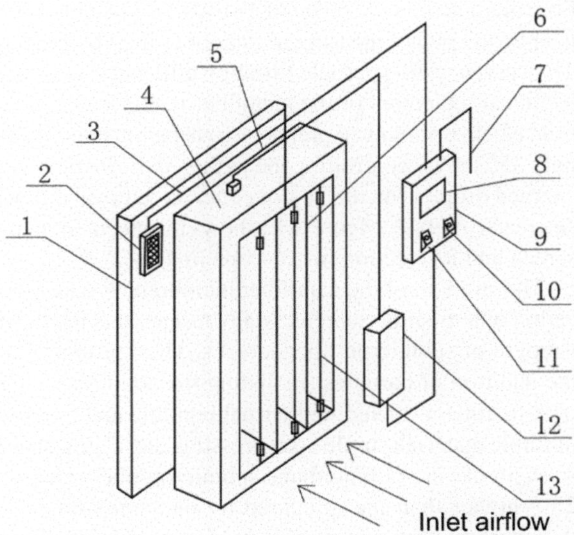

Fig. 9.10 Electrostatic window screen structure diagram [85] 1—Ordinary window screen; 2—Humidity detection probe; 3—Humidity detection probe lead; 4—Electrostatic dust removal device; 5—High voltage power line; 6—Dust collection plate; 7—External power line; 8—Humidity detection control device display screen; 9—Humidity detection control device; 10—Switch; 11—Humidity control range adjustment button; 12—Transformer; 13—Corona pole

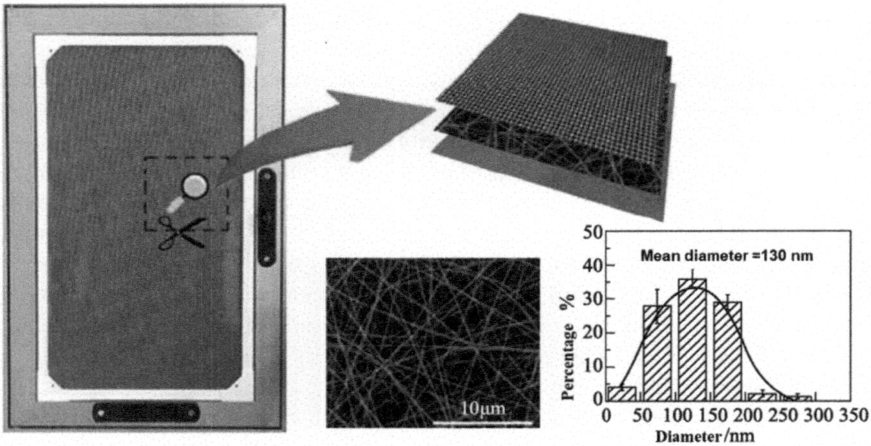

Fig. 9.11 Filter-type anti-smog screen [87]

9.4.1 Nanofiber Membrane Anti-Smog Screen

The nanofiber membrane anti-smog screen employs nanofiber membrane materials as core filtering, which are prepared by bidirectional stretching or electrospinning with fiber diameters in the range of 1–100 nm. The nanofiber membrane anti-smog screen is obtained by hot pressing of nanofiber membrane material with a nylon mesh [88]. Transparency, fine particle filtration efficiency, and air permeability are the three key technical indicators of the nanofiber membrane anti-smog screen. Zhang et al. [89] used electrospinning to prepare a transparent nanofiber membrane with a thickness of only 350 nm by controlling the jetting, deformation, and phase-separation process of charged droplets on the Taylor cone. Its removal efficiency for 0.07 μm fine particles can reach 99.98%. Figure 9.12 shows the electrospun nanofiber membrane anti-smog screen and its electron microscope image.

The strength of the nanofiber membrane is a key factor that limits the wide application of anti-smog screens. Many researchers have conducted a lot of studies on the strength of submicron fiber screens. Their proposed improvement methods include the addition of particles (including the addition of oxides and nanomaterials, etc.), post-treatment (including thermal bonding and modification treatment), blending of multiple materials, multi-axis pressing, etc. Tijing et al. [90] showed that the breaking strength of pure polyurethane submicron fiber screens prepared by biaxial stretch is 25% higher than that prepared by uniaxial stretch. Wang et al. [91] showed that the appropriate proportion of oxides, such as Fe_3O_4 and SiO_2 would multiply the strength of the submicron fiber screen. Ding et al. [92] found that the breaking stress of polyvinylidene fluoride-hexafluoropropylene electrospun submicron fiber screens gradually decreases with the increase of temperature in the range of 25–75 °C. Lou

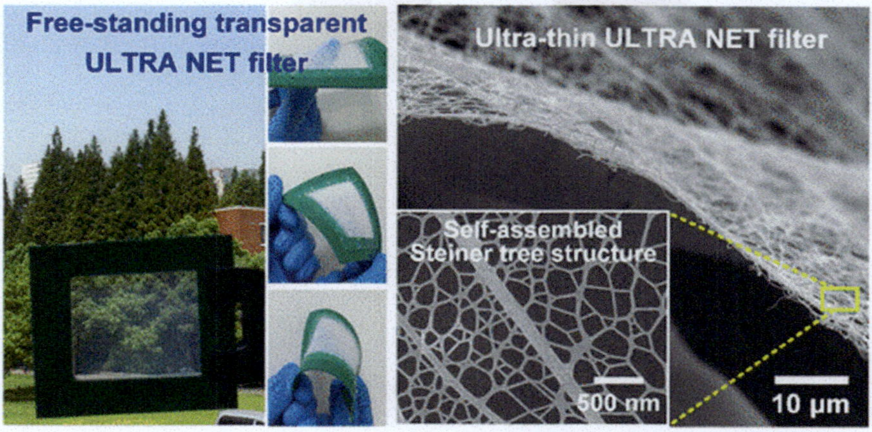

Fig. 9.12 **a** Electrospun nanofiber membrane anti-smog screen and **b** its SEM image [89]

Lihua [93] found that thermal rolling bonding can significantly improve the strength of the submicron fiber screen.

Polyacrylonitrile (PAN) is a commonly used electrospinning material, and the nanofibers produced from it have the characteristics of uniform thickness and high porosity [94]. Wang et al. [95] prepared PVA/PAN submicron fiber composite membranes by electrospinning, and improved the filtration performance of PVA through steam and chemical crosslinking. Cao et al. [96] added nanocrystals of jute fiber to the double-layer PAN membrane to improve the mechanical properties of the submicron fiber window screen. Polyvinylidene fluoride (PVDF) is an excellent membrane material with good stability and radiation resistance [97]. Zárate et al. [98] determined the optimal concentration of the PVDF/DMAc and PVDF/DMF solution systems used to prepare PVDF membranes. Bottino et al. [99–101] believe that mixed solvents can improve the pore size and porosity of PVDF membranes.

9.4.2 Nuclear Pore Membrane Anti-Smog Window Screen

The thermal neutrons in the nuclear reactor induce the fission of uranium 235. The fragments produced by the fission penetrate the organic polymer thin film, leaving a narrow radiation damage channel on the path of the fission fragments. After the channel is oxidized and etched with an appropriate chemical reagent, the channel on the thin film can be turned into a cylindrical micropore. By controlling the irradiation conditions of the nuclear reaction and the etching conditions, nuclear pore membranes with different pore densities and sizes can be obtained [102, 103]. The anti-smog window screen made by nuclear pore membrane realizes indoor-outdoor gas exchange by utilizing the diffusion principle of gas molecules from high concentration to low concentration. However, it has poor breathability and is difficult to meet the requirements for good light transmission [104].

Currently, the key research issue of the nuclear pore membrane anti-smog window screen is to improve the porosity and increase the breathability. Etching technology is the key technology for the preparation of nuclear pore membranes. This technology is also the main research field of nuclear pore membrane anti-smog window screens [105, 106]. There are two important parameters in the etching process: the track etching rate and the bulk etching rate. The track etching rate refers to the reaction rate of the etching solution with the irradiation products along the latent track direction, and the bulk etching rate refers to the reaction rate of the etching solution with the non-irradiated damaged area.

The track etching rate is affected by many factors, such as the physical and chemical properties of thin film material, irradiation conditions (types of ions and ion energy, etc.), post-irradiation treatment (sensitization), and etching conditions (concentration, temperature, etc.). The bulk etching rate is mainly affected by temperature and etching solution concentration.

The size of the micropores depends on the bulk etching rate and the etching time. The opening angle size of the micropores is determined by the ratio of the track etching rate to the bulk etching rate. When the track etching rate is much greater than the bulk etching rate, micropores close to cylindrical shape are formed [107]. The team of Professor Qu from Nanhua University obtained an understanding of the relationship between the etching rate and the etching conditions by etching at different etching solution concentrations and temperatures so they could control the pore size and pore shape of the nuclear pore membrane by adjusting the temperature and etching concentration. They have prepared nuclear pore membranes with micropores closer to cylindrical shape [108]. Figure 9.13 is a structural diagram of the nuclear pore membrane anti-smog window screen.

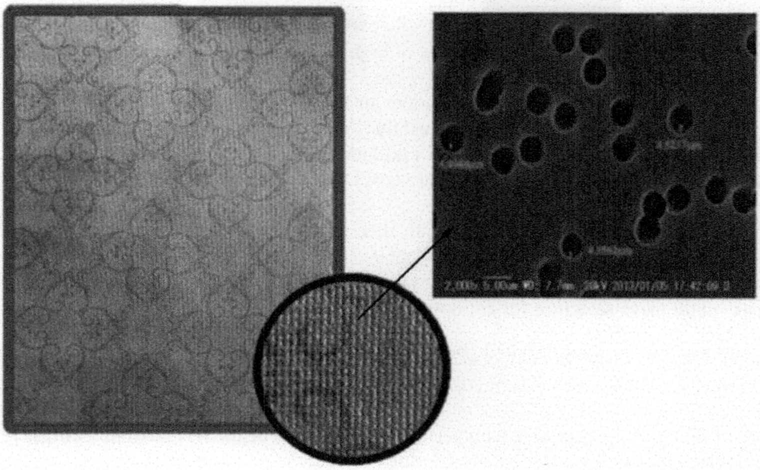

Fig. 9.13 Nuclear pore membrane anti-smog window screen [108]

References

1. Han Y, Bai ZP, Xi ZG. Indoor air pollution and prevention (Second Edition). Beijing: Chemical Industry Press; 2020.
2. Wang Y, Wild O, Chen H, et al. Acute and chronic health impacts of $PM_{2.5}$ in China and the influence of interannual meteorological variability. Atmos Environ. 2020;229:117397.
3. Che FX. Modern air microbiology and sampling inspection technology. Beijing: Military Medical Science Press; 2010.
4. Hang M. Microbiology. Hangzhou: Zhejiang University Press; 2019.
5. Zhong S, Zhang L, Jiang X, et al. Comparison of chemical composition and airborne bacterial community structure in $PM_{2.5}$ during haze and non-haze days in the winter in Guilin, China. Sci Total Environ. 2019;(655):202–10.
6. Zhang Y, Wu D, Kong Q, et al. Exposure level and distribution of airborne bacteria and fungi in an urban utility tunnel: a case study. Tunn Underground Space Technol. 2020;96:103215.1–103215.11.
7. Hou L. Invisible indoor air pollution. Beijing: China Building Materials Industry Press; 2020.
8. Lv Y, Lu Z. Indoor air pollution transmission and control. Beijing: Machinery Industry Press; 2014.
9. Zhao G, Zou J, Zhang T, et al. Recent progress on removal of indoor air pollutants by catalytic oxidation. Rev Environ Health. 2020;35(4):311–21.
10. Bai Y, Huo L, Zhang Y, et al. A spatial fractional diffusion model for predicting the characteristics of VOCs emission in porous dry building material. Sci Total Environ. 2020;704:135342.1–135342.6.
11. Becerra JA, Lizana J, Gil M, et al. Identification of potential indoor air pollutants in schools. J Cleaner Prod. 2020;242:118420.1–118420.19.
12. Basu AK, Byambasuren T, Chau NH, et al. Cooking fuel choice, indoor air quality and child mortality in India. Social Science Electronic Publishing; 2020.
13. Angelova RA, Stankov P, Markov D, et al. Human as a physiological source of deterioration of the air quality and comfort conditions indoors. In: CBU international conference proceedings. ISE Research Institute; 2019.

14. Kosmider L, Cox S, Zaciera M, et al. Daily exposure to formaldehyde and acetaldehyde and potential health risk associated with use of high and low nicotine e-liquid concentrations. Sci Rep. 2020;1(10):6546.
15. Schachter EN, Rohr A, Habre R, et al. Indoor air pollution and respiratory health effects in inner city children with moderate to severe asthma. Air Qual Atmos Health. 2020;2(13):247–57.
16. Mathew J, Goyal R, Taneja KK, et al. Air pollution and respiratory health of school children in industrial, commercial and residential areas of Delhi. Air Qual Atmos Health. 2015;4(8):421–7.
17. Hirsch T, Weiland SK, Mutius EV, et al. Inner city air pollution and respiratory health and atopy in children. Eur Respir J. 2010;14:669–77.
18. Park JK, Xiao Y, Ramuta MD, et al. Pre-existing immunity to influenza virus hemagglutinin stalk might drive selection for antibody-escape mutant viruses in a human challenge model. Nat Med. 2020;26:1–7.
19. Safa F, Chaiton M, Mahmud I, et al. The association between exposure to second-hand smoke and sleep disturbances: a systematic review and meta-analysis. Sleep Health. 2020;6(5):702–14.
20. Bi C, Chen Y, Zhao Z, et al. Characteristics, sources and health risks of toxic species (PCDD/Fs, PAHs and heavy metals) in $PM_{2.5}$ during fall and winter in an industrial area. Chemosphere. 2020;(238):124620.
21. Javier G, Richardus KNJ, Cristina P, et al. A state-of-the-art review on indoor air pollution and strategies for indoor air pollution control. Chemosphere. 2020;262: 128376.
22. Polley JR, Winkler CA, Nicholls RVV. Studies on the formation of hexamine from formaldehyde and ammonium salts in aqueous solution. Can J Res. 1947;25:525–34.
23. Association JAC. Principles and practical techniques of indoor air purification. Beijing: China Machine Press; 2016.
24. Jian XP, Xu W, Hou XL, et al. Progress in activated carbon modification technology. Biomass Chem Eng. 2020;54(5):66–72.
25. Han L, An X, Zhang P, et al. The removal of formaldehyde via visible photocatalysis using the black TiO_2 nanoparticles with mesoporous. ChemistrySelect. 2020;1(5):97–103.
26. An Y, Fu Q, Zhang D, et al. Performance evaluation of activated carbon with different pore sizes and functional groups for VOC adsorption by molecular simulation. Chemosphere. 2019;227:9–16.
27. Rangkooy HA, Jahani F, Ahangar AS. Photocatalytic removal of xylene as a pollutant in the air using ZnO-activated carbon, TiO_2-activated carbon, and TiO_2/ZnO activated carbon nanocomposites. Environ Health Eng Manage. 2020;7(1):41–7.
28. Filho BMDC, Silva GV, Boaventura RR, et al. Ozonation and ozone-enhanced photocatalysis for VOC removal from air streams: Process optimization, synergy and mechanism assessment. Sci Total Environ. 2019;687:1357–68.
29. Liang WJ, Li JX, Zhu Tao. Low temperature plasma air pollution control technology and application. Beijing: Chemical Industry Press; 2016.
30. Oda T, Kumada A, Tanaka K, et al. Low temperature atmospheric pressure discharge plasma processing for volatile organic compounds. J Electrostat. 1995;35:93–101.
31. Yamamoto T, Asada S, Iida T, et al. Nobel NO_x and VOC treatment using concentration and plasma decomposition. Industry Applications Society Meeting; 2010.
32. Wu ZC. Research on plasma-catalytic synergistic treatment of organic waste gas. Hangzhou: Zhejiang University; 2001.
33. Yan NQ, Wu ZC, Tan TE. Research on the treatment of organic waste gas by pulsed corona discharge—discharge reactor structure. Shanghai Environ Sci. 2000;6:278–81.
34. Ouyang P. Application of low temperature ion discharge technology in indoor air purification. Sci Technol Econ Guide. 2020;28:56–7.
35. Ren PK, Chang J, Mu GH. Structure design and innovation of air purifier. Light Ind Sci Techno. 2020;36(10):76–8.
36. Peng C, Ni P, Xi G. Evaluation of particle penetration factors based on indoor $PM_{2.5}$ removal by an air cleaner. Environ Sci Pollut Res. 2020;27(27):8395–405.

37. Sharma R, Balasubramanian R. Evaluation of the effectiveness of a portable air cleaner in mitigating indoor human exposure to cooking-derived airborne particles. Environ Res. 2020;183: 109192.
38. Ji YB. Research on the performance of filter-type air purifiers. Jinan: Shandong Jianzhu University; 2020.
39. Xu ZL, Feng X, Zhang YZ, et al. Factors affecting natural decay in air purifier environmental test cabin—discussion on several issues of air purifiers (3). HVAC. 2017;47(8):11–3.
40. Dongliang S. Air purifier structural design. China New Technol New Products. 2020;6:67–9.
41. Zhang L. Evaluation study of common indoor air purification technologies. Guangdong Chem Ind. 2020;47:140–4.
42. Yoda Y, Tamura K, Adachi S, et al. Effects of the use of air purifier on indoor environment and respiratory system among healthy adults. Int J Environ Res Public Health. 2020;17(10):3687.
43. Standardization Administration of the People's Republic of China. Air purifier: GB/T 18801-2015. Beijing: China Standards Publishing House; 2015.
44. People's Republic of China Building Industry Standard. Air purifier pollutant purification performance measurement: JG/T 294-2010. Beijing: Ministry of Housing and Urban-Rural Development of the People's Republic of China; 2010.
45. Xu ZL. Principles of air cleaning technology. Beijing: Science Press; 2014.
46. Osborne MW, Gail L, Ruiter P, et al. Applied membrane air filtration technology for best energy savings and enhanced performance of critical processes. AAF International. 2012.
47. Burton NC, Grinshpun SA, Reponen T. Physical collection efficiency of filter materials for bacteria and viruses. Ann Occup Hyg. 2007;51:143–510.
48. Jiang D, Zhang W, Liu J, et al. Filtration and regeneration behavior of polytetrafluoroethylene membrane for dusty gas treatment. Korean J Chem Eng. 2008;25:744–53.
49. Feng S, Li D, Low ZX, et al. ALD-seeded hydrothermally-grown Ag/ZnO nanorod PTFE membrane as efficient indoor air filter efficient indoor air filter. J Membr Sci. 2017;531:86–93.
50. Choi YH, Lee J, Khang DY. A reusable, isoporous through-hole membrane filter for airborne particulate matter removal. J Membr Sci. 2020;612: 118474.
51. Chen YH, Wang F, Chen S, et al. Application of PTFE microporous membrane filter material and its dust collector in mineral dust removal system. Modern Min. 2019;35:244–51.
52. Ding B, Yu JY. Electrospinning and nanotechnology. Beijing: China Textile Press; 2011.
53. Liu ZJ, Li H. Application and research of domestic membrane filtration equipment. Heilongjiang Med. 2004;4:306.
54. Liddament MW. A review of ventilation and the quality of ventilation air. Indoor Air. 2000;10:193–9.
55. Xie W, Qi DY. Discussion on the design of fresh air system based on controlling indoor $PM_{2.5}$. Clean Air Conditioning Technol. 2018;(4):76–8.
56. Ye L, Chen WJ. Application of bidirectional flow fresh air system in gymnasium. Build Thermal Energy Vent Air Conditioning. 2018;37(10):67–8.
57. Chao CYH, Wan MP, Cheng ECK. Penetration coefficient and deposition rate as a function of particle size in non-smoking naturally ventilated residences. Atmos Environ. 2003;37:4233–41.
58. Wang J. Indoor ventilation and purification technology. Beijing: China Architecture & Building Press; 2020.
59. Tang ZH. Ventilation dust removal and purification. Beijing: China Architecture & Building Press; 2020.
60. Wang JL, Wang QQ, Lin CQ. Distribution characteristics of indoor particulate matter mass concentration in office buildings in Beijing. HVAC. 2017;47:113–8.
61. Halek F, Boghozian SNA. Simultaneous size distribution study of indoor and outdoor particulates in Tehran. J Aerosol Sci. 1990;21:361–4.
62. Lange CDR. Particle size specific indoor/outdoor measurement. J Aerosol Sci. 1995;26:519–20.
63. Chen JB, Liu YY, Wang HT. Diffusion and distribution of indoor particulate matter in displacement ventilation systems. Build Energy Effi. 2019;47(5):11–7.

64. Li Y. Study on the distribution characteristics of pollutants in air-conditioned rooms. Harbin: Harbin Engineering University; 2008.
65. Hu ST, Ye BC, Wang G, et al. Research on the joint application of variable frequency multi-unit machine and combined ventilation system. In: Academic exchange conference of the building thermal energy power branch of the China Architectural Society; 2007.
66. Yan W, Tung TCW, Niu JL. On-site measurement of tracer gas transmission between horizontal adjacent flats in residential building and cross-infection risk assessment. Build Environ. 2016;99:13–22.
67. Yaghoubi MA, Knappmiller KD, Kirkpatrick AT. Three-dimensional numerical simulation of air contamination dispersal in a room. American Society of Heating. Refrigerating and Air-Conditioning Engineers, Inc., Atlanta. GA1995.
68. Yuan BB. Research on indoor $PM_{2.5}$ purification strategy in office buildings. Nanjing: Nanjing University of Science and Technology; 2017.
69. Mundt E. Non-buoyant pollutant sources and particles in displacement ventilation. Build Environ. 2001;36(7):829–36.
70. Zhao B, Zhang Z, Li X. Numerical study of the transport of droplets or particles generated by respiratory system indoors. Build Environ. 2005;40:1032–9.
71. Lin Z, Chow TT, Fong KF, et al. Comparison of performances of displacement and mixing ventilations. Part II: indoor air quality. Int J Refrig. 2005;28(2):288–305.
72. Zhu QQ. Simulation experiment study on indoor airflow organization and particle distribution under different air supply modes. Jinan: Shandong Jianzhu University; 2019.
73. Yit JE, Chew BT, Yau YH. A review of air filter test standards for particulate matter of general ventilation. Build Service Eng. 2020;41(6):758–71.
74. Park DH, Joe YH, Piri A, et al. Determination of air filter anti-viral efficiency against an airborne infectious virus. J Hazard Mater. 2020;396: 122640.
75. Pei J, Ji L. Secondary VOCs emission from used fibrous filters in portable air cleaners and ventilation systems. Build Environ. 2018;142:464–71.
76. Wang C, Yan K, Wang J, et al. Electrospun polyacrylonitrile/polyvinyl pyrrolidone composite nanofibrous membranes with high-efficiency $PM_{2.5}$ filter. J Polym Eng. 2020;40:487–93.
77. Zheng YY, Cai WL, Wang X, et al. Adhesive-free hot-pressed polytetrafluoroethylene coated high-temperature filter material. J Textile Res. 2013;34(8):22–6.
78. Kim MW, An S, Seok H, et al. Transparent metallized microfibers as recyclable electrostatic air filters with ionization. ACS Appl Mater Interfaces. 2020;22(12):25266–75.
79. Cao M, Gu F, Rao C, et al. Improving the electrospinning process of fabricating nanofibrous membranes to filter $PM_{2.5}$. Sci Total Environ. 2019;666:1011–21.
80. Zheng JW, Wu HY, Li Q, et al. Study on the filtration performance of anti-smog window screens for fine particles. Ind Safety Environ Protection. 2019;45(2):54–8.
81. Huang SC, Zhao J, Chen S. A kind of anti-smog air purification window screen: CN211448465U. 2020-09-08.
82. Yu M, Long YZ, Ramakrishna S. Anti-smog window screen for vehicles: CN111605386A. 2019-12-06.
83. Hu W, Wan L, Sun CC. A kind of anti-smog window screen and anti-smog window screen group: CN210598768U. 2020-05-22.
84. Zhang W, Liu B, Yang BY, et al. Dust removal test and efficiency analysis of electrostatic dust removal anti-smog window screen. J Xi'an Univ Technol. 2016;36(6):456–60.
85. Han DY, Li Z, Song BX, et al. Application of electrostatic technology in dust-proof window screens. Sci Technol Inf. 2014;12:1–2.
86. Hu BJ. Preparation of polyurethane polar nanofibers and their application in anti-smog window screens. Zhengzhou: Zhongyuan Institute of Technology; 2019.
87. Gao J. Preparation and performance study of photochromic anti-smog submicron fiber window screens. Shanghai: Donghua University; 2018.
88. Jeong S, Cho H, Han S, et al. High efficiency, transparent, reusable, and active $PM_{2.5}$ filters by hierarchical Ag nanowire percolation network. Nano Lett. 2017;7(17):4339–46.

89. Zhang S, Liu H, Tang N, et al. Highly efficient, transparent, and multifunctional air filters using self-assembled 2D nanoarchitectured fibrous networks. ACS Nano. 2019;11(13):13501–12.
90. Tijing LD, Choi W, Jiang Z, et al. Two-nozzle electrospinning of (MWNT/PU)/PU nanofibrous composite mat with improved mechanical and thermal properties. Curr Appl Phys. 2013;13:1247–55.
91. Wang S, Liu Q, Zhang Y, et al. Preparation of a multifunctional material with super hydrophobicity, super paramagnetism, mechanical stability and acids-bases resistance by electrospinning. Appl Surf Sci. 2013;279:150–8.
92. Ding Y, Di W, Jiang Y, et al. The morphological evolution, mechanical properties and ionic conductivities of electrospinning P(VDF-HFP) membranes at various temperatures. Ionics. 2009;15:731–4.
93. Lou LH. Study on the preparation and performance of high efficiency low resistance PAN electrospun micro/nano filtration membrane. Shanghai: Donghua University; 2016.
94. Wu QX. Preparation of polyacrylonitrile-based nanofiber yarn by electrospinning and its pre-oxidation. Tianjin: Tianjin Polytechnic University; 2014.
95. Wang X, Zhang K, Yang Y, et al. Development of hydrophilic barrier layer on nanofibrous substrate as composite membrane via a facile route. J Membr Sci. 2010;356:110–6.
96. Cao X, Huang M, Ding B, et al. Robust polyacrylonitrile nanofibrous membrane reinforced with jute cellulose nanowhiskers for water purification. Desalination. 2013;316:120–6.
97. Cai XH, Xiao TH, Chen SM. Study on the rheological properties of PVDF membrane forming system by thermal induced phase separation. Membr Sci Technol. 2015;35(3):7–14.
98. Zárate JMOD, Pen AL, Mengual JI. Characterization of membrane distillation membranes prepared by phase inversion. Desalination. 1995;100:139–48.
99. Bottino A, Camera-Roda G, Capannelli G, et al. The formation of microporous polyvinylidene difluoride membranes by phase separation. J Membr Sci. 1991;57:1–20.
100. Bottino A, Capannelli G, Monticelli O. Poly (vinylidene fluoride) with improved functionalization for membrane production. J Membr Sci. 2000;166(1):23–9.
101. Bottino A, Capannelli G, Munari S, et al. High performance ultrafiltration membranes cast from LiCl doped solutions. Desalination. 1988;68:167–77.
102. Zhang L. Study on the preparation of nuclear pore membrane and its application in tea water filtration. Lanzhou: University of Chinese Academy of Sciences (Institute of Modern Physics, Chinese Academy of Sciences); 2018.
103. Lv XL, Liu X, Fang WM, et al. High efficiency haze removal nuclear pore membrane filter piece and high efficiency haze removal nuclear pore membrane mask: CN103830980A. 2014-06-04.
104. Yang WJ. Breathable functional nuclear pore membrane. Textile Sci Res. 2015;9:32–3.
105. Wang Y, Qu H, Mo D. Study on the pore size measurement and filtration effect of nuclear pore membrane. Nucl Technol. 2016;1(39):38–42.
106. Wu X, Li H, Hao X, et al. Manufacture of polyimide micron-hole separation membrane by electron beam etching. J Radiat Res Radiat Process. 2007;25:345–9.
107. Zhou ML. Study on the preparation of nano silver and nano palladium with nuclear pore membrane. Shanghai: East China University of Technology; 2016.
108. Zuo ZZ. Study on the etching factors of PET nuclear pore membrane. Hengyang: Nanhua University; 2014.

Chapter 10
Application of Gas Cleaning Membranes in Medical Field

10.1 Overview

The development of membrane technology has transformed the concept, as well as the procedure of conventional chemical separation. It has also progressively advanced and implemented the technology in several nascent sectors, including the medical and healthcare industry.

Medical membrane is a term used to describe membranes utilized in the medical and healthcare industry. Various medical membranes exist, performing different functions: gas cleaning, selective gas permeation, medical microporous filtration (e.g. transfusion and infusion), and artificial organ applications. The artificial organ membrane has developed relatively quickly and is in the spotlight of medical membrane technology. It primarily encompasses membranes for blood dialysis, artificial heart valves, artificial skin, and blood oxygen supply [1, 2].

Gas cleaning membrane is one type of medical membranes that is aimed at medical equipment and environment purification. Most porous membranes used in gas cleaning for medical applications have higher criteria than other types of membranes. In addition to the typical chemical requirements, for instance, acidity, heavy metal content and leaching, these membrane materials also need to be free of cytotoxicity and have good biocompatibility, cell compatibility and a hemolysis rate of less than 5%.

The commonly employed materials for medical membranes are displayed in Table 10.1.

Z. Zhong and W. Xing, *Membrane Technology for Air Cleaning*,
https://doi.org/10.1007/978-981-96-1144-7_10

Table 10.1 Types and examples of medical membranes [3]

Type		Example
Cellulose Membrane	Regenerated (Unmodified) Modified Surface Coated Synthetic Modified	Copper imitation membrane, bio-flow membrane, regenerated cellulose membrane, saponified cellulose membrane, cellulose ester type Acetate cellulose membrane, diacetate cellulose membrane, triacetate cellulose membrane Bio-cellulose membrane, polyethylene glycol cellulose membrane Blood imitation membrane, SMC
Polymer Membrane	Natural Hydrophilic Process Hydrophilic Mixed Hydrophilic Disposal Hydrophilic	Polyvinyl alcohol membrane Polyacrylonitrile membrane, polymethyl methacrylate membrane, polycarbonate membrane Polyamide, polysulfone, polyethersulfone Polysulfone PSF-K, polysulfone PU-S

10.2 Medical Gas Cleaning Membranes (Transfusion and Infusion)

Intravenous infusion therapy holds a significant role in modern clinical treatment. Statistics indicate that a significant majority of hospitalized patients, ranging from 90 to 95%, require intravenous infusion therapy. Furthermore, the consumption of disposable infusion devices in China is remarkably high, reaching five billion sets per year. Research has indicated that contaminants such as dust, bacteria and fibers in the wards can enter the liquid through the exhaust pipe during the infusion process, leading to pollution and resulting in vascular embolism, phlebitis, allergic reactions, and other complications [4]. During the 1930s, researchers were already looking into the detrimental effects of particles in infusion. By the 1950s, the size of particles could be identified using an electron microscope. From the 1960s to 1970s, academics attained a consensus on the clinical harm caused by infusion particles.

The therapeutic effect can be influenced by varying the infusion velocity during the treatment procedure. Empirical studies indicate that the optimal rate of infusion for adults should range from 40 to 60 drops per min, with a maximum limit of 100 drops per min. According to the national standard GB 8368-2018 for disposable infusion sets [5], it is required that the infusion set should deliver a minimum of 1000 mL of 0.9% sodium chloride solution when subjected to a static pressure of 1 m height of water for a duration of 10 min. To maintain the equilibrium of pressure in and outside the infusion bottle and ensure a regular infusion rate, it is essential for air to enter the infusion set. A precision infusion set is needed to effectively remove bacterial contamination and suspended particles from the air in liquid medicine.

The precision filter infusion set typically comprises a venous needle, an infusion tube, a liquid medicine filter, a flow regulator, a drip bottle, and an air filter for the air intake pipe (Fig. 10.1). The Chinese pharmaceutical industry standard YY 0770.2-2009 [6] defines the specifications for the air filter used in the air intake pipe.

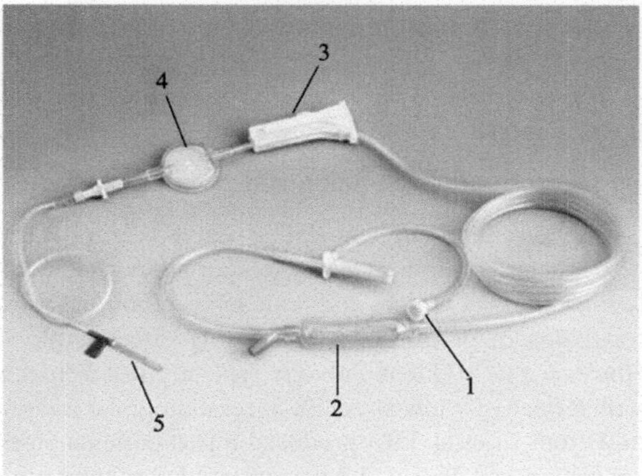

Fig. 10.1 Medical Infusion Filter. 1—Air filter for air intake pipe; 2—Drip bottle; 3—Flow regulator; 4—Liquid medicine filter; 5—Venous needle

Particularly, it pertains to the air filter material used in medical infusion and syringe filters. The precise criteria for infusion air filters are shown in Table 10.2.

At present, air cleaning effectiveness in infusion devices in China mainly relies on glass fiber filter material. Glass fiber cloth is produced by exposing glass to high-temperature melting and drafting operations. It possesses desirable characteristics with a good resistance to temperature and humidity. However, glass fiber possesses a hydrophilic nature, rendering it susceptible to water absorption and pore obstruction, resulting in reduced air permeability. A water-resistant treatment is necessary for the glass fiber cloth when it is used as a functional material for infusion systems. Fluorinated polyester is a widely employed and efficient hydrophobic agent for surfaces. A hydrophobic film is created by spraying a fluorinated polyester emulsion evenly onto the material surface and subjecting it to high-temperature heat treatment, resulting in the perpendicular alignment of fluorocarbon chains. Glass fiber non-woven fabric, as treated with fluorinated polyester, can maintain highly hydrophobic, effectively meeting the requirement for air permeability and preventing the passage of medication solutions. But glass fiber has constraints as brittleness and low resistance to bending, making it prone to fiber shedding, which might lead to contamination of the

Table 10.2 Requirements for air filter material

Performance	Indicator
Filtration efficiency	The filtration efficiency of particles above 0.5 μm in the air is ≥ 95%
Outflow per unit area	Water resistance > 15 kPa
Hydrophobic performance	The outflow per square millimeter for 10 min is not less than the nominal value
Air pressure transmission	The time required to transmit 10 kPa pressured air ≤ 3 s

medication solution during medical processing. Hence, alternative materials need to be developed as replacement materials to glass fiber cloth [7–9].

The polymeric materials listed below are often used as glass fiber substitutes in medical infusion air filters.

(1) Polyester/Polytetrafluoroethylene

Polytetrafluoroethylene (PTFE) exhibits excellent hydrophobic properties, as well as resilience to high and low temperatures, acids, and alkalis. It also possesses a low friction coefficient and may achieve optimal filtration effectiveness by precise regulation of pore size. It is a highly effective air-filtering material for infusion. However, the PTFE membrane typically has a thickness of less than 10 μm, thus with a relatively low strength, necessitating the use of supporting substrates such as polyester. Li et al. [10] produced a two-component composite membrane material consisting of PTFE and polyester/polytetrafluoroethylene (PET/PP). This material has exceptional filtration performance, achieving a filtration efficiency of 99.95% for particles with a size of 0.3 μm. Additionally, the membrane has a low filtration pressure drop of only 350 Pa.

(2) Polypropylene/Polyethersulfone

Polyethersulfone (PES) is a membrane material known for exceptional performance thanks to the molecule's ether group flexibility, benzene ring stiffness, and conjugation unit structure with sulfone group. In recent years, it has gained attention as an air filtration material because of its mechanical properties and film-forming characteristics. In addition, PES has high blood compatibility and holds significant promise in the field of artificial organs. With a filtering effectiveness of up to 99.99% for particles sized 0.33 μm, a composite membrane was prepared by Wang et al. as embedding PES microspheres in polypropylene (PP) fibers [11]. Sun et al. [12] attempted to use the additives of ZnO and SiO_2 in the polypropylene/polyethersulfone (PP/PES) membrane to increase its filtration performance. They discovered that adding SiO_2 was an effective way to meet the goal.

(3) Polypropylene/Polyvinylidene Fluoride

Polyvinylidene fluoride (PVDF) possesses superior hydrophobic characteristics, along with commendable resistance to corrosion and heat. Consequently, PVDF materials have gained considerable interest in air cleaning studies in recent years. The PVDF membrane materials utilized in infusion vessels feature a homogeneous pore size range, spanning from 0.5 to 2 μm, and offer excellent air permeability. A particular technological company's air filter PVDF membrane used in infusion devices is compared with conventional glass fiber membranes in Table 10.3 based on technical parameters. Table 10.3 demonstrates that polypropylene/polyvinylidene fluoride (PP/PVDF) membranes provide a distinct advantage in air flux volume when compared to conventional glass fiber membranes.

Nano-scale coating of fluorine ester is formed on the thermal-resistant and hydrophobic non-woven substrate by submerging and post-heating, according to Zhu et al. [7] It can perform as the air filter membrane for the medical infusion system as

Table 10.3 Comparison of technical parameters between PVDF membrane and glass fiber membrane for medical infusion air filter

Technical parameter index	Polypropylene/Polyvinylidene fluoride	Glass fiber
Pore size/μm	0.5–2	0.7–1.9
Air flow rate (0.01 MPa)/[mL/(m^2 h)]	4800	4200
Membrane thickness/μm	100–200	100–200
Resistance/kPa	22	23

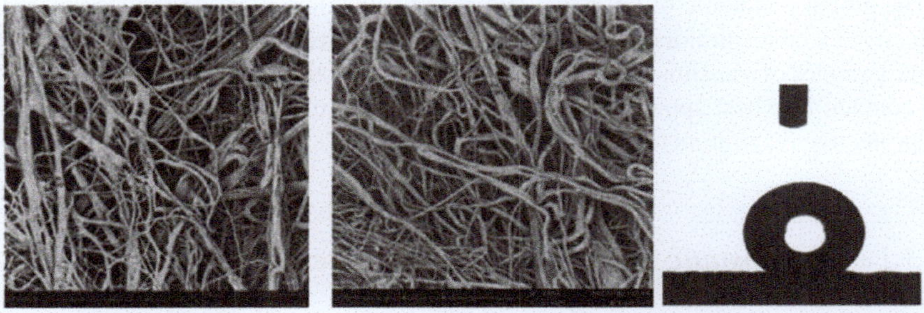

Fig. 10.2 Comparison of microscopic structure before and after the formation of fluorine ester coating

shown in Fig. 10.2. This membrane material exhibits exceptional hydrophobic properties and high gas permeability, with water contact angle exceeding 150°, water rolling angle below 4° and water resistance pressure up to 32 kPa; and a porosity up to 80% and air flow rate of 12 L/h after pressurization. Through the thermal treatment in the preparation procedure, a covalent bond forms between the fluorine coating and polyester substrate, guaranteeing the stability of the membrane layer and eliminating the membrane peeling. The fluorine ester-coated polyester membrane was tested at the Zhejiang Food and Drug Testing Institute, and the results confirmed the full satisfaction of the infusion air filter criteria.

10.3 Respirator Membranes

The naturally occurring viruses exhibit a wide range of shapes and sizes, often ranging in dimension from 10 to 300 nm. Large viruses are sized from 200 to 300 nm, as the smallpox virus; medium viruses include the influenza virus, sized around 100 nm and coronaviruses, sized from 80 to 130 nm; small viruses include the porcine circovirus having dimensions about 17 nm. In order to propagate, viruses must cling to a living carrier because they are unable to exist on their own. Airborne transmission of viruses mostly occurs through two methods: droplet transmission and aerosol transmission. Bioaerosols are solid or liquid airborne particles composed of bacteria, fungi, viruses, and other biological byproducts. International academic standards have not yet been established for the precise particle size threshold to differentiate between droplets and aerosols. According to the World Health Organization (WHO) and the US Centers for Disease Control and Prevention (CDC), particles that have a diameter greater than 5 μm are categorized as droplets, whereas particles having a diameter equal to or less than 5 μm are classified as aerosols. Aerosols can enter the human body through the respiratory system and lead to illness and sickness [13, 14].

The global pandemic of the novel coronavirus pneumonia in 2020 has rendered protective facial respirators an essential safeguard in individuals' everyday lives. The primary modes of transmission for the new coronavirus are through droplets, aerosol and contact. The best approach for preventing and controlling epidemics now is to use protective respirators to block the new coronavirus's pathways of transmission.

10.3.1 Electrostatic Melt-Blown Fabric Filter Materials

The primary protection effect of medical surgical respirators and N95 respirators is achieved by a melt-blown fabric filter layer. This intermediate layer in the respirators comprises of a loosely spun film of polypropylene fiber. The fibers have diameters ranging from 0.5 to 10 μm, resulting in a film having small pore size and high porosity. This membrane has the ability to capture droplets and aerosols that transport viruses by several mechanisms, including diffusion, inertial collision, deposition interception, and electrostatic adsorption [15–23].

The melt-blown process entails the extrusion of the polymer via spinneret hole of the casting head and forms a fine stream of molten polymer. The air channel on both sides of the casting spinneret hole expels heated air at a high velocity to elongate the polymer melt into a fine stream (Fig. 10.3).

A cooler is positioned beneath the casting head to facilitate the cooling and crystallizing of the fibers. Although melt-blown technology boasts a short process flow and good production efficiency, the nanofiber preparation still remains a significant problem, whilst the typical melt-blown fiber having a minimum diameter of 1–2 μm [24, 25].

Researchers have employed diverse techniques to decrease the diameter of melt-blown fibers in recent years. The study by Deng et al. [26] fabricated nanofiber membrane by employing a one-step melt-blown technique, utilizing polypropylene (PP) and polystyrene (PS) as the basis materials (Fig. 10.4). This material is micro and nano-structured with fibers having the lowest diameter of 300 nm and fiber size differential of 30 times. The multi-scale structure of the PP/PS membrane guarantees both the high filtration efficiency (99.87%) and the exceptionally low filtration pressure drop (37.73 Pa). This study provides a reference for upcoming research on low-resistance and high-performance melt-blown fiber materials.

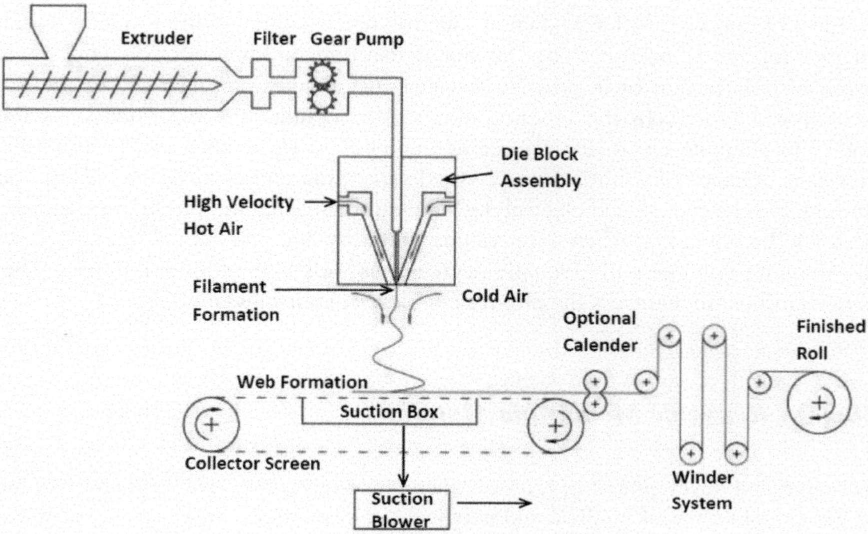

Fig. 10.3 Melt-blown process flow-chart [25]

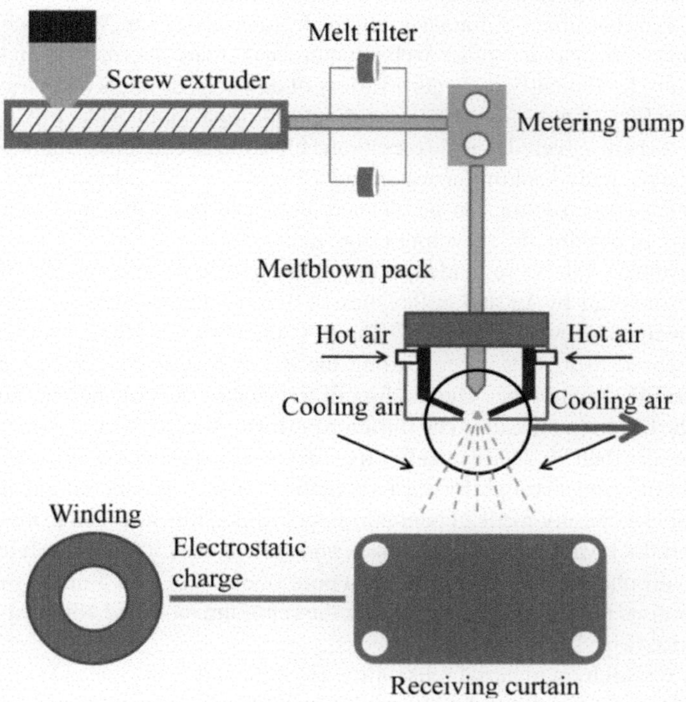

Fig. 10.4 One-step melt-blown process for preparing PP/PS nanofiber membrane [26]

Electret materials, often known as electrolytes, have the ability to maintain the charges for an extended period. The charge can refer to immobilized charges as a result of polarization or to positive and negative charges confined on the surface or within a "trap". In the absence of an electric field, the electrolyte has the ability to generate electrostatic force autonomously. The respirator filtration efficiency is enhanced by combining physical separation and electrostatic adsorption through polypropylene and electrolytes, without increasing the filtration resistance. It should be noted that when temperature, humidity and operational time increase, the amount of charge will gradually decrease, as well as the filtration impact. This issue significantly hampers the progress of electret melt-blown fabric.

10.3.2 Reusable Membrane Materials

Research and development of reusable membrane materials have been spurred due to the global deficit of medical and personal-use respirators. Zhong et al. produced a spider-web-like polytetrafluoroethylene membrane material [27, 28] of a unique microporous structure, which has advantages such as waterproofness, breathableness, and corrosion resistance. It has a strong ability to inhibit the infiltration of microbes. This membrane has fibers of diameter ranging from 0.1 to 0.2 μm, which intertwine to create a three-dimensional spider-web structure and create microspores in size from 0.1 to 0.5 μm. It efficiently prevents a variety of aerosols, droplets and viruses while filtering over 99% of non-oily particles without additional input of electrostatic effect. The researchers discovered that the membrane filtration effectiveness remained above 99% after repeatedly soaking in water and 75% alcohol. It indicates that the fiber membrane-based respirator can be cleaned and sterilized with water and alcohol, significantly improving the recycling usage rate.

ZIF-8 nanocrystals were loaded onto polypropylene/polycarbonate (PPC) fiber membrane material by Li and colleagues [29] using the melt-blown process and in-situ growth, as shown in Fig. 10.5. The PPC membrane's filtration efficiency for $PM_{2.5}$ increased to 91.68% after loading the ZIF-8 particles, 32.83% higher than the pristine PPC membrane. The loading of ZIF-8 nanocrystals had no discernible effect on the filtration pressure. The filtration performance of this membrane to $PM_{2.5}$ remains greater than 80.77% even after five recycling experiments by water washing and drying, offering a straightforward alternative for reusable membrane materials.

A reusable respirator made of nanoporous polyimide film attached to the commercial N95 product was generated by Nazek and colleagues [30]. This polyimide film exhibits hydrophobic characteristics and a pore size of less than 5 nm. By replacing this nanoporous polyimide layer, it allows the reuse the modified N95 respirator for everyday use (Fig. 10.6).

For the research on material reusability, Horvath and colleagues [31] discovered that bacteria and viruses can be eliminated through photocatalysis by adding TiO_2

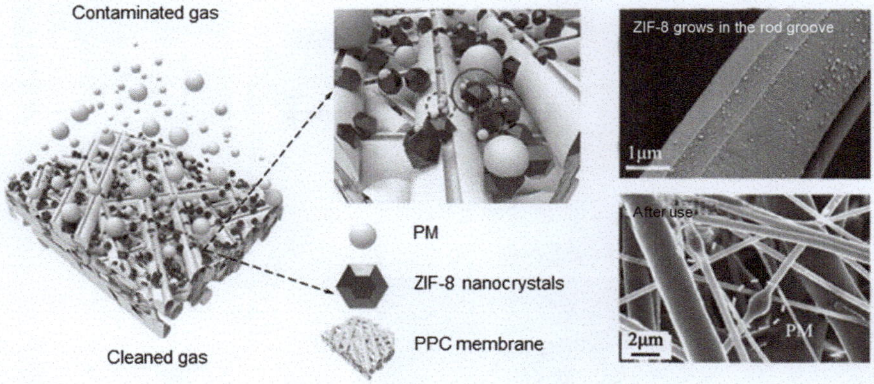

Fig. 10.5 Schematic diagram of polypropylene/polycarbonate (PPC) fiber membrane [29]

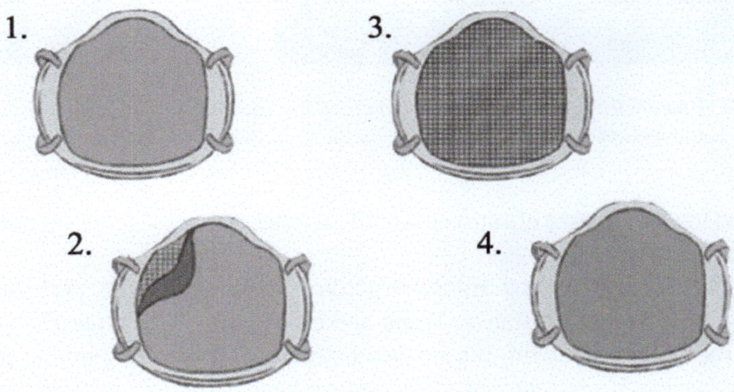

Fig. 10.6 Schematic diagram of reusable respirator replacement [30]

nanowires to the membrane layer to ensure hygiene and reuse. The TiO_2 nanowire-based respirator is capable of being reused over 1000 times, and its preliminary application has already been carried out, as presented in Fig. 10.7.

10.3.3 Electrospun Nanofiber Membranes

Electrospinning technology is a method that uses electric field force to stretch the spinning solution or melt jet situated in a high-voltage field. The resultant nanofibers create a layer structure with a large number of micropores when they are deposited in an erratic state on the receiving device. The electrospun-induced fibers can have diameter as low as 5 nm, and the corresponding fiber-based membrane can achieve

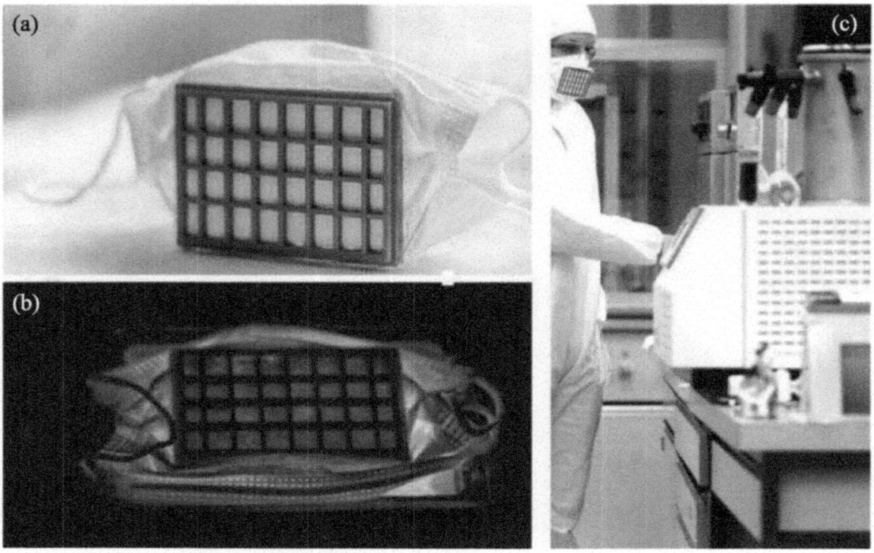

Fig. 10.7 TiO_2 nanowire respirator: **a** photograph of the product, **b** photocatalytic sterilization process, and **c** application situation [31]

efficient and low-resistance filtration to small particles due to the large specific surface area, and uniform pore size.

Researchers create unique micro-structures to enhance the performance of nanofiber membrane. For instance, Wang and colleagues [32] prepared a polylactic acid nanofiber membrane with porous bead structure by electrospinning polylactic acid (PLA) as the basis material. The porous bead structure in fiber network is presented in Fig. 10.8. With a pressure drop of just 165.3 Pa, this membrane material exhibits exceptional filtration efficiency of up to 99.997% for aerosol particles of 260 nm.

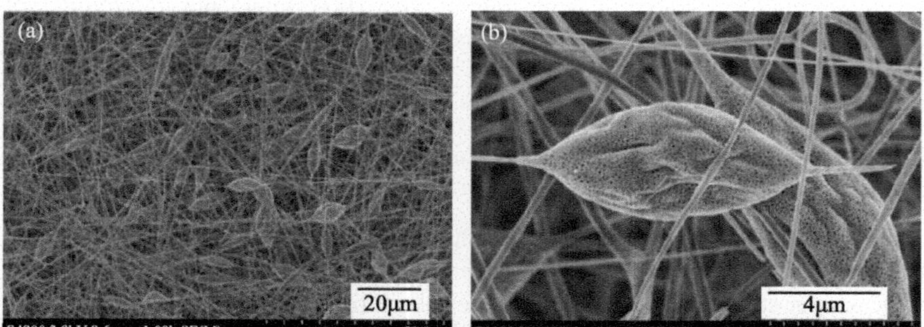

Fig. 10.8 Microstructure of porous bead in polylactic acid nanofiber membranes [32]

Fig. 10.9 Microstructure of PVDF dendritic nanofiber membranes [33]

Li and colleagues [33] have successfully employed electrospinning technology to produce a novel form of polyvinylidene fluoride (PVDF) nanofiber membrane. Figure 10.9 illustrates the distinctive dendritic nanofiber network topology of this material. The coarse fibers support the membrane skeleton with good mechanical properties. The finer fibers build an intra-connecting structure, effectively reducing the pore size to a minimum of 5–10 nm, which increases the van der Waals force between the fiber and dust particles and improves the filtration performance. The filtration efficiency to 0.26 μm-sized sodium chloride particles is as high as 99.999%, and the filtration pressure drop is just 124.2 Pa.

Some researchers have discovered that graphene or nanocrystals can help enhance the performance of fiber membranes in addition to constructing specific microstructures. For instance, by adding graphene oxide (GO), Zhang et al. [34] enhanced the performance of membrane materials. Using graphene as an incorporating additive into polyacrylonitrile (PAN), they electrospun to create PAN/GO nanofiber membrane (Fig. 10.10). While the filtration efficiency and pressure drop of pristine PAN membrane materials were 88.3% and 462 Pa, respectively, the PAN/GO nanofiber membrane achieved a 99.6% filtration efficiency to $PM_{2.5}$ at just 117 Pa filtration pressure drop. It is evident that graphene modification has significantly improved the filtering effectiveness and pressure resistance of PAN membrane materials. The stability of the PAN/GO membrane was examined in a 100-h application: the filtration efficiency to $PM_{2.5}$ was maintained at 99.1%, with a slight decrease of 0.5%. Relevant research has pointed out that the loading of graphene oxide also improved the antibacterial effect of protective respirators.

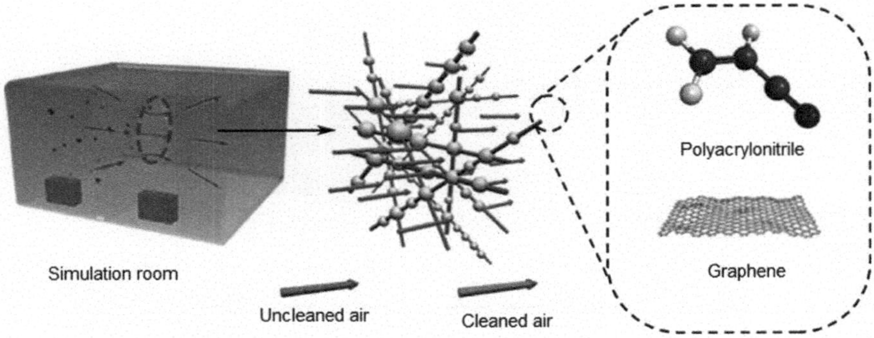

Fig. 10.10 Schematic diagram of PAN/GO nanofiber membranes [34]

Huang et al. [35] developed laser-induced graphene (LIG) technology to prepare the antibacterial respirator, whose disinfection rate increased to 81% compared to the unmodified respirators. With the additional solar radiation treatment of 0.75 kW/m^2, the sterilization rate increased to 99.998% in 10 min owing to the photothermic effect. The LIG surface properties can be controlled by precisely adjusting the laser parameters, thus making a moisture-powered generator with breathing as the driving force. The adhesion of bacteria or atmospheric-suspended particles will change the LIG surface characteristics and affect the moisture-induced potential, consequently offering feedback information on the respirator's pollution status.

Lee et al. [36] utilized polyacrylonitrile (PAN) and conductive metal-organic framework material (MOF, Ni-CAT-1) to prepare an MOF-coated nanofiber membrane with the hydrothermal reaction (Fig. 10.11). Since the MOF has a large specific surface area and specific functions, the membrane can perform a high filtration rate to $PM_{2.5}$ (> 99%) and is capable to applicable in oil fume treatment. Because of MOF's conductivity, breathable detection can be realized for this membrane as the material's resistance correlates with the breathing air flux.

With the advancement of nanotechnology, significant progress in the electrospinning technique has been achieved, with numerous research institutions in China leading the way. Professor Ding's group at Donghua University has developed the electrospun membrane of spider-web structure, which has high filtration efficiency for ultrafine particles. The MOFilter technology has been developed by the group of Professor Wang Bo at Beijing Institute of Technology, effectively deactivating bacteria, fungi and viruses by catalyzing atmosphere oxygen to superoxide radicals [37]. The electrospun nanofiber material developed by the group of Professor Wang Ce at Jilin University has a demonstrative application in the high-temperature gas filtration process. In addition, some companies, including Chongqing Zhongna Technology Company and Xiamen Zhongke Bestar Company, have implemented the electrospun technique to fabricate commercial medical respirators.

In addition to the two applications mentioned above, gas cleaning membranes can also be facilitated in medical environmental air purifiers and surgical ventilator

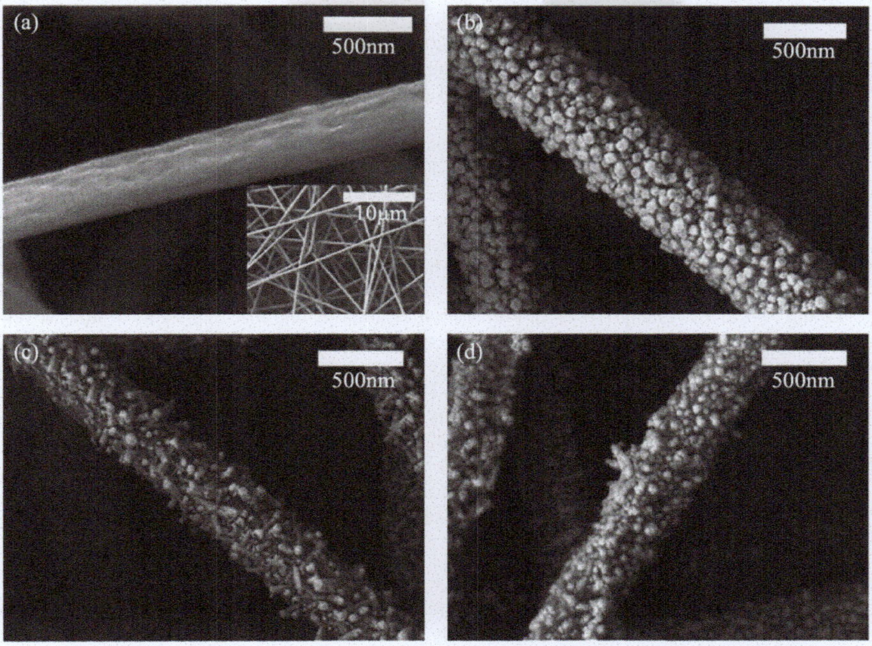

Fig. 10.11 Microstructure of PAN/MOF nanofiber membranes [36]

air filters. With the development of material and technology, there will be a largely expanded application scope for gas cleaning membranes in the medical and healthcare field.

References

1. Xing WH, Gu XH. High-performance membrane materials and membrane technology. Beijing: Chemical Industry Press; 2017.
2. Wang XS. Gas membrane technology. Beijing: Chemical Industry Press; 2010.
3. Pan F, Duan YF. Application and prospect of membrane technology in artificial organs. Ind Textiles. 2003;2:21–4.
4. Jiang Y, Zhan SG. Principle and application development of precision filter infusion set. Med Health Equip. 2012;33(6):80–3.
5. Disposable infusion set gravity infusion type: GB 8368—2018. 2018.
6. Filter material for medical infusion and injection devices: YY 0770. 2—2009. 2009.
7. Zhu BK, Yin X, Liang ZY. Ultra-hydrophobic air purification material for infusion set. In: Proceedings of the 2016 China-EU medical biofilm science and technology symposium; 2016. pp 120–4.
8. Zhang YH, Yang SY, Zhu BK, et al. Medical microporous filter membrane material and device material. In: The 7th National membrane and membrane process academic report conference; 2011. p. 213.

9. Zhang L. Preparation and performance study of medical water-repellent breathable air membrane. Shanghai: Donghua University; 2018.
10. Li M, Dai ZJ, Huang C, et al. Preparation and filtration performance of PTFE microporous membrane/bicomponent meltblown material composite air filter material. J Donghua Univ. 2018;44(2):174–81.
11. Wang JN, Ma LC, Li L, et al. Research on electrospun PES microsphere/fiber low-resistance composite air filter membrane. J Polym Sci. 2014;11:1479–785.
12. Sun LY, Qian JH, Shen WJ, et al. Study on the structure and performance of polyethersulfone composite membrane with inorganic additives. Modern Textile Technol. 2020;28(1):11–7.
13. Yang SS, Zhou JC. Microbiology. Beijing: Science Press; 2004.
14. Chen XQ, Li S, Zhang MS, et al. Progress in biological aerosol research. J Nanjing Med Univ. 2020;40(6):783–8.
15. Zhang X, Liu JX, Zhang HF, et al. Preparation technology and research status of non-woven filter material for protective respirators. J Textile Res. 2020;41(3):168–74.
16. Chen FX, Zhai LS, Liu KS, et al. Research progress and development trend of protective respirators. J Xi'an Univ Eng. 2020;34(2):1–12.
17. Han L, Hu MY, Ma YB, et al. Research status of medical non-woven respirator materials and their new technologies. J Xi'an Univ Eng. 2020;34(2):20–6.
18. Zuo SY, Chen YH, Zeng C, et al. Introduction to the application range and related standards of respirators in various countries. Chin J Infection Control. 2020;19(2):109–16.
19. Respiratory protection self-priming filter anti-particulate respirator: GB 2626—2019. 2019.
20. Disposable medical respirators: YY/T 0969—2013. 2013.
21. Medical surgical respirators: YY 0469—2011. 2011.
22. Technical requirements for medical protective respirators: GB 19083—2010. 2010.
23. Technical specifications for daily protective respirators: GB/T 32610—2016. 2016.
24. Hu JY. Prevention of structural molding study 2. Shanghai: Donghua University Press; 2016.
25. Drabek J, Zatloukal M. Melt-blown technology for production of polymeric microfibers nanofibers: a review. Phys Rev Fluids. 2019;31(091301):1–26.
26. Deng N, He H, Yan J, et al. One-step melt-blowing of multi-scale micro/nano fabric membrane for advanced air-filtration. Polymer. 2019;165:174–9.
27. Zhong ZX. The membrane technology achievements of Nantong University's Membrane Science Institute take up the mission of fighting the "epidemic." Membr Sci Technol. 2020;1:63.
28. Zhong ZX. Nantong University developed new type of nano spider web bionic membrane respirator. Membr Sci Technol. 2020;41:64.
29. Li T, Cen X, Ren H, et al. Zeolitic imidazolate framework-8/Polypropylene-Polycarbonate barklike meltblown fibrous membranes by a facile in situ growth method for efficient $PM_{2.5}$ capture. ACS Appl Mater Interfaces. 2020;12:8730–9.
30. Nazek E, Nadeem Q, Husa B, et al. Flexible nanoporous template for the design and development of reusable anti-COVID-19 hydrophobic face respirators air-filtration. ACS Nano. 2020;14:7659–65.
31. Horvath E, Rossi L, Mercier C, et al. Photocatalytic nanowires-based air filter: towards reusable protective respirators. Adv Func Mater. 2020;30(2004615):1–8.
32. Wang Z, Zhao C, Pan Z. Porous bead-on-string poly (lactic acid) fibrous membranes for air filtration. J Colloid Interface Sci. 2015;441:121–9.
33. Li Z, Kang W, Zhao H, et al. Fabrication of a polyvinylidene fluoride tree-like nanofiber web for ultra high performance air filtration. RSC Adv. 2016;6:91243–9.
34. Zhang C, Yao L, Yang Z, et al. Graphene oxide-modified polyacrylonitrile nanofibrous membranes for efficient air filtration. ACS Appl Nano Mater. 2019;2:3916–24.
35. Huang LB, Xu SY, Wang ZY. Self-reporting and photothermally enhanced rapid bacterial killing on a laser-induced graphene respirator. ACS Nano. 2020;14:12045–53.
36. Lee H, Jeon S. Polyacrylonitrile nanofiber membranes modified with Ni-based conductive metal organic frameworks for air filtration and respiration monitoring. ACS Appl Nano Mater. 2020;3:8192–8.
37. Li P, Li J, Feng X, et al. Metal-organic frameworks with photocatalytic bactericidal activity for integrated air cleaning. Nat Commun. 2019;10(21771):1–10.

Chapter 11
Application of Membrane Technology in Industrial Exhausts Purification

11.1 Application of Membrane Technology for Ultra-Low Emission of Dust

11.1.1 Application of Membrane Technology for Ultra-Low Emission of Dust in Coal-Fired Power Plants

The flue gas composition of coal-fired power plants is complex, containing not only a substantial quantity of solid dust, but also corrosive gases such as SO_2, NO_x, O_2, and water vapor [1]. Electrostatic precipitators are favored by many power plants due to their large flue gas treatment capacity and suitability for high temperature, high pressure, and high humidity conditions. However, as environmental awareness increases, there is a need for further improvement in dust removal technology. Bag dust collectors have become one of the most effective pieces of technical equipment for controlling industrial dust and smoke in China. Nevertheless, during the industrial implementation process, the traditional filter material gradually reveals some shortcomings. Its retention rate for small particle-size dust is low, and it cannot meet the latest emission standards [2].

When traditional filter materials are used to filter dust, the larger pores between the filter material fibers allow small particle size dust to easily penetrate the filter material, resulting in low dust removal efficiency. The filtration efficiency can be improved when a filter cake layer is formed, but this significantly increases the filtration resistance [2, 3]. Membrane materials change the deep filtration mode of traditional filter materials to the surface filtration mode. The dust filtration process is primarily achieved by the membrane layer. The membrane material is a three-dimensional network with a intertwined microporous structure and no through pores. The pore size ranges from 0.1 to 3.5 μm, resulting in higher dust collection efficiency. The principle of traditional filter material filtration and membrane filtration is illustrated in Fig. 11.1 [2].

Z. Zhong and W. Xing, *Membrane Technology for Air Cleaning*,
https://doi.org/10.1007/978-981-96-1144-7_11

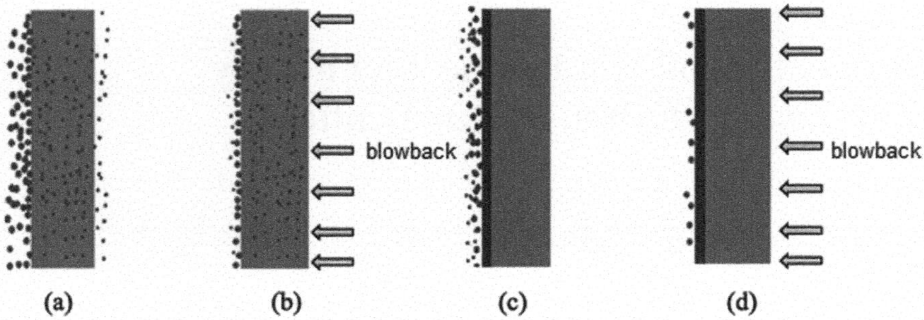

Fig. 11.1 Dust filtration principle of traditional filter materials (**a**, **b**) and membrane materials (**c**, **d**)

11.1.1.1 Coal-Fired Boiler Membrane Dust Collector

This project is a 210 t/h coal-fired boiler ultra-low emission renovation project in Changzhou, China. The filter material uses an organic composite filter membrane. The operational requirements are such that the dust emission concentration of the outlet flue gas is less than 5.0 mg/m^3, and the operating pressure difference of the dust collector is less than 1200 Pa.

(1) Process overview

The power plant process is shown in Fig. 11.2. Air, after being heated by the preheater, enters the boiler and mixes with coal powder for combustion. The combustion exhaust gas carries a large amount of dust particles and gaseous pollutants. After denitrification and heat exchange, it enters the dust collector. Inside the dust collector, dust particles are separated by the filter material, and the exhaust gas is purified. The dust-free exhaust gas then passes through the desulfurization tower to remove sulfides before being discharged through the chimney [4, 5]. In the process, the dust collector is the main equipment for separating dust particles, and its performance and stability are crucial to the operation of the entire system.

The particle size of the dust is an important basis for designing membrane materials. According to the particle size characteristics of the dust, the pore size of the filter membrane material can be designed. The dust in the process system is characterized by scanning electron microscopy and a particle size analyzer, and the results are shown in Fig. 11.3. It is evident that the dust exhibits agglomeration, uneven particle size distribution, and an approximately spherical morphology, with particle sizes typically ranging from 0.5 to 3 μm, indicating the presence of ultrafine powder.

(2) Process parameters design

The dust collector used in this process is a membrane dust collector. It is designed to operate for more than 3 years, with a filtration wind speed of 0.6–1.2 m/min. The dust removal approach employs an electromagnetic pulse valve for back-blowing, and includes both resistance control and time control working modes. The start value

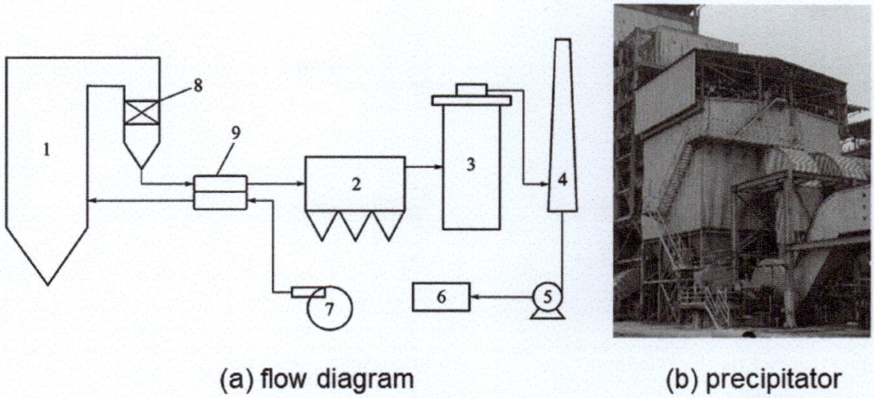

Fig. 11.2 The process flow diagram of coal-fired power plant (**a**) and dust collector appearance (**b**). 1—Boiler; 2—Dust collector; 3—Desulfurization tower; 4—Chimney; 5—Wastewater pump; 6—Wastewater treatment device; 7—Blower; 8—SCR denitrification; 9—Air preheater

Fig. 11.3 Dust characterization results

for resistance control is set to 1000 Pa, while the set value for time control is 50 min. The back-blow gas pressure is 0.4 MPa. The project process parameters are displayed in Table 11.1.

(3) Membrane product design

The filtration performance of the dust collector largely depends on the performance of the membrane product [16, 21]. In accordance with the project characteristics, the filter material of the dust collector uses a PPS organic composite membrane. The membrane pore size is designed to be 0.5 μm, and the membrane product specification is designed to be ϕ160 mm × 7500 mm. The performance parameters of the membrane product are designed as shown in Table 11.2.

Table 11.1 Process parameters

Parameter	Value
Flue gas temperature/°C	< 180
Oxygen content/%	< 10
Humidity/%	6
Flue gas volume/($\times 10^4$ m^3/h)	40–45
Filter wind speed/(m/min)	0.6–1.2
Operating pressure difference/Pa	< 1000
Lifespan/years	> 3.0
Filter material	Organic composite membrane material
Filter material specification/mm	ϕ 160 × 7500
Injection method	Offline

Table 11.2 Performance parameters of the membrane product

Parameter	Value
Membrane product material	PPS organic composite membrane material
Membrane product specification/mm	ϕ 160 × 7500
Thickness/mm	1.8–2
Membrane pore size/μm	0.5
Air permeability (200 Pa)/(m/min)	2–4

(4) Analysis of influencing factors

The parameters that affect the operation performance of the dust collector include flue gas characteristics (including gas temperature, humidity, corrosiveness, and explosiveness), dust characteristics (including dust particle size, adhesion, hygroscopicity, and explosiveness), membrane product performance (including membrane material, membrane pore size, air permeability, thickness), operating parameters (including dust removal method, dust removal cycle, compressed air pressure), and usage requirements (including service life, operating resistance, emission requirements, maintenance). Here, a few parameters are selected for impact analysis.

(1) Temperature

Temperature is the most important influencing factor when choosing a filter product. When choosing a membrane material, it is required that its continuous use temperature must be higher than the temperature during normal operation of the working condition. For working conditions with drastic fluctuations in gas temperature, a relatively large safety factor should be chosen, and the instantaneous peak temperature of the working condition must not exceed the upper limit of the membrane material usage temperature. The normal operating temperature of this coal-fired boiler process is below 180 °C, so PPS organic composite membrane material can be chosen.

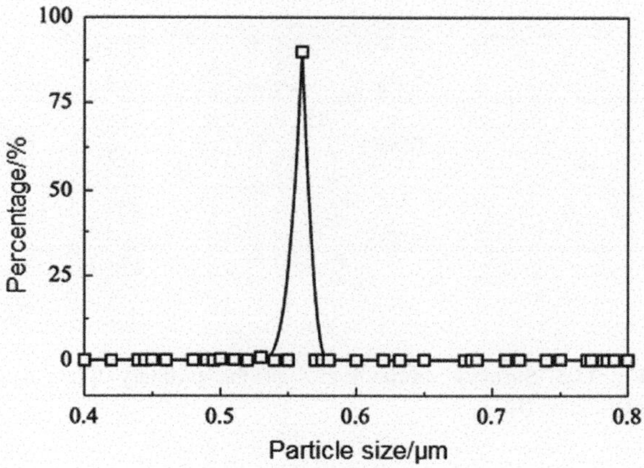

Fig. 11.4 Pore size distribution of organic composite membranes

(2) Membrane materials

The material and pore size of composite membrane materials have a significant impact on their filtration performance. Figure 11.4 shows the characterization results of the pore size of organic composite membrane materials. The most probable pore size of the membrane material is around 0.5 μm [6, 7], which is smaller than the average particle size of dust. Moreover, the distribution of pore sizes is narrow, indicating that the sizes of the pores are relatively uniform, and the performance of the membrane material is excellent.

Figure 11.5 shows the scanning electron microscope images of the surfaces of ordinary filter materials and organic composite membrane materials. By comparison, it can be seen that both ordinary filter materials and membrane materials have a fibrous structure, but the former has a larger fiber diameter. The pores formed by interweaving are uneven in size and are basically large pores, with pore diameters exceeding 50 μm. The fiber diameter of organic composite membrane materials is small, and the pores are denser and more uniform, with an average pore diameter of about 0.5 μm. Smaller and denser pores are more conducive to capturing dust on the material surface, preventing dust from entering the filter material and blocking the filtration channels, and are also beneficial for back-blowing dust removal.

Figure 11.6 presents the scanning electron microscope images of the ordinary filter materials and the organic composite membranes after 30 months of operation. From the figure, it is evident that the structure of the organic composite membrane material remains intact. The interior of the material is relatively clean, and there is no dust pollution. In contrast, the pore channels inside the ordinary filter material have been blocked by a significant amount of dust. Continued operation in this state will increase equipment resistance and operating energy consumption. Therefore, the advantages of organic composite membranes over ordinary filter materials are clearly demonstrated [8–10].

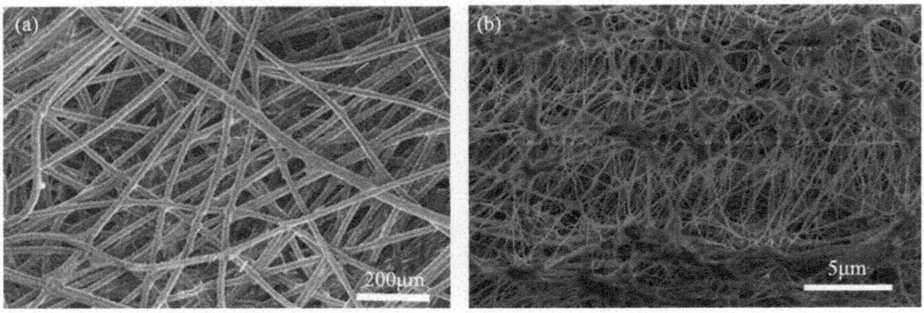

Fig. 11.5 Surficial morphologies of the ordinary filter material (**a**) and organic composite membrane (**b**)

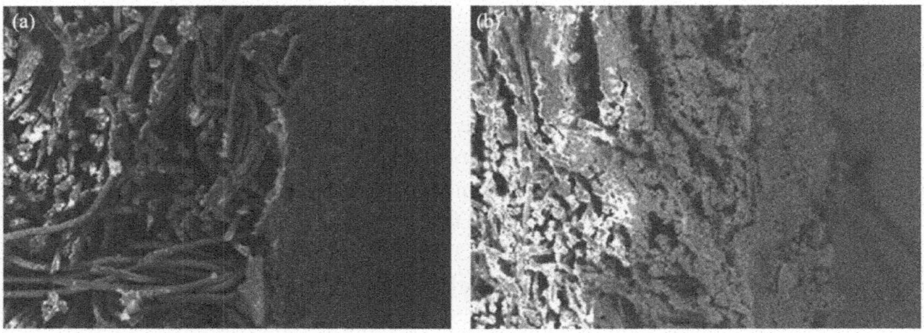

Fig. 11.6 Surficial morphologies of the ordinary filter material (**a**) and organic composite membrane (**b**) after filtration test

(3) Operating resistance

During the operation of the dust collector (taking 30 months as a study period), the operating pressure difference data of the dust collector of ordinary filter materials and organic composite membrane filter materials are compared at regular intervals, as shown in Fig. 11.7. It can be seen that at the beginning of the operation, the ordinary filter material dust collector has a slight advantage, and the operating pressure difference of the ordinary filter material dust collector is 50 Pa lower than that of the organic composite membrane filter material dust collector. However, after 2 months of operation, the organic composite membrane filter material showed an obvious advantage and maintained a higher advantage over the ordinary filter material dust collector in the following time. The operating pressure difference of the ordinary filter material dust collector has reached 1300 Pa after 30 months, while the organic composite membrane material still maintains around 850 Pa. Compared to ordinary filter materials, the pore size of the membrane material is smaller, and the resistance is greater at the beginning of operation. However, because it belongs to the

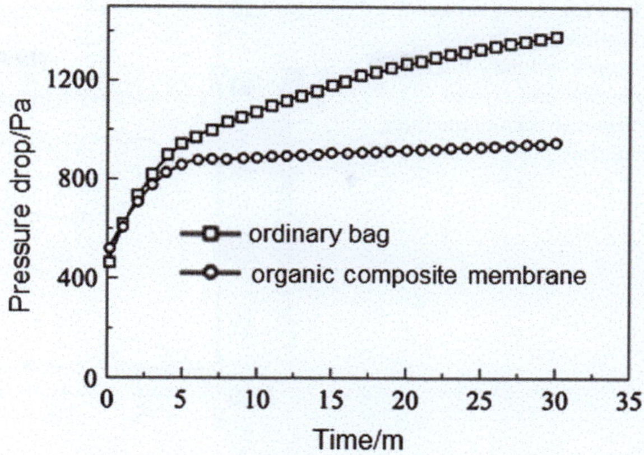

Fig. 11.7 Comparison of operating pressure differences between ordinary filter materials and organic composite membrane filter materials

surface filtration mechanism, the dust removal effect is better during operation, and the membrane filtration recovery is more thorough, so it can maintain a low energy consumption, high efficiency operation state for a longer time [11].

(4) Emission level

During the operation of the dust collector, at regular intervals (taking 3 months as a study period), dust emission concentration tests are conducted at the outlets of the dust collectors of ordinary filter materials and organic composite membrane filter materials, as shown in Fig. 11.8. At the beginning of the operation of the ordinary filter material dust collector, the dust emission concentration is relatively high. After 5 months of operation, the dust emission concentration tends to stabilize, maintaining at 20 mg/m^3. This is mainly because, in the early stages of operation, the pore size of ordinary filter materials is relatively large, and the dust interception efficiency is not high. As the filtration progresses, a large amount of dust enters the interior of the filter material, causing the filter pores to be blocked, thereby improving the filtration efficiency. Organic composite membranes effectively avoid the above problems, maintaining relatively low resistance from the start of filtration and their emission concentration is relatively stable, always below 5.0 mg/m^3, basically maintained at the level of 3.0 mg/m^3.

(5) Economic benefits

The organic composite membrane, with its dense microporous structure, operates on the principle of surface filtration and can retain most of the dust on the membrane surface, making dust removal more thorough. The dust removal efficiency can reach 99.99%, and the dust concentration at the flue gas outlet is less than 5.0 mg/m^3. The filtration pressure drop is low, and energy consumption is small, not only saving energy for enterprises but also extending the service life of the dust collector, which can improve the overall benefits of the enterprise. The membrane dust collector has a running resistance of more than 400 Pa lower than that of ordinary bag dust collectors,

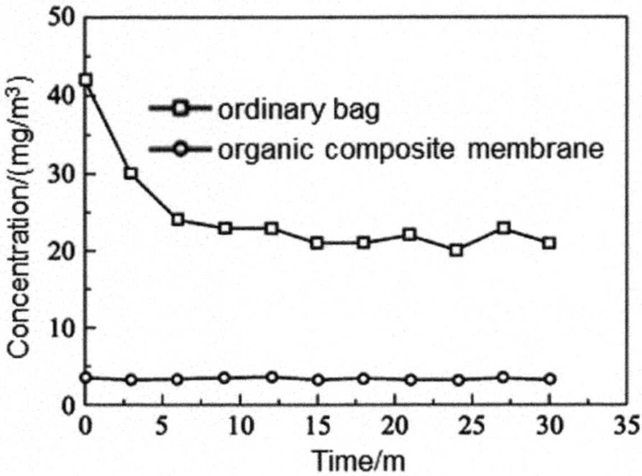

Fig. 11.8 Comparison of filtration performance between ordinary filter materials and organic composite membrane filter materials

which can reduce the power of the induced draft fan and has a long back-blowing cycle, reducing the use of gas. The annual cost savings is more than 1 million RMB. The membrane dust collector can increase the boiler capacity by more than 6%, and the annual increase in output value is more than 41 million RMB. The economic benefits are significant.

11.1.1.2 Ultra-Low Emission Transformation of Coal-Fired Flue Gas

This project is an ultra-low emission renovation project for a dust collector supporting the No. 3 boiler of a power plant in Lianyungang. The filter material used is an organic composite membrane. The following is an introduction to the operation of the membrane dust collector.

(1) Process overview

Figure 11.9 is a schematic diagram of the power plant boiler process system. This process is a conventional coal-fired power plant flue gas cleaning process, which includes SCR denitration, low and medium dust removal, and gypsum method desulfurization. Among them, the dust collector is the main equipment for purifying flue gas particulates. The flue gas volume is 3×10^5 m^3/h, the inlet dust concentration is 25–30 g/m^3, and the flue gas temperature is approximately 130 °C.

The distribution of flue gas dust particle size, as shown in Fig. 11.10, is broad. The most probable particle size is 30 μm. The distribution also includes ultrafine powder with a particle size of about 0.1 μm and large particles with a particle size close to 1 mm. This wide range indicates the complexity of managing and filtering such diverse particulate matter.

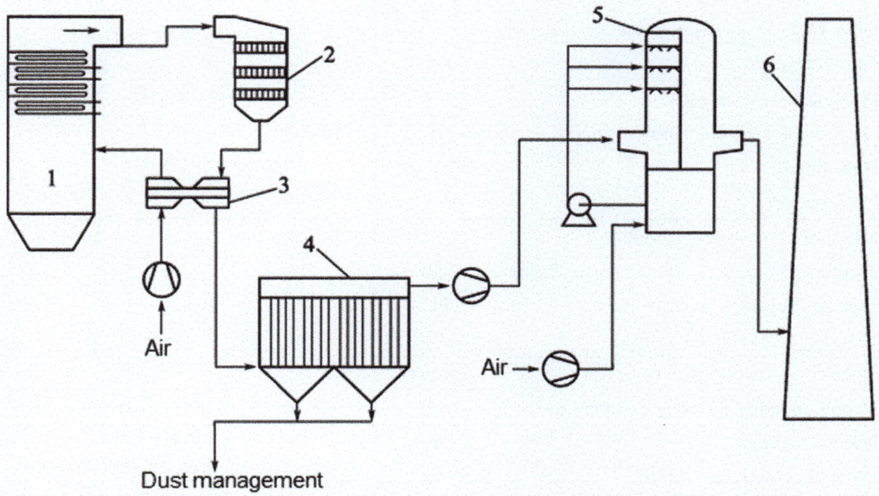

Fig. 11.9 Schematic diagram of the process flow of the coal-fired power plant boiler system. 1—Boiler; 2—SCR reactor; 3—Air preheater; 4—Dust collector; 5—Desulfurization tower; 6—Chimney

Fig. 11.10 Detection results of the size distribution of coal-fired flue gas

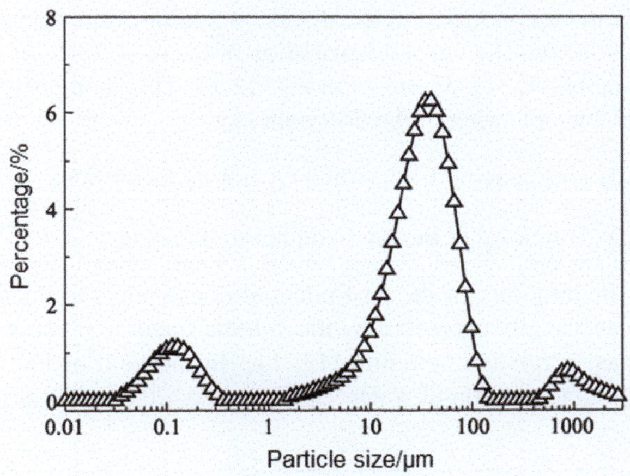

(2) Process design

The dust collector used in this process is a membrane dust collector, designed to operate for more than 3 years. The filtration wind speed is set between 0.7 and 1.0 m/min. The dust removal method employs an electromagnetic pulse valve for back-blowing, which includes both resistance control and time control modes. The start value for resistance control is set to 800 Pa, while the set value for time control is 30 min. The back-blowing pressure is 0.45 MPa and the project process parameters are shown in Table 11.3.

Table 11.3 Process parameters

Parameter	Value
Flue gas temperature/°C	130
Oxygen content/%	10
Humidity/%	11
Lifespan/year	> 3.0
Membrane product material	Modified PTFE composite membrane material
Membrane product surface Density/(g/m^2)	550–845
Flue gas volume/($\times 10^4$ m^3/h)	30
Filtering wind speed/(m/min)	0.7–1.0
Operating pressure difference/Pa	< 800
Membrane product usage temperature/°C	250
Injection mode	Offline
Back blowing interval/min	30

This system has a high concentration of flue gas dust. During the design of the dust collector, considerations were made for the uniform distribution of dust flue gas and the effective removal of dust. The flue was positioned in the middle, which is more conducive to the diversion and uniform distribution of flue gas in the dust collector. The weight distribution of the dust collector is reasonable, and it operates stably[31, 32], as shown in Fig. 11.11. This thoughtful design ensures efficient and stable operation of the dust collector.

(3) Influencing Factors and Operating Conditions

① Humidity of the dust-containing flue gas.

The humidity of the dust-containing gas refers to the amount of water vapor in the gas, usually expressed by the volume fraction of water vapor in the dust-containing gas or relative humidity [13, 14]. In the field of flue gas dust removal, when the volume fraction of water vapor is greater than 10% or the relative humidity exceeds

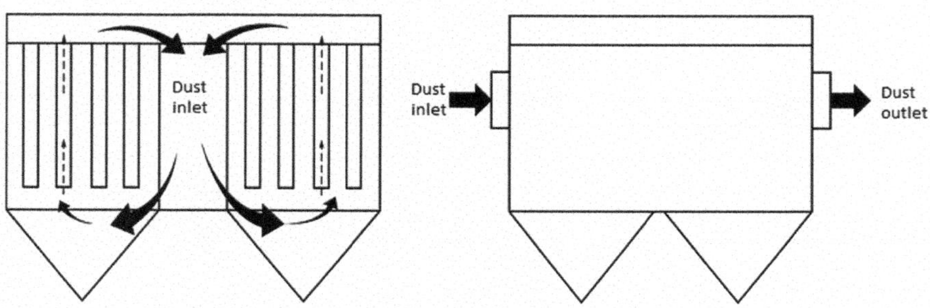

Fig. 11.11 Membrane-based dust collector

80%, it is referred to as wet dust-containing gas. Special attention should be paid to the selection of filter products and system design for wet dust-containing gas. This is because when water vapor and chemical gases coexist, they can affect the temperature resistance of the filter product and accelerate the hydrolysis of the filter product. At the same time, wet dust-containing gas can cause the dust on the surface of the filter product to become wet and sticky, especially for hygroscopic and deliquescent dust, which can even cause bag clogging.

During the operation of the dust collector in this project, the moisture content in the coal-fired flue gas remains above 10% for a long time. During the process operation, some of the moisture will condense on the surface of the filter material and combine with the dust, causing the performance of the filter material to decline or even be scrapped. The impact is particularly severe during fluctuations in working conditions and during start-up and shutdown. Traditional filter materials are mostly hydrophilic, which can easily cause dust to stick together, which is very detrimental to the operation of the dust collector. PTFE modified membrane products have strong hydrophobic characteristics, large operational flexibility, and strong adaptability to working conditions, and they will not cause bag clogging.

② The impact of oil pollution.

When the boiler of a coal-fired power plant is ignited, the common ignition methods include heavy oil (residual oil) gun ignition, light oil (diesel) gun ignition, and micro oil (less oil) ignition. Considering that the unburned oil pollution during the ignition process will cause serious damage to the filtering performance of the dust collector bag, which is irreversible and fatal, manufacturers generally avoid direct filtering of ignition flue gas with new bags, and the most common choice is to take a bypass [4]. With the increase of environmental protection requirements, new dust collectors are no longer allowed to design bypasses, which puts higher requirements on the performance of dust collector filter materials.

Modified PTFE membrane materials not only have strong hydrophobicity, but also have strong oleophobicity, which can resist organic pollution. The oil smoke produced by the ignition of the membrane dust collector in the furnace is directly filtered by the membrane product. In order to test the endurance of the membrane product, the ignition time was extended by 4 h, and then the performance of the membrane product was examined. The results showed that the membrane product still maintained good working performance.

Figures 11.12 are SEM photos of membrane products and ordinary filter materials after long-term operation of the coal-fired power plant. It can be seen that the dust does not enter the membrane, but is retained on the membrane surface, indicating that the membrane product has high resistance to pollution and will not be contaminated by dust.

③ Operational flexibility

In mid-July, the dust collector was debugged and stabilized. In November, the load of the No. 3 boiler was increased to 300 t/h, at which point the dust collector pressure

Fig. 11.12 SEM images of the membrane and ordinary filter materials after long-term operation. **a–c** membrane products, where **a** is the membrane surface diagram, **b** is the magnified diagram of the membrane surface, and **c** is the back diagram of the membrane product; **d–f** are ordinary products, where **d** is the product surface diagram, **e** is the magnified diagram of the product surface, and **f** is the back diagram of the product

difference was about 1200 Pa. However, when the boiler load was reduced to approximately 220 t/h, the dust collector pressure difference quickly dropped to below 600 Pa (Fig. 11.13). The decrease in the operating pressure difference of the dust collector with the boiler load indicates that the membrane product can effectively resist load impacts, thus demonstrating high operational flexibility.

④ Filter efficiency

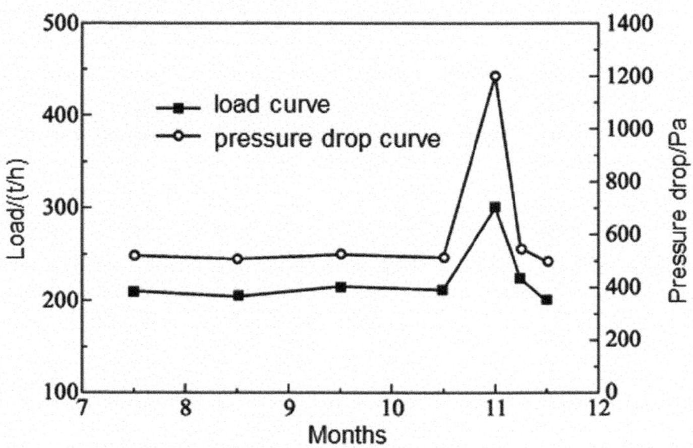

Fig. 11.13 Dust collector operation data using membrane material

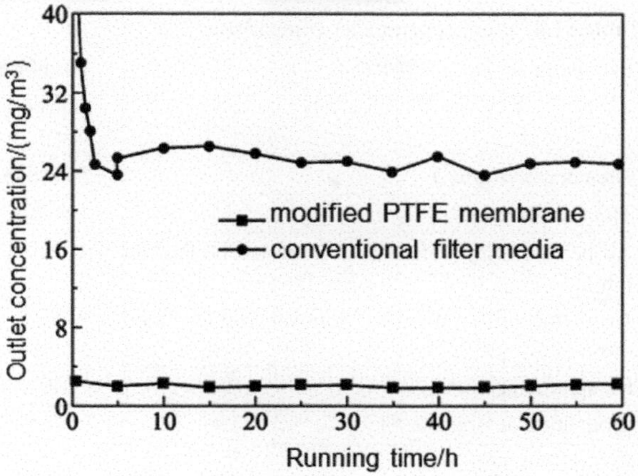

Fig. 11.14 Changes in dust concentration of outlet flue gas with operation time

Figure 11.14 presents the test results of the dust emission concentration at the dust collector outlet when using both modified PTFE composite membrane material and traditional filter material. It can be observed that the initial dust emission concentration is high when using traditional filter material, but after 5 h, it gradually stabilizes, maintaining around 24 mg/m^3. In contrast, when filtering with modified PTFE composite membrane material, the dust emission concentration essentially stabilizes at 2 mg/m^3 after the operation starts. This outcome is attributed to the differences in the properties, structure, and filtering mechanism of the two materials [10]. The modified PTFE membrane material, which is more suitable for flue gas filtration under harsh conditions, offers high dust filtration efficiency, resistance to acids and alkalis, high back-blowing efficiency, long service life, and good economic benefits [6, 12].

⑤ Operation stability

Table 11.4 compares the performance parameters of the membrane product and the ordinary dust collector filter material. It can be observed that after the membrane product undergoes dusting, all its performance indicators can be restored, demonstrating that the product performance is well-maintained and stable. In contrast, after the ordinary filter material undergoes dusting, its air permeability and elongation at break are already lower than the Chinese national standard. This indicates that the product has been severely polluted and damaged, rendering it unsuitable for further use.

⑥ Energy saving and consumption reduction effects and economic evaluation. Membrane products have a significant advantage in energy saving and consumption reduction compared to ordinary filter materials. Figure 11.15 illustrates the relationship between the pressure difference of the dust collector operation and the boiler load. In this figure, ‘A’ represents the operation data of the dust collector using ordinary filter materials, while ‘B’ represents the operation data of the dust collector using organic composite membrane products. As can be observed

Table 11.4 Product parameter comparison

Parameter	Membrane product		Ordinary filter bag	
	Before use	After use	Before use	After use
Area density/(g/m^2)	555–845	576.00	500–600	735.00
Thickness/(mm)	1.8–2.0	2.24	1.7	1.98
Air permeability (before dust removal, 200 Pa)/(m/min)	–	1.32	–	1.12
Air permeability (after dust removal, 200 Pa)/(m/min)	2–4	3.01	13	1.46
Elongation at Break *T*/%	≥ 20	28.31	30	19.41
Elongation at Break *W*/%	≥ 25	27.15	40	18.78
Tensile Strength *T*/N	≥ 1500	1676	≥ 1000	929
Tensile Strength *W*/N	≥ 1800	2505	≥ 1200	1262

from the figure, under essentially the same load, the operating resistance of the membrane product is about 400 Pa lower than that of the ordinary filter material. Figure 11.16 depicts the relationship between the induced draft fan current and the boiler load. As can be seen from the figure, the induced draft fan current exhibits a significant reduction under similar boiler loads.

Through data collection and analysis, the emission concentration of dust particles is reduced from the original 24–2 mg/m^3 after using the membrane dust collector. Calculated based on 8000 h of operation per year and a flue gas volume of 300,000

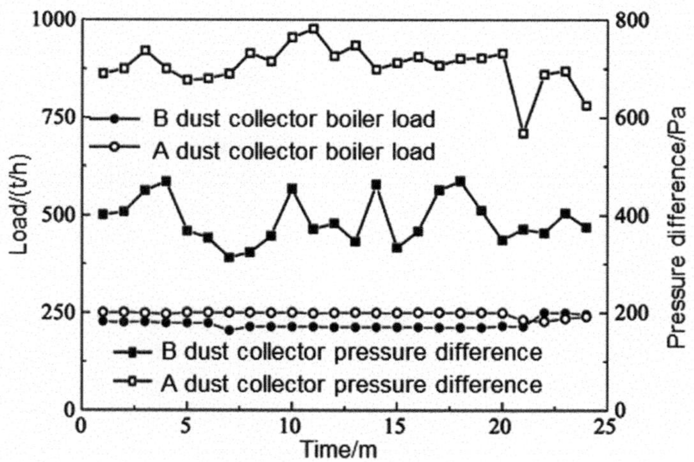

Fig. 11.15 Relationship diagram between loading and pressure. A—Using ordinary filter material; B—Using organic composite membrane

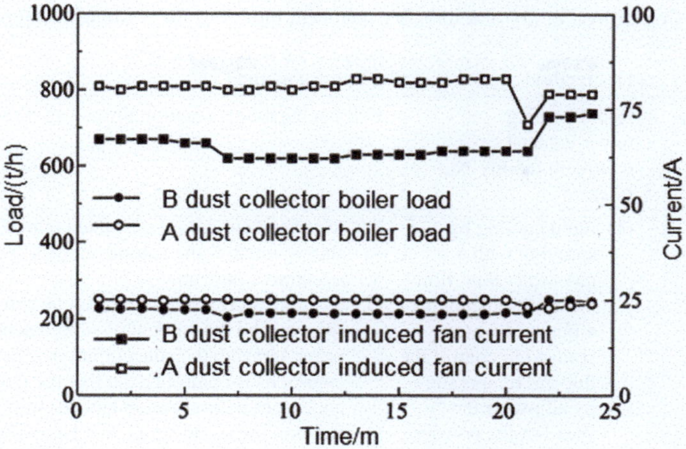

Fig. 11.16 Relationship diagram between loading and induced draft fan current. A—Using ordinary filter material; B—Using organic composite membrane

m^3/h, it can be concluded that more than 50 t of dust particles can be reduced each year, and the annual electricity saving is 4.98×10^5 kW h.

11.1.2 Membrane Technology Applied to Ultra-Low Emissions of Waste Flue Gas

With the rapid development of the global economy, the volume of various types of waste is increasing. If not handled properly, these wastes can cause significant harm. Therefore, waste treatment is receiving increasing attention. Waste treatment includes the treatment of household waste, hazardous waste, and medical waste, among others. Currently, waste incineration is one of the most effective methods for achieving harmless, reduced, and resource-utilized waste [13]. In developed countries worldwide (such as Japan and Switzerland), waste disposed of by incineration accounts for more than 70% of the total waste disposed of. Although waste incineration has clear advantages, it also has its limitations. For instance, the flue gas produced by incineration needs to be treated. However, the composition of the flue gas is complex, the parameters frequently change, there are many types of pollutants, and particulate matter is challenging to remove effectively. These are all problems faced by waste incineration technology [14]. At present, bag filters have become the mainstream technology for waste incineration flue gas cleaning. The flue gas cleaning process, with bag filters as the main component, includes a dry treatment process and a semi-dry treatment process. Their respective advantages, disadvantages, and comprehensive performance are shown in Table 11.5 [15].

Table 11.5 Comparison of dry and semi-dry treatment processes for waste incineration flue gas cleaning

Parameter	Dry method	Semi-dry method
Treatment process	Rapid cooling tower-exhaust gas treatment facility-bag filter	Pulping degassing reaction tower-bag filter
Principle	(1) rapid cooling by spraying with a gas–water dual-flow nozzle, adjusting the amount of water sprayed to control the flue gas temperature (2) by adjusting the speed of the powder machine, the quicklime powder is sprayed into the rapid cooling tower and the flue, and a certain amount of reaction aid is added as needed (3) the flue gas and the adsorbent undergo neutralization and adsorption in the flue and the dust layer of the dust collector (4) the adsorbent and dust are filtered and removed by the dust collector to achieve purification	(1) water and quicklime are mixed to make lime slurry, which is then atomized by a special structure atomizer (2) by controlling the amount of lime slurry sprayed and the amount of supplementary water sprayed, the concentration of acid gas emissions is controlled, so that the flue gas undergoes neutralization and adsorption in the reaction tower, flue, and dust layer of the dust collector (3) the large particle reaction products generated fall to the bottom of the tower and are discharged (4) finally, the dust collector filter layer is used for dust removal and purification
Removal efficiency/%	HCl: ≥ 90 SO_2: ≥ 80 Dust: ≥ 99.5	HCl: 90–95 SO_2: 85–90 Dust: ≥ 99.5
Emission concentration /(mg/m^3)	HCl: ≤ 60 SO_2: ≤ 200 NO_x: ≤ 300 Dust: ≤ 50 Dioxins (daily average): ≤ 0.1 (ng-TEG/m^3) Mercury (daily average): ≤ 0.2 Cadmium (daily average): ≤ 0.1 Lead (daily average): $\leq$ 0.6	HCl: ≤ 50 SO_2: ≤ 150 NO_x: ≤ 300 Dust: ≤ 50 Dioxins (daily average): ≤ 0.1 (ng-TEG/m^3) Mercury (daily average): ≤ 0.2 Cadmium (daily average): ≤ 0.1 Lead (daily average): ≤ 0.6

(continued)

Table 11.5 (continued)

Parameter	Dry method	Semi-dry method
Operation and control	(1) the preparation and transportation system of hydrated lime powder is simple. There is no equipment corrosion and pipeline scaling and blockage (2) the system and dust collector operate stably and reliably (3) the inlet flue gas temperature and acid gas emission concentration of the dust collector are two independent control systems which do not affect each other	(1) the preparation and transportation system of lime slurry is complex, the adjustment requirements are high, the nozzle is easy to wear, and the pipeline is easy to block (2) the moisture content in the lime slurry significantly impacts the flue gas temperature. When the concentration of acid gas components is high, it can cause the flue gas temperature to drop excessively, leading to the formation of condensation water. This can ultimately result in a 'bag sticking' phenomenon, which disrupts the normal operation of both the system and the dust collector
Maintenance	(1) the entire system is in a dry state environment, requiring fewer power equipment (2) the system has fewer equipment failure factors and a lower damage rate (3) management and maintenance are convenient, and maintenance is simple	(1) the system is in a semi-dry state, requiring more power equipment (2) the equipment failure rate is higher, and equipment parts are easily damaged (3) the maintenance management workload is large, and the requirements are high
Operating cost	(1) slightly lower acid removal rate, consumes a certain amount of compressed air (2) low power consumption (3) lower cost for spare parts and maintenance	(1) slightly higher acid removal rate, but due to an additional liquid production and transportation process, the amount of acid remover used is not saved (2) higher power and water consumption (3) higher cost for spare parts and maintenance
Land occupation	Less	More

Table 11.6 Characteristics of waste incineration fly ash

Parameter	Value
Particle size/μm	20–30
Bulk density/(g/cm^3)	0.3–0.5
Dust concentration/(g/cm^3)	1.5–25
Moisture content/%	30–60

The dust generated during the waste incineration process is commonly referred to as fly ash, with its characteristics outlined in Table 11.6. As indicated in the table, the flue gas has a high moisture content and the fly ash is lightweight, which makes it challenging to manage. Experience has demonstrated that bag dust collectors are the most practical choice. The filter material utilized in the dust collector is an organic composite membrane material, known for its robust anti-pollution capability and superior flue gas cleaning performance [12, 16].

The waste incineration plant you mentioned employs a dry process for flue gas cleaning. As depicted in Fig. 11.17, the flue gas first enters the dust collector after being cooled by the quenching tower. Here, it reacts with the injected adsorbent. The dust collector retains the adsorbent, and the purified exhaust gas is then discharged from the chimney.

The dust collector is designed to handle a flue gas volume of 4 × 10^4 m^3/h, operating at a temperature of 220 °C, with a filter material of PTFE composite membrane material. The specific system parameters can be found in Table 11.7.

Key points of system design:

(1) The flue gas generated by waste incineration can reach temperatures above 600 °C. Organic filter materials cannot withstand this temperature, so a quench tower is set up before the dust collector. The quench tower uses spray cooling to strictly control the flue gas temperature entering the dust collector to below 170 °C, thereby preventing dioxin polymerization.

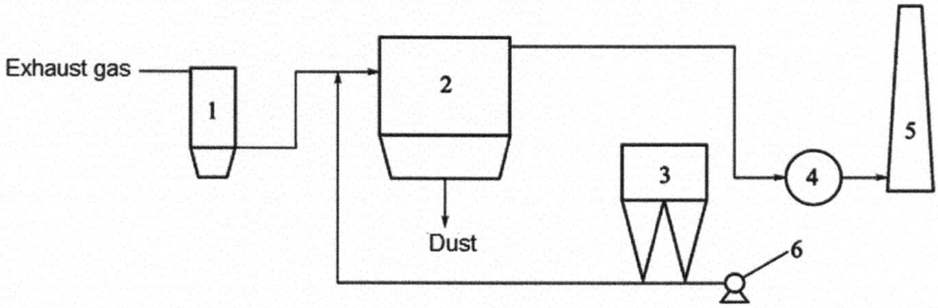

Fig. 11.17 Dry treatment process flow for flue gas cleaning from garbage incineration. 1—Quench tower; 2—Dust collector; 3—Adsorbent storage tank; 4—Induced draft fan; 5—Chimney; 6—Feeding pump

Table 11.7 Design parameters of the dry treatment system for flue gas cleaning in waste incineration furnaces

Parameter	Value and equipment selection	Remarks
Boiler exhaust volume/(m^3/h)	4×10^4	Residual heat boiler at the tail
Flue gas temperature/°C	220	–
Quench tower spray volume/(m^3/h)	1032	–
Outlet flue gas temperature/°C	170	–
Lime injection volume/(kg/h)	104	–
Reaction aid addition volume/(kg/h)	20	Granular chemical
Dust collector type	Composite membrane pulse bag type dust collector	–
Filter material	PTFE composite membrane	–
Equipment resistance/Pa	< 1000	–

(2) The spray equipment is controlled by a water–gas two-phase atomizing nozzle and temperature and humidity control devices. The reaction tower is equipped with insulation and heating devices to control the temperature and humidity of the flue gas at the dust collector inlet.
(3) The system uses quicklime powder as an adsorbent to remove acidic gases, dioxins, and heavy metals. Reaction aids are added to the quicklime powder to improve the removal efficiency of target pollutants. The combustion of 1 t of garbage produces flue gas that requires about 7–10 kg of quicklime powder, and the consumption of reaction aids is 1.5–5 kg. The system uses quicklime and reaction aids instead of activated carbon, achieving energy saving, consumption reduction, and waste reduction. Quicklime forms an adsorption layer on the surface of the organic composite membrane, which not only fully absorbs the pollutants in the purified flue gas but also improves the filtration efficiency [17, 18].
(4) The dust collector uses an electromagnetic pulse valve to control compressed air backblowing, resulting in high removal efficiency and stable system operation.
(5) Using an organic composite membrane, the flue gas cleaning is more thorough, the energy saving and consumption reduction are significant, and the adaptability is strong [19].

11.1.3 Application of Membrane Technology in Ultra-Low Emission of Cement Kiln

The production process of cement kiln is complicated, and there are many pollution points that emit dust. During the production process of cement products, there are nearly 40 pollution sources from raw material preparation to finished product delivery. Table 11.8 shows the equipment that will produce dust pollution during the cement production process and its flue gas characteristics [15].

The new dry process cement production usually has 4–5 stages of cyclone devices. The dry raw material is usually added in the passage between the first and second stages, and then it is carried into the cyclone device by the high-speed airflow, where it contacts the high-temperature flue gas from the kiln tail for heat transfer, and the temperature is raised. It is then successively heated by other cyclone devices, and

Table 11.8 Flue gas characteristics of some equipment in cement production process

Equipment		Dust concentration/ (g/m^3)	Gas temperature/ °C	Moisture content (volume fraction)/ %	Dew point/ °C	Dust particle size/ %	
						< 20 μm	< 88 μm
Suspended preheater kiln		30–80	350–400	6–8	35–40	95	100
External decomposition kiln		30–80	300–400	6–8	35–40	95	100
Clinker cooler		2–20	100–250	–	–	10	30
Rotary Dryer	Clay	40–150	70–130	20–25	50–65	25	45
	Slag	10–70				–	–
	Coal	10–50				60	–
	Gravity discharge drying mill	50–150					
Raw material mill	Air swept mill	300–500	60–95	10	45	50	95
	Vertical mill	300–800					
Separator		800–1200	70–100	–	–	–	–
Cement mill	Mechanical ventilation mill	20–120	90–120	–	–	50	95
Coal mill	Ball mill (air swept)	250–500	60–90	8–15	40–50	–	–
	Vertical mill						
Crusher	Jaw type	10–15					
	Hammer type	30–120	–	–	–	–	–
	Impact type	40–100					
Packaging machine		20–30	–	–	–	–	–

when the temperature rises to 700 °C, it enters the kiln to be calcined into clinker. The flue gas temperature at the kiln tail is about 350–400 °C, and after heat exchange with the grinding flue gas and cooling in the humidification tower, the temperature is about 120 °C, and then it is discharged through the chimney after dust removal by the dust collector. Figure 11.18 is a flow chart of the pre-decomposition rotary kiln and dust removal process.

The properties of the flue gas at the tail of the dry kiln are shown in Table 11.9.

A certain cement production line returns the kiln tail flue gas to the front end to preheat the raw material gas, making use of the heat [20]. The dust collector uses an organic composite membrane pulse dust collector, and its flue dust parameters and dust collector design operation parameters are shown in Tables 11.10, 11.11 and 11.12.

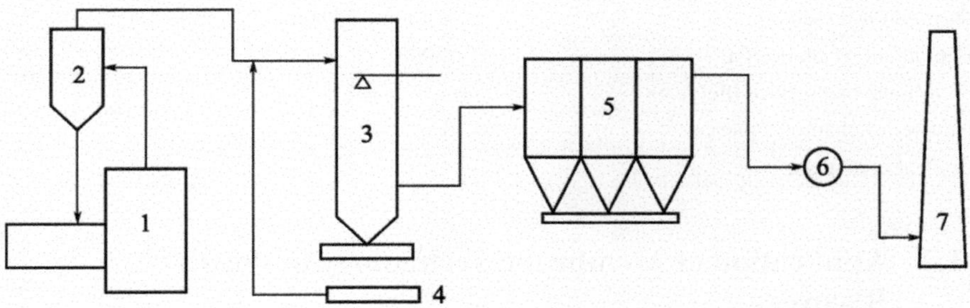

Fig. 11.18 Process flow of pre decomposition rotary kiln tail and raw material grinding. 1—Rotary kiln; 2—Preheater; 3—Humidification tower; 4—Raw material mill; 5—Dust collector; 6—Fan; 7—Chimney

Table 11.9 Properties of flue gas at the tail of the dry kiln

	Flue gas volume/ (m^3/kg clinker)	Temperature/ °C	Humidity/ %	Dew point/ °C	Dust concentration/ (g/m^3)	Chemical composition/%		
						CO_2	O_2	N_2
General	1.6–2.5	350–400	4–6	40	40–130	–	–	–
Actual measurement	2.5–3.0	300–350	7	40	60–80	12	12	76

Table 11.10 Cement kiln flue dust chemical composition test data

	SiO_2	Al_2O_3	Fe_2O_3	CaO	MgO	Na_2O	K_2O
Mass fraction/%	13.2	4.7	2.3	42	1.1	0.2	0.5

Table 11.11 Cement kiln flue dust particle size distribution

Particle size/μm	< 15	15–20	20–30	30–40	40–88	> 88
Ratio/%	94	2	2	1	1	0

Table 11.12 Specifications and parameters of membrane dust removers

Parameter	Value
Flue gas treatment volume/(m^3/h)	23,000
Temperature/°C	$\leq$ 240
Filter bag material	Organic Composite Membrane
Membrane bag specifications/mm	ϕ 200 × 8000
Number of membrane bags/pieces	960
Dust removal method	Dust removal method
Equipment resistance/Pa	< 2000
Membrane bag service life/years	6
Number of chambers/pieces	10
Control system/set	1

11.2 Application of Membrane Technology in Powder Recovery

11.2.1 Application of Membrane Technology in Molecular Sieve Catalyst Recovery

Molecular sieve catalysts are a type of porous material with a specific surface area of up to several hundred square meters per gram. They exhibit high activity, strong adsorption capacity, and good pressure resistance, making them commonly used as adsorbents, catalysts, and catalyst carriers in petroleum refining and petrochemical industries. However, molecular sieve catalysts are prone to loss during production, leading to not only resource waste but also serious environmental pollution. Therefore, their production process requires strict separation and purification to improve the recovery rate of catalysts. Compared to traditional separation methods such as sedimentation, plate and frame filtration, and centrifugal separation, membrane technology offers high separation efficiency, low energy consumption, minimal pollution, and a high degree of automation. As a result, it is receiving increasing attention. The following is an introduction based on actual engineering applications.

11.2.1.1 Engineering Overview

In the context of an actual engineering application, a certain catalyst company produces molecular sieve catalysts. After the catalysts pass through the dryer, their concentration is 30–50 g/m^3. If the catalysts are not further recovered, it will result in a significant loss of molecular sieve catalysts and the exhaust gas emissions will not meet the standard, leading to air pollution. Traditional cyclone separators, ordinary bag dust collectors, or water washing absorption processes used for material recovery not only have limited material recovery capacity but also produce a large amount of new pollutants such as wastewater. Furthermore, catalysts with small particle sizes are more likely to block the pores of the ordinary bag, causing the filtration resistance to rise sharply. Frequent back blowing can cause a large number of ordinary bags to break, with a service life of less than 6 months, resulting in insignificant economic benefits. The process flow is depicted in Fig. 11.19.

The particle size distribution and surface morphology of the molecular sieve catalysts were characterized using a Malvern particle size analyzer and a scanning electron microscope. Additionally, the bulk density of the molecular sieve catalysts was tested. The results are as follows:

(1) Particle size distribution of molecular sieve catalysts

As shown in Fig. 11.20, the molecular sieve catalysts have an average particle size of 3 μm. Particles with a size of less than 1 μm are present, and agglomeration is observed.

(2) Surface morphology of molecular sieve catalysts

Figure 11.21 presents a scanning electron microscope image of the surface morphology of the molecular sieve catalysts. As depicted in the figure, the ultrafine powder of the molecular sieve catalysts agglomerates into nearly spherical particles. Most of the particle sizes range between 3 and 8 μm, which is consistent with the particle size analysis data. Both dispersed ultrafine dust and agglomerated large particle dust are present. The bulk density is small, with a measured value of 0.42 g/cm^3. This information provides valuable insights into the properties and behavior of the molecular sieve catalysts.

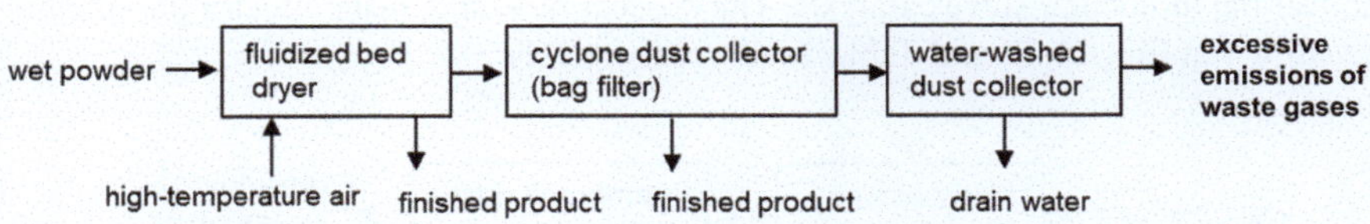

Fig. 11.19 Existing molecular sieve catalyst recovery process

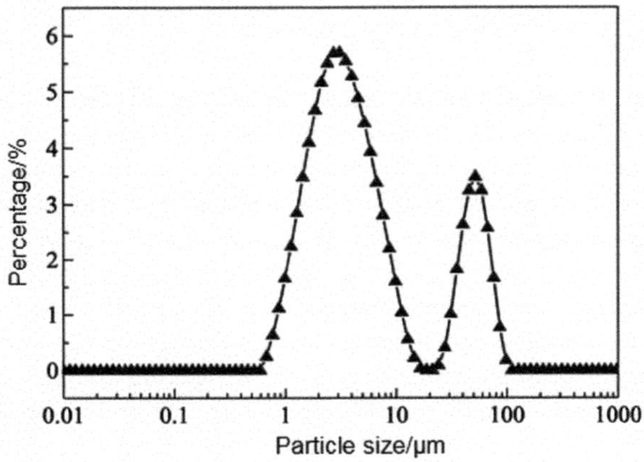

Fig. 11.20 Particle size distribution of the molecular sieve catalysts

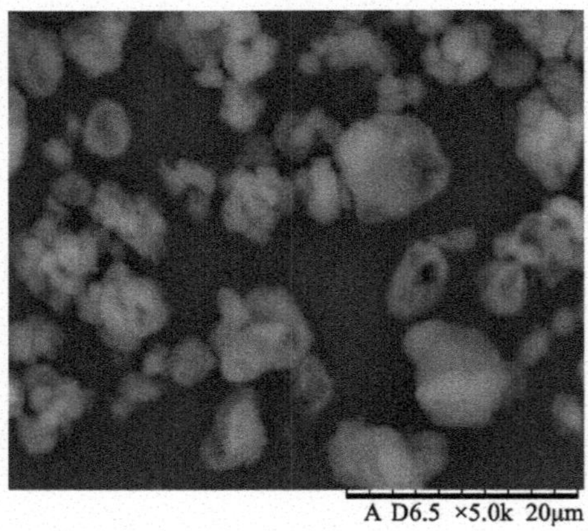

Fig. 11.21 SEM image of the molecular sieve catalysts

11.2.1.2 Scheme Design

In view of the characteristics of existing process, the manufacturer, after investigation and understanding, finally adopted an organic composite membrane dust collector to recover the molecular sieve catalysts. The process flow and dust collector parameters are shown in Fig. 11.22 and Table 11.13, respectively.

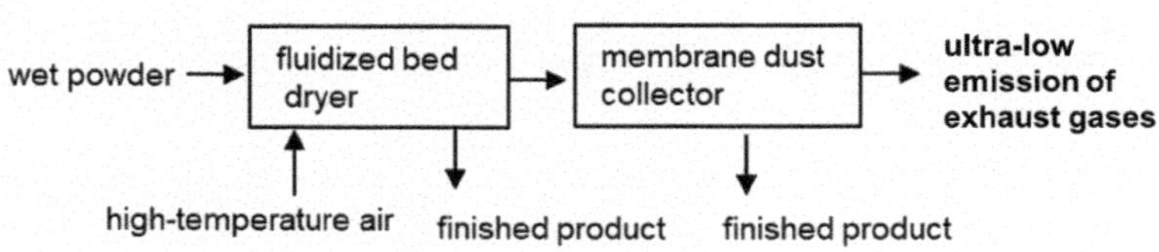

Fig. 11.22 Membrane method for molecular sieve catalyst recovery process

Table 11.13 Parameters of membrane dust collector for molecular sieve catalysts

No.	Parameter	Value
1	Dust collector material	304 stainless steel
2	Inlet temperature/°C	120–200
3	Inlet dust concentration/(g/m^3)	30–50
4	Gas handling volume/(m^3/h)	8000–12,000
5	Membrane bag material	Organic composite membrane
6	Membrane bag size/mm	130 × 3000
7	Number of membrane bags	144
8	Filter area/m^2	176
9	Filtration speed/(m/min)	0.76–0.95
10	Number of pulse valves	40
11	Cage material	Stainless steel
12	Cage length/mm	2850
13	Cage circumference/mm	Top 38 cm, middle 40 cm, bottom 38 cm
14	Number of cage ribs	8 longitudinal ribs, 13 reinforcement rings

11.2.1.3 Membrane Product Design

The performance of the dust collector is influenced by several factors, including the dust removal material, the dust collector equipment, the nature of the dust being treated, and the operating conditions [21]. With a design goal of achieving an emission concentration of less than 10 mg/m^3, the existing conditions of the dust collector and the properties of the recovered molecular sieve catalyst are taken into consideration. The gas cleaning membrane material is customized accordingly, with adjustments made to the pore size, porosity, and thickness of the membrane material to match the particle size, surface morphology, and packing density of the catalyst. This approach aims to achieve the best catalyst recovery effect. The performance parameters of the membrane material are detailed in Table 11.14.

11.2.1.4 Operating Results

(1) The emission concentration of molecular sieve catalyst powder is less than 10 mg/m^3;
(2) The service life of the membrane material is more than 24 months;
(3) The operating pressure drop of the dust collector is less than 1200 Pa.

Table 11.14 Performance parameters of membrane material

Parameter	Value
Average pore diameter/μm	3
Porosity	> 85%
Finished surface density/(g/m^2)	580
Thickness/mm	2
Air permeability (200 Pa)/(m/min)	3
Operating temperature/°C	160
Instantaneous temperature/°C	220
Radial strength/(N/50 mm)	≥ 1000
Weft strength/(N/50 mm)	≥ 1200

11.2.2 Application of Membrane Technology in Dye Recovery

Traditional dye product recovery processes have problems such as insufficient dye recovery, serious dye leakage, and non-compliant dust emission from exhaust gas, which are not conducive to environmental protection and affect production. The membrane dust collector is expected to become the most effective solution.

11.2.2.1 Project Overview

A certain dye company uses cyclone separation and spray for exhaust gas cleaning in the dye production process. This exhaust gas cleaning method can only remove some large dye particles, the dye collection is not sufficient, and the dye exhaust gas cools down after passing through the water washing tower. The tiny dye molecules in the gas will regenerate large gel particles, which is very detrimental to the operation of the equipment.

Samples of dye particles produced during the entire process were taken for testing, and the particle size distribution was tested. The results are shown in Fig. 11.23.

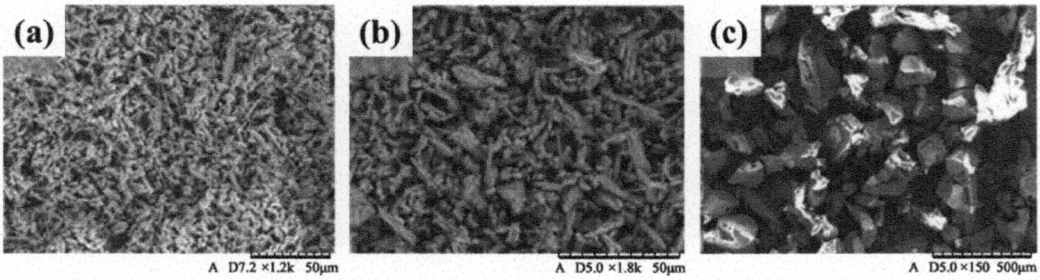

Fig. 11.23 SEM images of the dye particles. **a** Dye in the exhaust gas at the outlet of the spray tower; **b** dye collected by the cyclone separator; **c** dye in the exhaust gas at the outlet of the cyclone separator

From the test results of the above three samples, it can be seen that the dye particles are agglomerated, with a particle size ranging from 10 to 100 μm. The particle size of dyes A and C is concentrated at 10–20 μm, but they contain some dye powder with a smaller particle size (about 1–3 μm). This portion of the dye powder has a small particle size and strong penetration. The particle size of dye B is larger, and it contains almost no small particle size dye. The relative density of dyes A and C is smaller, which is the main reason for the difficulty in dye recovery.

11.2.2.2 Flue Gas Cleaning Requirements

The emission index of the purified flue gas (standard state, dry basis, reference oxygen content) is: dust concentration less than 10 mg/m^3.

11.2.2.3 Scheme Design

By comparison, the membrane method for dye recovery is ultimately adopted, effectively addressing issues such as high dust emission concentration, significant resistance during equipment operation, and powder bridging. The process of the membrane dust collector is depicted in Fig. 11.24. After the dye waste gas is processed by the cyclone dryer and cyclone separator, the waste gas is dried, and a portion of the large particle dye is recovered. The remaining ultrafine powder dye and waste gas pass through the membrane dust collector. Under the influence of the specific membrane material (or product), the dye powder is intercepted and recovered. The concentration of dye in the purified tail gas is less than 10 mg/m^3, and it is discharged through the chimney.

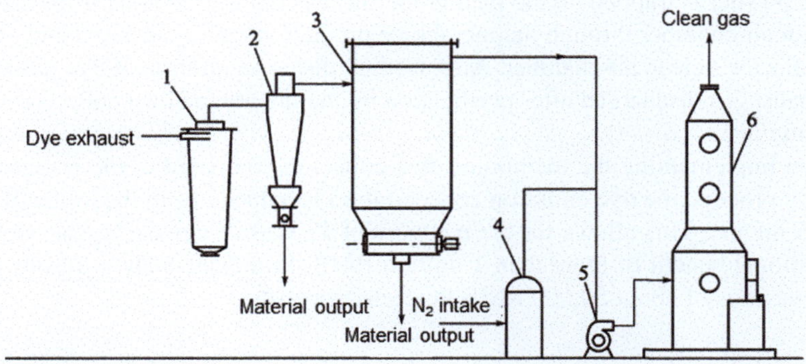

Fig. 11.24 Membrane based dye recovery process. 1—cyclone dryer; 2—cyclone separator; 3—membrane dust collector; 4—nitrogen buffer tank; 5—induced draft fan; 6—spray tower

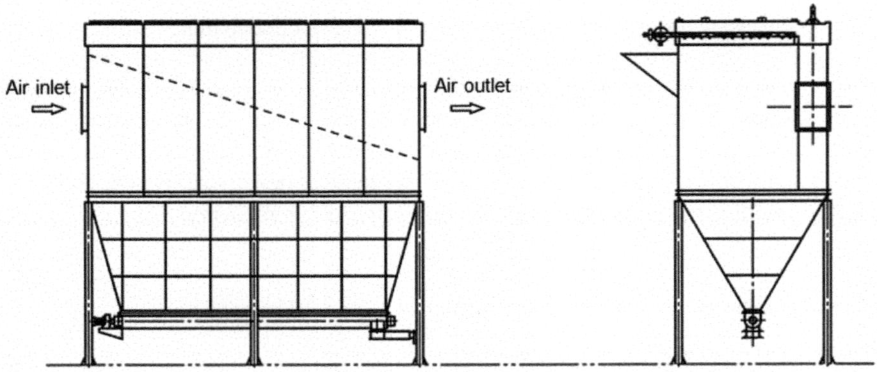

Fig. 11.25 Design of dust collector for dye recycling process

11.2.2.4 Dust Collector Design

The dust collector has been designed, and the specific layout is shown in Fig. 11.25:

The process specification design of the dust collector is shown in Table 11.15.

11.2.2.5 Operating Results and Economic Analysis

After the membrane dust collector has been operating stably for one year, the effectiveness of the membrane product compared to the ordinary filter material is illustrated in Fig. 11.26. As depicted in the figure, the ordinary filter material leads to significant pollution in the clean air chamber of the dust collector due to powder leakage. However, when the membrane product is utilized for exhaust gas cleaning, the dye powder is trapped on the membrane surface and is prevented from entering the clean air chamber through the membrane product, resulting in a relatively clean dust collector's clean air chamber. After testing, the concentration of dye powder in the exhaust gas discharged after purification by the membrane dust collector is less than 5 mg/m^3.

Upon implementing the membrane dust collector, compared to the original dye recovery process, the dye emission concentration is reduced from the initial 50 mg/m^3 to 5 mg/m^3. This allows for the recovery of 12 tons of dye each year, yielding an additional benefit of more than 1 million RMB. Simultaneously, it results in an annual savings of about 50,000 RMB in electricity costs.

Table 11.15 Design of process specifications for dye recovery dust collector

No.	Component name	Model specifications and technical parameters	Quantity	Unit	Remarks
1	Steel structure	Section steel q235	1	Set	Includes columns and air ducts
2	Steel structure	Stainless steel	1	Set	–
3	Spray system	Stainless steel spray pipe	1	Set	Air bag is carbon steel
4	Pulse valve	Submerged, explosion-proof	1	Batch	–
5	Offline valve	Stainless steel piston rod	1	Batch	With magnetic switch
6	Organic composite membrane	130 mm × 3050 mm	1	Batch	–
7	Cage frame	Carbon steel, organic silicone coating	1	Batch	–
8	Spiral machine	s400, L = 6 m	1	Unit	Stainless steel shell and shaft
9	Ash discharge valve	DN400	1	Unit	Stainless steel impeller, stainless steel shell
10	Sealing strip	–	1	–	–
11	Gas storage tank	1 m^3	1	Unit	–
12	Pneumatic system valve	–	1	Set	–
13	Differential pressure transmitter	–	1	Piece	–
14	Level meter	–	2	Pieces	–
15	Electrical control instrument	–	1	Unit	–
16	Thermistor	–	2	Pieces	Air inlet
17	Cable tray	–	1	Set	–
18	Transportation	–	1	Set	–
19	Installation and debugging	–	1	Set	–
20	Insulation	–	1	Set	Color steel tile + rock wool

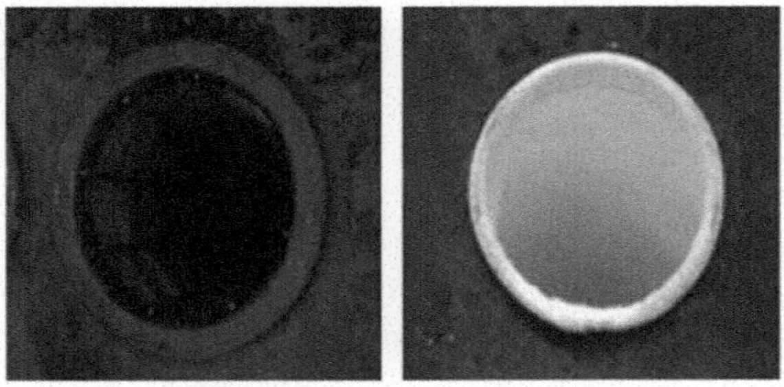

Fig. 11.26 Internal pollution situation diagram of dye recovery filter material

11.3 Application of Ceramic Membrane Technology in the Field of High Temperature Flue Gas Cleaning

With the advancement of the energy industry and energy-saving technology, more sophisticated requirements have been established for the development of high-temperature dust removal technology. The high-temperature dust removal technology currently under study involves the purification of high-temperature gases above several hundred degrees Celsius without the need for cooling, with attempts to recover and reuse the heat from clean high-temperature gases [22–25]. In recent years, numerous countries have dedicated their efforts towards the research and development of combined cycle power generation and coal gasification processes, which necessitate the resolution of high-efficiency dust removal technology under conditions of high temperature and high pressure. This section exemplifies the application of ceramic membrane high-temperature filters in coal-fired power plants.

11.3.1 Ceramic Membrane Technology Applied in Coal-Fired Power Plants

11.3.1.1 Project Overview

The flue gas process flow of a coal-fired power plant in Jiangsu is depicted in Fig. 11.27. The flue gas, post-combustion, traverses through the heater and economizer for cooling before entering the SCR denitration. The flue gas temperature reaching the SCR is about 350 °C, and the dust concentration is about 50 g/m^3. A large amount of dust enters the SCR denitration device with the flue gas, which will cover and block the pores of the catalyst, affecting the performance of the catalyst. Therefore, if a high-temperature filtering device can be installed before the SCR device to remove the dust, it will greatly improve the SCR performance.

The flue gas temperature before the SCR is usually higher than 350 °C. Under these conditions, ordinary organic materials cannot function properly. Ceramic membrane

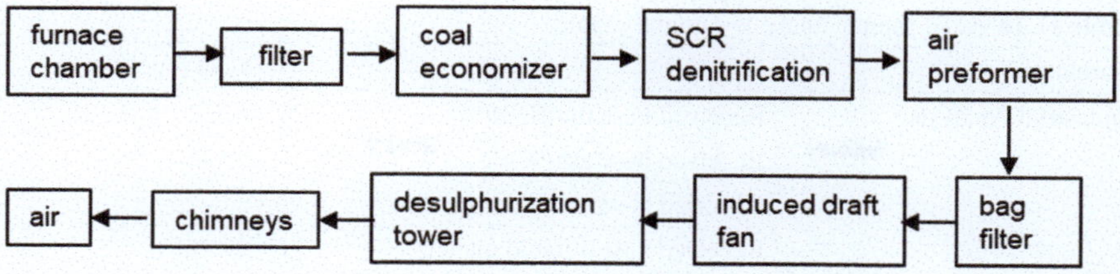

Fig. 11.27 Flue gas cleaning system of power plant

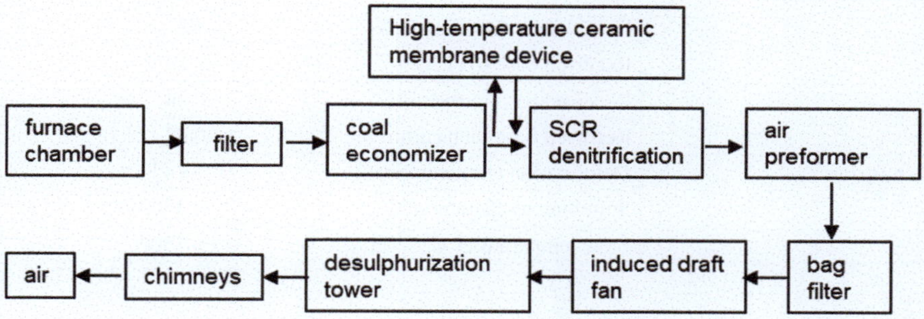

Fig. 11.28 Location of high temperature ceramic membrane equipment

materials, owing to their superior high-temperature resistance, corrosion resistance, and high filtration precision, emerge as the optimal choice. As illustrated in Fig. 11.28, a flue gas pipeline is extended adjacent to the denitration equipment to connect to the ceramic membrane equipment. The flue gas, after undergoing high-temperature filtration, enters the SCR equipment. This process reduces the dust pollution in the SCR and enhances the efficiency of the equipment.

11.3.1.2 Process Parameters

Before the design, a detailed analysis of the dust from boiler combustion was carried out, and the results are shown in Table 11.16. The coal type burned by the boiler is bituminous coal, which contains 0.32% sulfur and 0.84% nitrogen. After combustion, nitrogen oxides and sulfides will be produced in the flue gas.

The particle size of the flue gas dust was analyzed, and the results are depicted in Fig. 11.29. As illustrated in the figure, the dust particle size is primarily distributed between 10 and 100 μm, but there are also some fine particles smaller than 10 μm. Therefore, if the minimum pore size of the ceramic membrane is less than 10 μm, it can effectively ensure a high dust retention rate.

Figure 11.30 is a morphology diagram of the dust (SEM image). As can be seen from the figure, the fly ash is predominantly spherical, the coal powder combustion is relatively complete, and there are many small particle dusts in the powder.

Table 11.16 Coal sample analysis summary table

Parameter		Coal type
Industrial Analysis	Received base moisture/%	6.5
	Received base ash/%	29.45
	Air dried base moisture/%	2.0
	Received base volatile matter/%	26.46
Received base low calorific value/(MJ/kg)		20.3
Hardgrove grindability index		60
Elemental analysis	Received base carbon/%	51.2
	Received base hydrogen/%	3.5
	Received base oxygen/%	8.19
	Received base nitrogen/%	0.84
	Received base total sulfur/%	0.32
Ash fusion	Deformation temperature/°C	> 1450
	Softening temperature/°C	> 1500
	Flow temperature/°C	> 1500

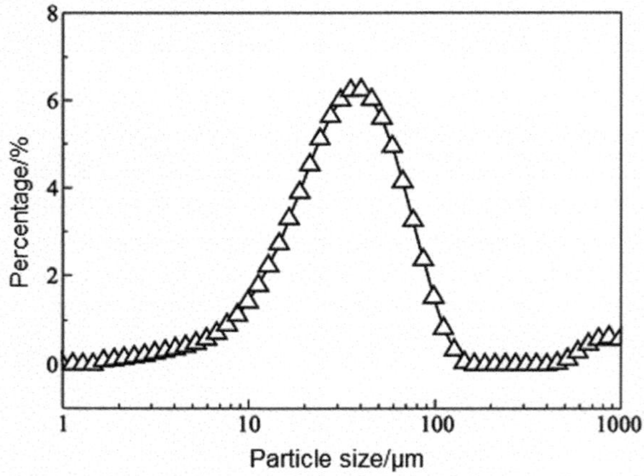

Fig. 11.29 Size distribution of the dust in power plant flue gas

11.3.1.3 Design of Ceramic Membrane Filter

(1) Design basis and parameters

The design basis for the high-temperature ceramic membrane filter of this project is shown in Table 11.17, and the design parameters are shown in Table 11.18.

(2) Tube plate design and tube bundle distribution

The ceramic membrane filter system comprises a traditional tube plate-based container, which separates the lower dusty side from the upper clean gas, while

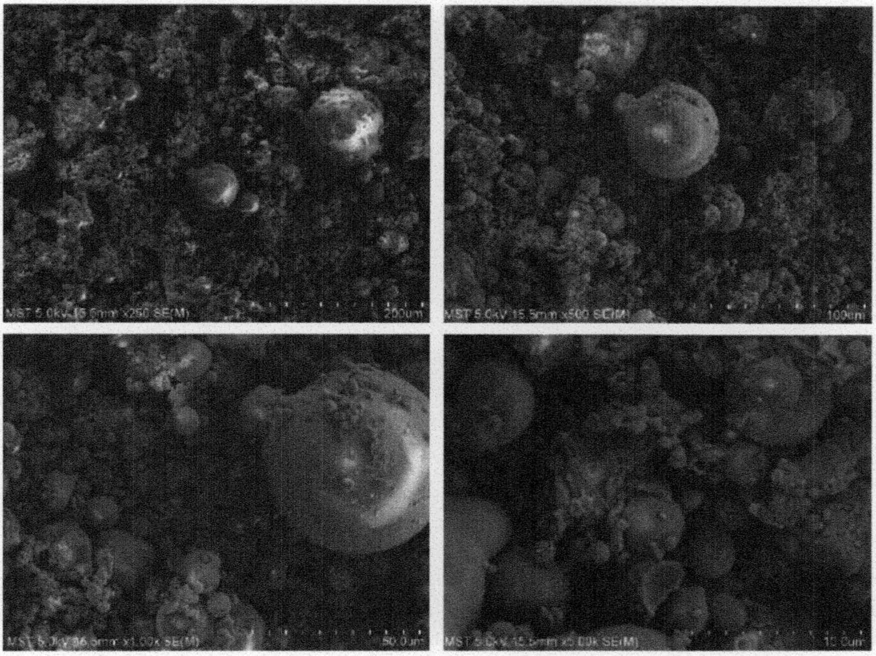

Fig. 11.30 Surficial morphology of the dust

Table 11.17 Design basis

Parameter	Value
Flue gas treatment volume/(m^3/h)	1000
Flue gas temperature/°C	305–460
Inlet dust concentration/(g/m^3)	< 50
Project name	Value
Inlet SO_2 concentration/(mg/m^3)	400–700
Inlet NO_x concentration/(mg/m^3)	350–500

the ceramic tubular membrane is suspended on the tube plate. The entire tube plate, also known as the main tube plate, is designed to accommodate the installation of 100 ceramic tubular membranes. The main tube plate employs a flange cover design, with a wall thickness of no less than 32 mm, and is fixed in the container using bolts and flanges.

In ceramic membrane filtration devices, ceramic tubular membranes are typically arranged in groups, with a specific requirement for the distance between each group. If the distance is too large, the volume of the filter increases, thereby decreasing its economic efficiency. If the distance is too small, it can easily lead to bridging phenomena. Bridging phenomena can severely impact the filtration performance of

Table 11.18 Design parameters

Parameter	Value
Flue gas treatment volume/(m^3/h)	1500
Ceramic membrane specification/mm	ϕ 60/40 × 1500
Design temperature/°C	450
Outlet concentration/(mg/m)	≤ 5
Spray gas	N_2
Spray gas source temperature/°C	250
Item	Parameter
Number of ceramic membranes	100
Number of pulse valves	10
Design pressure difference/kPa	5.0–8.0
Maximum fan pressure/kPa	10.0
Spray gas source pressure/MPa	0.3–0.6
Spray gas source heating	Electric heating

the ceramic membrane, potentially leading to the breakage of the ceramic tubular membrane, and ultimately causing the system to shut down. Therefore, selecting the appropriate distance between ceramic tubular membranes is crucial. The project designs the distance between the ceramic tubular membranes to be 60 mm. The arrangement of the tube bundle is depicted in Fig. 11.31a. Figure 11.31b illustrates the velocity and pressure distribution cloud diagram of the ceramic tubular membrane. As shown in the figure, the overall flow field distribution is relatively uniform, but there is a phenomenon of concentrated flow velocity in some areas.

(3) Operating conditions

The crux of the ceramic membrane filtration system lies in ensuring the system's long-term operational stability [24, 25]. Figure 11.32 illustrates the curve of system

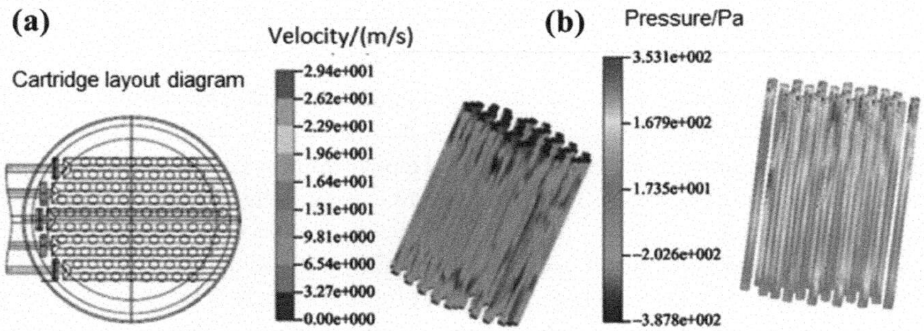

Fig. 11.31 Arrangement of tube bundles and cloud map of velocity and pressure distribution inside tubular membranes

pressure drop with respect to flow rate and time. As depicted in the figure, the system pressure drop exhibits minor fluctuations with the variation in flow rate. The system flow rate range is 1000 to 1500 m^3/h, and the system pressure drop is stable at 2300–2500 Pa. Figure 11.33 presents the diagram of the high-temperature filtration equipment of ceramic membrane.

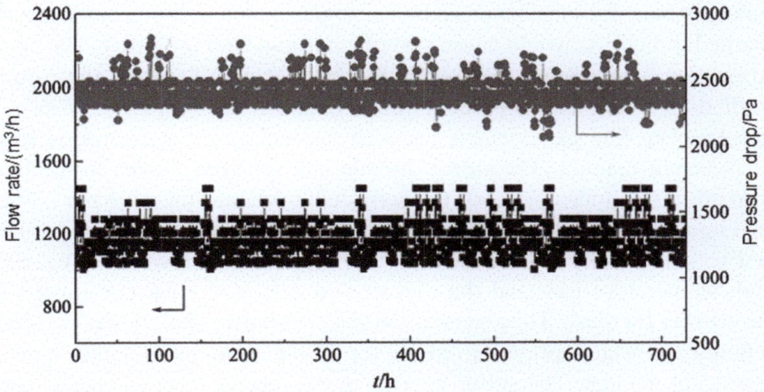

Fig. 11.32 Relationship curve between pressure drop, flow rate, and time

Fig. 11.33 Digital photo of the high-temperature filtration equipment of ceramic membrane

11.3.2 Application of Ceramic Membrane Technology in Coke Oven Flue Gas Treatment

11.3.2.1 Project Overview

The coke oven flue gas from a steel plant in Anhui Province, China, contains a significant amount of harmful substances such as nitrogen oxides, sulfides, and particulates. If discharged directly, it would cause severe environmental pollution. The activated coke adsorption-regeneration method for flue gas cleaning is one of the commonly used methods [26–28], and its process flow is depicted in Fig. 11.34.

The coke oven flue gas contacts the activated coke in the adsorption tower in a counter-current manner. The sulfur dioxide and nitrogen oxides in the flue gas are adsorbed in the micro-pores of the activated coke, and the nitrogen oxides are catalyzed into nitrogen by the catalyst in the micro-pores of the activated coke. The saturated activated coke is transferred to the regeneration tower for regeneration with high-temperature gas, and the regenerated activated coke is returned to the adsorption tower for reuse. The regeneration gas, post the regeneration tower, contacts the ammonia and water sprayed from the top of the ammonium sulfate tower in a counter-current manner. The sulfur dioxide in the regeneration gas reacts with the ammonia and water to form ammonium bisulfite, which is then aerated to produce ammonium sulfate product. The purified regeneration gas and the purified gas from the adsorption tower proceed to the next process together.

In this process, because the regeneration gas exiting the regeneration tower carries a certain amount of activated coke, the purity and color of the produced ammonium sulfate product are relatively poor. To enhance the purity of the ammonium sulfate product, it is necessary to purify the regeneration gas. However, the regeneration gas exhibits characteristics such as high temperature, strong corrosiveness, and small particle size, and conventional removal methods cannot meet these requirements.

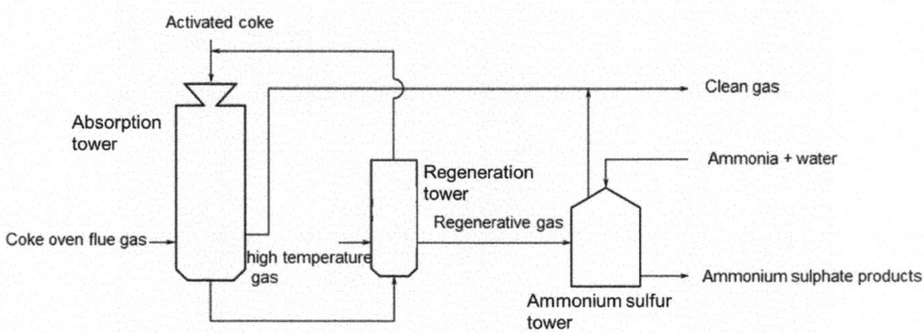

Fig. 11.34 Process flow of adsorption regeneration method for purifying coke oven flue gas

11.3.2.2 Process Design

The design of the regeneration gas cleaning process is depicted in Fig. 11.35. After the regeneration gas is purified through the ceramic membrane filter, the active coke within it is filtered by the ceramic membrane, and the purified regeneration gas enters the ammonium sulfate tower for the production of subsequent products. The flow rate of the regeneration gas is 140 m^3/h (normal operation gas volume, SO_2 concentration 5.5%), the flow rate during high concentration maintenance is 170 m^3/h, the flue gas temperature is < 420 °C, the flue gas pressure is ~ 6000 to ~ 3000 Pa, and the main components of the gas are shown in Table 11.19.

Through on-site testing of the dust particle size, it was discovered that the particle size of the active coke is mainly concentrated between 10 and 100 μm, accounting

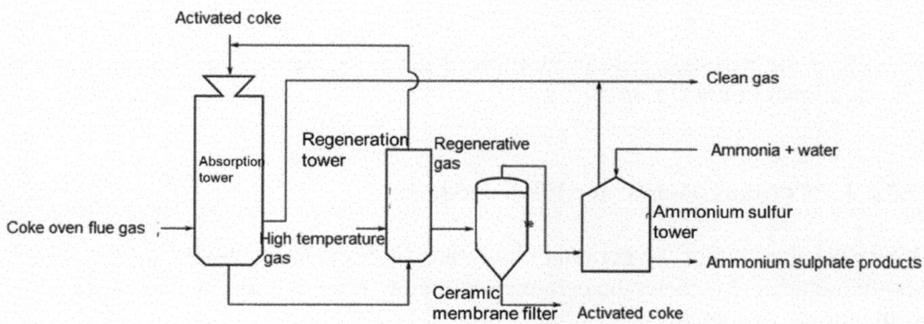

Fig. 11.35 Schematic diagram of the process flow for purifying regenerated gas using ceramic membrane filters

Table 11.19 Main components of flue gas

Component	Proportion
SO_2/%	5–9
NH_3/%	1–5
H_2O/%	16–30
N_2/%	74–55.8
CO_2/%	3–4.75
Component	Proportion
O_2/%	0
CO/%	0.05
SO_3/%	0.09
Dust/(g/m^3)	2

Note The dust primarily consists of active coke, and the design of the ceramic membrane filter needs to take into account the prevention of ammonium sulfate and sulfite crystallization to avoid blockage

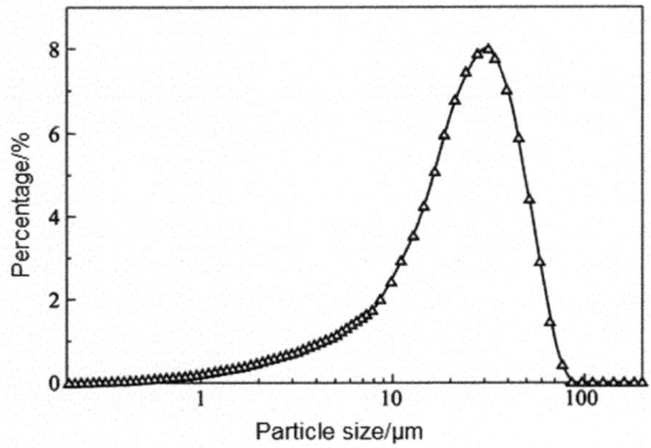

Fig. 11.36 Size distribution of the dust in regeneration tail gas

for 90% of the particles. The proportion of particles smaller than 10 μm is about 10%, as shown in Fig. 11.36.

11.3.2.3 Ceramic Membrane Filter Design

The design diagram of the ceramic membrane filter device is shown in Figs. 11.37, mainly includes two parts: the ceramic membrane filter cylinder and the back-blow purification component.

The design parameters of the ceramic membrane filter are shown in Table 11.20.

11.3.2.4 Features of the Filtration System

(1) The filter element employs a high-performance, high-temperature silicon carbide ceramic membrane. This ceramic membrane boasts advantages such as high filtration precision, low air resistance, and high temperature and pressure resistance, making it especially suitable for the removal of high-temperature dust (hot gas). The highest filtration precision can reach 0.1 μm, and the removal efficiency can reach 99.9%.
(2) The regeneration of the ceramic membrane utilizes online high-pressure pulse back-blowing technology, with nitrogen serving as the back-blowing medium, and employs electronic timing and pressure difference control. The ceramic membrane inside the filter is divided into several groups, with each group being back-blown in sequence, thereby completing the online back-blowing regeneration of the ceramic membrane.

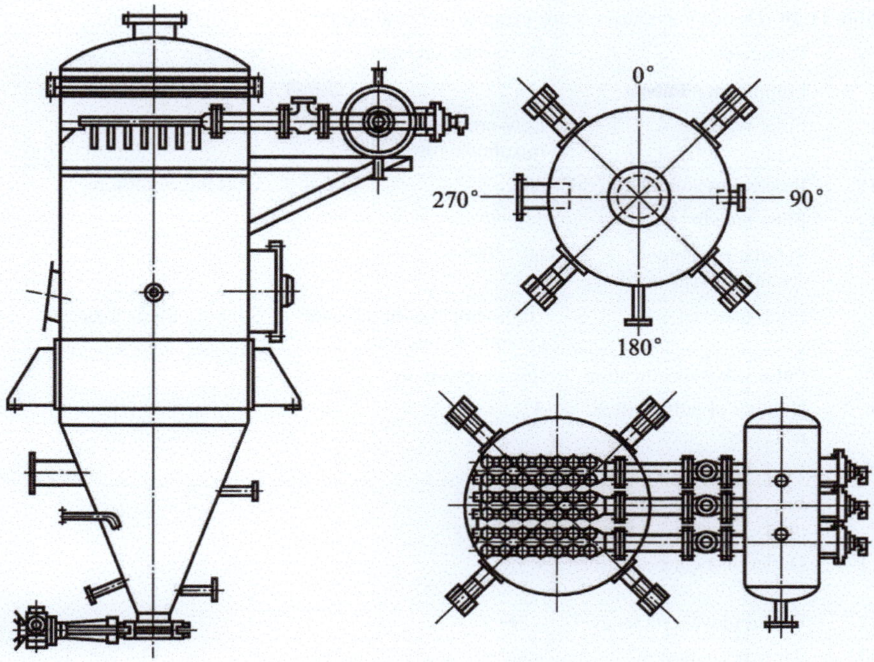

Fig. 11.37 Design diagram of the ceramic membrane filter device

11.3.2.5 Operating Results

The installation of the ceramic tubular membrane in the ceramic membrane device is shown in Fig. 11.38. Since the successful commissioning, the ceramic membrane device has been running stably, and all parameters such as filtration precision and operating resistance meet the design standards.

(1) Outlet emission concentration $\leq$ 10 mg/m^3.
(2) Dust removal efficiency $\leq$ 99.9%.
(3) Air leakage rate $\leq$ 2%.
(4) Operating resistance $\leq$ 1500 Pa.

Table 11.20 Design parameters of the ceramic membrane filter

No.	Parameter	Performance/Quantity	Remarks
1	Equipment number	–	Non-standard production
2	Filter type	Low-pressure pulse ceramic membrane filter	–
3	Processing volume/(m^3/h)	355	Working condition
4	Filter area/m^2	11	–
5	Tubular membrane specifications/(mm, pcs)	60/1500, 40	–
6	Filter material	High-temperature resistant ceramic membrane	Silicon carbide material
7	Pulse valve specification	Submerged type	–
8	Number of pulse valves/pcs	3	–
9	Back-blow gas source	Nitrogen	–
10	Inlet dust concentration/(g/m^3)	< 2.0	–
11	Outlet dust concentration/(mg/m^3)	< 10	–
12	Dust removal efficiency/%	≥ 99.9	–
13	Air leakage rate/%	≤ 2	–
14	Maximum operating resistance/pa	≤ 1500	–
15	Operating temperature/°c	≤ 450	–
16	Filter pressure resistance/pa	− 10,000	–

Fig. 11.38 Device diagram of ceramic membrane

11.3.3 Application of Ceramic Membrane Technology in the Treatment of Yellow Phosphorus Tail Gas

11.3.3.1 Project Overview

A specific chemical company in Guizhou Province, China, generates a substantial amount of phosphorus-containing furnace gas during the production of yellow phosphorus. The current furnace gas treatment process is depicted in Fig. 11.39. The furnace gas produced by the phosphorus furnace directly enters the scrubbing tower, where hot water is sprayed, and yellow phosphorus transitions from gas to liquid under hot water conditions. The dust in the furnace gas enters the water phase and then settles and stratifies. The crude phosphorus enters the phosphorus receiving trough, and after further settling, the phosphorus-containing water phase enters the refining pot. The refining pot is fed with hot water for additional refining, and the wastewater containing dust enters the sedimentation tank post-settling. This process results in a large amount of phosphorus sludge, causing environmental pollution, and some phosphorus products enter the phosphorus sludge, leading to resource wastage [29]. To conserve energy, enhance economic benefits, and protect the environment, ceramic membrane filtration technology can effectively improve the treatment process, based on the principle of technical feasibility and economic savings.

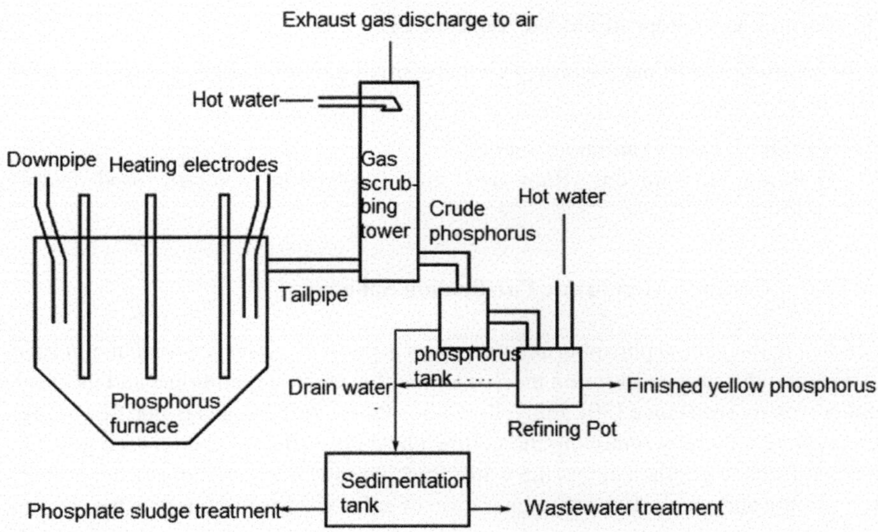

Fig. 11.39 The existing treatment process for yellow phosphorus tail gas

Table 11.21 Furnace gas composition

No.	Composition	Content (volume fraction)/%
1	CO	55–60
2	CO_2	2
3	O_2	0.3
4	N_2	1
5	H_2	11
6	HF	1000 mg/m^3
7	P_4	25–30
8	H_2S	680–2100 mg/m^3
9	PH_3	500–1100 mg/m^3

11.3.3.2 Design Basis

The gas parameters of the phosphorus furnace are as follows.

(1) Raw gas: Fully enclosed high-temperature gas from yellow phosphorus furnace.
(2) Furnace gas generation: 6000 m^3/h (maximum furnace gas generation).
(3) Dust concentration: 50–200 g/m^3.
(4) Outlet gas temperature of electric furnace: Normally 120–180 °C; maximum 340 °C.
(5) Furnace gas composition: See Table 11.21.

In addition, the furnace gas also contains small amounts of tar, phosphoric acid, and other components.

(6) Working mode: Continuous mode
(7) Working environment: Rain and snow protection, semi-enclosed factory building.

11.3.3.3 Ceramic Membrane Purification Scheme

The dust in the yellow phosphorus furnace gas is purified using a ceramic membrane. Leveraging the surface filtration mechanism of the ceramic membrane and its excellent corrosion resistance [30], the yellow phosphorus furnace gas is purified at high temperatures, thereby removing the yellow phosphorus dust in the furnace gas. The purified furnace gas then enters the water washing tower, which reduces the production of phosphorus mud and the generation of solid waste, ultimately achieving the goal of clean production.

The ceramic membrane yellow phosphorus furnace gas cleaning scheme is depicted in Fig. 11.40. The primary equipment is a ceramic membrane filter, and the auxiliary systems encompass a back-blowing gas generation system, a back-blowing system, and a filter dust transport system, among others.

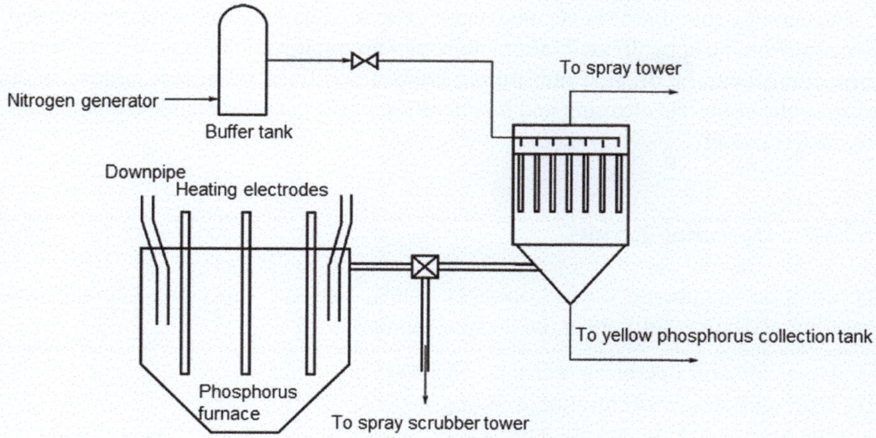

Fig. 11.40 Ceramic membrane technology for treating yellow phosphorus tail gas treatment process

11.3.3.4 Design Parameters

The design parameters of the ceramic membrane yellow phosphorus furnace gas cleaning equipment are presented in Table 11.22. The main equipment should implement suitable insulation measures to prevent significant temperature fluctuations, ensuring that the temperature inside the equipment remains above 230 °C, making the yellow phosphorus gaseous.

Table 11.22 Design parameters

No.	Parameter	Value
1	Processing air volume/(m^3/h)	6000
2	Flue gas temperature processing/°C	230–340
3	Inlet dust concentration/(g/m^3)	50–200
4	Outlet dust concentration/(mg/m^3)	< 10
5	Filter area/m^2	201
6	Ceramic membrane material	Silicon carbide
7	Pulse valve specification	Right angle type
8	Number of pulse valves/pieces	12
9	Injection air source pressure/MPa	0.2–0.4
10	Body leakage rate/%	< 2
11	Dust collector operating resistance/Pa	< 2000
12	Main system operating time/h	8400

The ceramic membrane is cleaned and regenerated by back-blowing compressed nitrogen, which can be divided into manual and automatic control modes. The automatic control can be divided into timing control and fixed resistance control, thus ensuring the complete cleaning and regeneration of the ceramic membrane filter and avoiding accidents.

11.3.3.5 Operating Results

The ceramic membrane device operates stably, and the filtration accuracy and operating resistance fully meet the design standards.

(1) Dust collector operating resistance < 2000 Pa.
(2) Dust emission concentration < 10 mg/m^3.

References

1. Xiong ZH. Atmospheric pollution prevention technology and engineering applications. Beijing: Mechanical Industry Press; 2003.
2. Pan JS, Wang GH, Zhang F, et al. Application of membrane dust removal technology with ultra-high efficiency in ultra-low emission of coal-fired power plants. China Salt Industry. 2018;305(2):42–7.
3. Wang GH, Chen LP, Zhang F, et al. Application of membrane technology in dust removal in coal-fired power plants. Salt Ind Chem Ind. 2015;305(2):50–3.
4. Ji JW, Chen AQ. Bag-type dust removal technology for boiler flue gas. Beijing: China Electric Power Press; 2006.
5. Chen YL. Desulfurization, denitrification, and flue gas dust removal technology for power plant boilers. China Equip Eng. 2020;7(13):196–7.
6. Wu YY, Liu Y, Yang GH, et al. Membrane material for controlling $PM_{2.5}$ particles in industrial emissions. Chem Fiber Textile Technol 2013; 42(3):30–2.
7. Zhong ZX, Zhang F, Wu JW. A membrane treatment system and process for the flue gas of high-concentration salt-containing organic wastewater incineration: CN107255288B. 2019-05-21.
8. Xu T, Fei CJ. The development of bag dust removal technology and the application of domestic membrane materials. Cem Technol. 2018;5:96–100.
9. Yin YH, Han J, Yu B. The influence of membrane covering treatment of filter material on pressure loss of bag filter. Modern Textile Technol. 2016;5:43–5.
10. Zhou GC, Liu JF, Xu H, et al. Comparative study of the performance of coated filter material and ordinary filter material. Light Ind Technol. 2016;1:93–4.
11. Tian W. Research on the dust cleaning of pulse jet bag filter. Xi'an: Xi'an University of Architecture and Technology; 2005.
12. Zhang F, Zhong ZX, Xu ZL, et al. A kind of waste liquid incineration flue gas cleaning equipment: CN207169346U. 2018-04-03.
13. Nie YF, Jin YY, Liu FQ. Solid waste treatment engineering technology manual. Beijing: Chemical Industry Press; 2012.
14. Wang C, Zhang DY, Wang HT, et al. Waste gas treatment engineering technology manual. Beijing: Chemical Industry Press; 2012.
15. Chen LS, Tao H. Bag-type dust removal technology manual. Beijing: Machinery Industry Press; 2010.

16. Yang D, Xu H, Liu JF, et al. Comparative study and application of different filter material properties. Chem Fiber Textile Technol. 2018;47(1):6–10.
17. Zhou YC, Hua YL, Ma KW, et al. Municipal solid waste incineration flue gas cleaning treatment technology. Chem Eng Equip. 2020;10:277–9.
18. Mao ZJ. Discussion on the technology of ultra-low emission transformation of garbage incineration flue gas. Sci Innov. 2020;3:100–1.
19. Mu LY, Wu G, Wang J, et al. Application of filter material in waste incineration flue gas treatment system. Technol Eng Appl. 2013;3:47–9.
20. Duan YG, A XY. Technical transformation of the dust collector of the kiln tail flue gas of Tianshan Cement Company. In: Proceedings of the China cement technology annual meeting and the 11th National Cement Technology exchange conference, Chongqing; 2009.
21. Sun WG. Application of bag dust removal technology in flue gas treatment in coal-fired power plants. Equip Manage Maintenance. 2019;10(20):107–8.
22. Zhong YS. Application review of the high-temperature filter material in bag filters. China Environ Protection Ind. 2020;8:57–61.
23. Liu L, Ji Z, Luan X, et al. Multi-objective optimization model of high-temperature ceramic filter. Korean J Chem Eng. 2020;37:883–90.
24. Gu YL. Research on key technologies of high temperature smoke bag dust removal system. Shenyang: Shenyang University of Technology; 2015.
25. Liu W, Jin J, Ma F. Progress of high-temperature resistant dust removal filter material. Resource Conserv Environ Protection. 2018;3:13–4.
26. Bao DS. Exploration of coke oven flue gas emissions and influencing factors. Yunnan Chem. 2020;47(8):67–71.
27. Zhou YL. Analysis of sintered pellet flue gas cleaning technology by activated coke approach. Chem Ind Manage. 2016;5:205.
28. Xiong YW, Li YF, Sun ZC. Low-temperature denitration process and reactor design of coking flue gas with activated coke. Coal Eng. 2020;52(7):174–81.
29. Gao HP. Research on high temperature corrosion behavior and combustion characteristics of gas equipment caused by yellow phosphorus tail gas. Kunming: Kunming University of Science and Technology; 2015.
30. Li D, Wang H, Wang K, et al. CFD-DEM simulation of gas-solid two-phase flow characteristics in ceramic membrane. Membr Sci Technol. 2019;39(2):51–7.
31. Weng HM. Flow field analysis and structure optimization of multi-chamber bag filter. Hangzhou: Hangzhou University of Electronic Science and Technology; 2013.
32. Yu Y, Chen LF, Cong DS, et al. Numerical simulation and optimization design of the dust collector's front flue. J Changchun Inst Technol (Natural Science Edition). 2018;2:39–42.

The manufacturer's authorised representative in the EU is Springer Nature Customer Service Centre GmbH, Europaplatz 3, 69115 Heidelberg, Germany. If you have any concerns regarding our products, please contact ProductSafety@springernature.com

Printed and bound by CPI Group (UK) Ltd, Croydon, CR0 4YY
30/09/2026
02225302-0013